D0769469

**Principles of Applied Climatology**

# Principles of Applied Climatology

**Keith Smith**
*Senior Lecturer in Geography*
*The University of Strathclyde, Scotland*

A HALSTED PRESS BOOK

John Wiley and Sons
New York

Published in the U.S.A.
by **Halsted Press**
A Division of **John Wiley and Sons, Inc., New York**

**Library of Congress Cataloging in Publication Data**

Smith, Keith, 1938–
Principles of applied climatology.

'A Halsted Press book.'
Includes bibliographies and index.
1. Climatology. I. Title.
QC981.S62 1975 551.6 74–20976

ISBN 0–470–80169–7



Text set in 10/12 pt Monophoto Times New Roman
Printed in Great Britain by
William Clowes & Sons, Limited
London, Beccles and Colchester.

To my wife, Muriel

# Contents

# Preface

This book has its origins in my efforts over several years to introduce university students of geography to some of the relationships between the atmospheric environment and human activity. These associations have always existed but it is only within the last ten to fifteen years that the continuing evaluation of the nature of both the earth and social sciences has tended to erode many of the traditional divisions between the various disciplines involved. Among other things, this trend has produced a more man-oriented type of physical geography with an increasingly deliberate emphasis on the physical basis of the natural resource systems of the world.

In no branch of the earth sciences has this development been more appropriate than in the study of the atmosphere, which represents what is, arguably, man's most fundamental natural resource. At the same time, the widespread concern for environmental matters has encouraged social scientists, including human geographers, to explore the consequences of the resource use of the atmosphere by the community as a whole. Essentially, therefore, this book seeks to explain some aspects of atmospheric behaviour, which are especially relevant to its utilization as a resource, and to clarify certain interactions between the climatic environment and contemporary socio-economic systems.

At the present stage of evolution, it is not easy to make a rigid definition of applied climatology. To some extent the workings of the atmosphere are of such immediate importance to mankind that it could be claimed that all climatology is applied climatology. On the other hand, until recently, a great deal of climatological information has been only potentially applicable, since we have lacked a detailed knowledge of the ways in which the atmosphere and man are linked. Thus, further progress is likely to depend on a better understanding of the effects of the atmosphere on man through an awareness of the weather-sensitivity of various human activities, together with a more realistic appraisal of man's capacity to modify his atmospheric environment deliberately or otherwise. Above all, the atmosphere must be viewed as a finite resource, which is ultimately in need of comprehensive management, and it is with these principal themes that this volume is mainly concerned.

The structure of the book reflects its philosophy. Following an introductory chapter which outlines the antecedents and scope of applied climatology, chapter 2 deals with the physical processes operating near the earth's surface in a dominantly natural or rural setting. Conversely, chapter 3 describes the inadvertent modifications imposed on climate by the construction of towns and then goes on to consider the wider implications of the release of pollutants into the air. The middle part of the book details the impact of climate on selected economic enterprises, chapter 4 contains the climatic framework of agricultural production, chapter 5 is concerned with the climatology of water and power resources and chapter 6 discusses the role played by the atmosphere in the major forms of transportation. In chapter 7 rather more weight is given to some of the broader social issues of human health, building climatology, and atmospheric facets of employment and recreation. Finally, weather forecasting and weather modification techniques are reviewed in chapter 8 in terms of their total significance for man.

Any effective treatment of applied climatology must be adequately rooted in the parent discipline and some previous knowledge of elementary meteorology or physical climatology will certainly not hinder the interpretation of this text. On the other hand, it is a major premise of the book that future advances in applied climatology will depend on a genuinely inter-disciplinary approach to climatic resources and, therefore, a determined effort has been made to present the material so that it will be understood by readers with little formal background in the more traditional aspects of atmospheric science.

Where possible, the book leads from a study of physical processes through to an assessment of atmospheric management policies and, given such a broad field, it will be apparent that it is virtually impossible to provide a uniformly adequate treatment, even at the introductory level attempted here. Since completing the manuscript in June 1973, I have become increasingly aware of shortcomings in the analysis, but I hope that a reasonably acceptable balance has been struck between the physical science and the social science aspects. Considerable thought has also been given to the problem of the enormous and rapidly expanding wealth of literature in the field of applied climatology. No single volume can do more than sample the published work now available and it has not been my intention to be encyclopaedic in any way. I have tried to illustrate general principles by examples drawn from all parts of the world but, perhaps inevitably, I have concentrated specifically on the English language material, which is mainly relevant to Britain and North America and is to be found in the more accessible international journals. Nevertheless, I have still found it necessary, even with careful selection, to include a total bibliography of over 1300 items. This has been done not only to make explicit my personal reliance on the work of others, but also to provide a reliable guide for those who may wish to pursue any topic further.

This book has been written primarily for undergraduate courses in geography and for other environmental sciences which embrace a study of climatology such as agriculture, horticulture, forestry, ecology, hydrology, civil engineering, and architecture. Similarly, it is felt that much of the book will serve the needs of social scientists such as economists, sociologists, marketing experts, and planners who have an interest in atmospheric resources. The comprehensive approach adopted should also appeal to Sixth Form students and to those in Colleges of Education. In addition, while the wide scope of the book precludes a fully definitive treatment of any one particular theme, the fairly extensive bibliography should prove a convenient and up-to-date source of reference for the intending specialist. It is certainly appreciated that there may well be readers whose interests lie almost exclusively in one or more of the topics covered, and perhaps it should be stated that they ought to find the chapters sufficiently self-contained to be read independently.

It is with real pleasure that I record my thanks to the people who have helped and advised me during the preparation of this book. My most obvious debt is to the various authors, editors, and publishers who gave me permission to reproduce published material. Every attempt has been made to identify the original sources of the illustrations and to make a suitable acknowledgement on the figure captions. I am also very grateful to Professor S. Gregory, Dr R. C. Ward, and Dr G. E. Jones, respectively of the Departments of Geography at the Universities of Sheffield, Hull, and Strathclyde, for reading and commenting on parts of an earlier draft. All the faults which remain are, of course, unquestionably my own responsibility. The diagrams were specially drawn for this volume by Miss D. C. Evans, Miss L. M. Gilchrist, and Mr R. Jolly of the cartographic staff of the Department of Geography in the University of Strathclyde. The entire manuscript was typed by Mrs J. Simpson. Finally, I dedicate this book to my wife, whose unfailing support—both moral and practical—has done so much to transform an idea into reality.

Keith Smith
June 1974.

# 1. Introduction

## 1.1 Some evolutionary trends in climatology

An association between climatology and geography can be traced back to the emergence of both disciplines in ancient Greece. Originally, the word *climate* meant *slope* and was used to describe the significant regional differences noted by pre-Christian scholars either as they travelled northwards from the Mediterranean shores into colder and wetter conditions or as they ventured into the increasingly hot, dry environment which lay southwards up the Nile and into Egypt (Thornthwaite, 1962). The Greeks concluded that the earth must slope up to the sun in the south and slope down away from the source of heat towards the north, and this theory was eventually rationalized into a latitudinal division of the earth's surface into Torrid, Temperate, and Frigid Zones. This interpretation of climate remained valid for some 2000 years, and it was not until the great voyages of discovery in the fifteenth and sixteenth centuries that the real complexity of regional climatic distributions began to be apparent. This important phase of geographical exploration coincided with the reawakening of scientific thought that followed the Renaissance, and the new emphasis on experiment and observation was greatly facilitated by the subsequent development of meteorological instruments such as the thermometer, constructed in 1593 by Gallileo, and the mercury barometer invented by his pupil, Torricelli, in 1643. For the first time, these instruments made possible a reasonably reliable, quantitative comparison of climatic elements for different times and places. In Britain, weather observations were encouraged by the Royal Society from 1666 onwards, while the first steps towards the establishment of national meteorological networks were taken in France during the 1770s and in Prussia by 1817 (Lamb, 1959). These observations soon revealed large local variations in the weather elements and, partly in an attempt to minimize such differences, considerable stress was placed on the standardization of instruments and their exposure. Gradually, during the nineteenth century, climatology became increasingly identified with the collection and analysis of meteorological data recorded at standard sites over long periods of time.

As indicated by Leighly (1949), too much of the work done in climatology from 1800 down to the middle of the twentieth century was dominated by regional climatic description and classification, often based on the arithmetic mean, and too little attention was paid to the physical interpretation of the recorded data. At first, the large-scale descriptive approach did have value not only in providing an overall global framework, but also because it assisted geographers around the turn of the century towards the concept of the natural region first outlined by Herbertson (1905). Unfortunately, the growing appreciation of the physical world tempted many geographers of the time into an attempted explanation of human activity largely on the basis of environmental factors.

The specific case for climatic determinism was most vigorously propounded by Huntington (1907 and 1945), who viewed much of human history, as well as the vicissitudes of contemporary society, as a more or less direct response to climatic fluctuations on various scales of time. Such extreme and over-simplified theories became unacceptable to most geographers by the late 1920s and the inevitable back-swing of the philosophical pendulum during the following years tended to stifle any further progress in unravelling the interrelationships between man and his climatic environment. Indeed, climatology as a whole suffered a severe recession in the first half of the present century. Part of this recession may be attributed to relative isolation from current trends in geography, but, more importantly, climatology languished because the initiative in atmospheric understanding was now firmly grasped by the rapidly expanding science of meteorology. As meteorology began to evolve into a specialized branch of applied mathematics and physics, climatology found itself without a well-defined role and became relegated to fill a somewhat sterile, bookkeeping function for meteorology. In turn, geographers were unable to relate meteorological advances in upper-atmosphere physics to the need for better understanding of atmospheric conditions nearer the ground. Not surprisingly, there was a marked decline of interest in climatology and Sewell *et al.* (1968a) have noted that the proportion of articles dealing with weather and climate, and the impact of the atmosphere on human affairs, published in the major American geographical journals fell from more than a third in 1916 to less than a twentieth in 1966.

Climatology still remains something of a poor relation with respect to both geography and meteorology but, within recent years, there have been signs of a modest improvement in the activity within climatology as a whole. For our purposes, however, two of the most significant trends have been a growing interest in atmospheric processes within the lower parts of the planetary boundary layer, where the characteristics of the air are closely dependent upon the properties of the earth's surface, and an increasing awareness of the complex relationships between the atmosphere and the economic and social framework of the modern world. These developments have been due to the cumulative efforts of numerous workers but, in retrospect, perhaps most of the changes can be summarized best in terms of *environmental climatology* and the separate, but linked, rationalization of the idea of *atmospheric resources*. Together, these two concepts represent the foundation of a great deal of the current work in applied climatology.

## 1.2 Emergence of applied climatology

Much of the new direction in climatology was foreseen by Thornthwaite (1953) when he introduced the term *topoclimatology* as a means of focusing attention on the qualities of the land surface that influence the exchanges of heat, moisture, and momentum in the lower atmosphere. In Thornthwaite's view, the micrometeorologist had already undertaken reliable observations of the vertical distributions of temperature, humidity, and wind in the air layers near the ground, and the time was ripe for an analysis of the horizontal variation of these vertical profiles as controlled by the composite mosaic of the countless local climates found in the valley and the forest, on the hilltop, on the south-facing slope, within the cornfield, or above the meadow. These suggestions were expanded further by Thornthwaite (1961) and adopted enthusiastically by other geographers such as Miller (1957 and 1965a) and Tweedie (1967), who saw the developments in the context of a new outlook for the whole of physical geography, whereby climatology could become the central core of all the environmental sciences. The same notion was taken up by Hare (1966), who recognized ecological, physiological, and hydrological forms of environmental climatology and pointed out that the establishment, in 1966, of the Environmental Science Services Administration in the USA and the Natural Environment Research Council in Britain could be cited as indicative of the fact that an integrated approach to the physical environment was accepted in government as well as academic circles.

The environmental interpretation of climatology, comprehensively documented by Geiger (1965) and Miller (1965b), has a two-fold significance for the emergence of applied climatology. In the first place, a knowledge of topoclimates was seen to be of direct relevance to the work of farmers, foresters, water engineers, transport engineers, builders, or any occupation which depended to some extent on local climatic conditions and much of this book is concerned with the elaboration of this theme. Second, it was also apparent that, since much of the earth's surface had already been altered through the centuries by the actions of man, many of the observed climates reflected a modified environment. This implication was recognized by Thornthwaite (1956), especially with respect to irrigation, which is one of the main ways in which a farmer adapts to his local climate and is also one of the major methods by which the surface moisture budget may be deliberately transformed. In expressing concern for the deliberate and the inadvertent modification of rural topoclimates, Thornthwaite echoed the much earlier writings of Marsh (1864 and 1874), who dealt at length with what he believed to be the widespread modification of the atmosphere by interference with the vegetation cover of the earth, especially through forest clearance, or the alteration of hydrologic regimes brought about by draining lakes or the introduction of irrigation. Marsh's essentially circumstantial evidence led to exaggerated claims and it is now known that changes in the surface characteristics of the land in rural areas are usually less significant than the climatic consequences of urbanization. Thus, the importance of the city lies not only in the highly distinctive urban climate which is produced internally, but also in the fact that extensive built-up areas comprise the main source of atmospheric pollution. Through the emission of airborne effluent from both residential and industrial areas, cities have modified the climate on regional as well as local scales and, for some pollutants, the effects have been monitored on a global scale.

Within the last two or three decades, it has become progressively clear that, largely as a result of the accelerating growth of population and the pace of technological innovation, man is no longer restricted to the entirely passive role with respect to the atmosphere that characterized the life of his forbears. At the same time, it has become equally clear that the relationships between climate and man have inevitably developed in a highly complex way. For example, atmospheric pollution is the most striking illustration of man's capability for unconscious adverse modification of the atmosphere and also indicates that the atmosphere does not possess an unlimited potential for waste disposal. In an attempt to preserve environmental quality, society has had to make economic, technical, and legal decisions designed to improve the standard of pollution emissions.

On the other hand, man is still subject to the age-old, natural variations of the atmosphere. Many parts of the world remain too dry or too cold for effective settlement, and major atmospheric hazards, such as hurricanes, continue to create disasters. Advances in weather forecasting have accomplished a great deal in the prediction of severe storms, but, despite the investment of much capital and research effort, the impact of atmospheric hazards shows that the sophisticated economic and social fabric of the modern world is very dependent on climatic conditions.

During the early 1950s, in an era of misplaced confidence, it began to seem that man could achieve deliberate weather control beyond the microscale for the first time in history. Once again, however, capital investment and scientific technology have been unable to make this dream a reliable reality. On the contrary, subsequent research on weather modification has tended to raise almost as many problems as it has solved, not just in the technical field but more particularly in connection with issues of public policy such as the extent of weather modification experiments that can, or ought to be, permitted with the present inadequate understanding of the possible consequences.

The culmination of these trends and evaluations has been the gradual realization that the atmosphere is a basic natural resource, which is subject to spatial and temporal variations, is used and polluted, and is ultimately in need of conservation and management in much the same way as, for example, water resources. Miller (1956) noted that, like other natural endowments, the resources of the atmosphere are not infinite and, while elements such as rainfall and sunshine are renewable, there is a definite limit to their availability in any one area. Climatic resources may be utilized in a variety of ways, some of which are more obvious than others. Thus, air pollution is a constant reminder of man's prolonged use of the atmosphere for waste disposal, despite the unfortunate effects on the 17 kg of air we each breathe daily (Chandler, 1970). The farmer clearly makes use of a climatic income in order to support his agricultural production; in fact, agricultural climatology is one of the more important areas of expansion within applied climatology and Chang (1968) has reported that the number of articles on the relationship between climate and crop growth published in agronomy journals increased from about two per year before 1920 to as many as fifteen per year during the 1960s.

The concept of a resource implies the application of economic methods of analysis and a dominantly economic view of the atmosphere has been expressed by Curry (1952), Perry (1971) and, in an important book, by Maunder (1970). According to Crutchfield and Sewell (1968), an economic appraisal of weather and climate involves an appreciation of the use of the atmosphere for production purposes, the losses of property and income due to severe weather events, and also the uncertainties in decision-making which arise from the unpredictability of weather variations. In order to understand the detailed relationships, it is first necessary to know which economic activities are weather-sensitive in either a direct or indirect sense and, so far, most progress has been achieved in agriculture, transportation, and the construction industry. Once the weather-sensitive mechanisms have been accurately defined, it should be possible to assess the gains and losses likely to result from each of the technically feasible adjustments to the atmosphere. The adjustments available depend a great deal on the type of activity under consideration, but Crutchfield and Sewell have recognized three basic categories.

First, there is the development of techniques designed to reduce the direct impact of the weather, and these might include responses as diverse as air-conditioning, drought-resistant crops, insurance, or the evacuation of an area from the path of a hurricane (Sewell, 1968a). Second comes from a reduction of uncertainty about the vagaries of the atmosphere through the improvement of weather information. Again, various methods are possible, but this adjustment depends largely on progress in weather forecasting and Lamb (1969), for example, has outlined the potential value of forecasts of the probable climatic trend over the next few years, or even decades. Finally, the ultimate adjustment possible is that of deliberate weather modification.

At the present time a comprehensive economic appraisal of atmospheric resources is restricted because the weather sensitivity of most activities is imperfectly understood, especially in quantitative terms, and, in turn, this has set a limitation on the immediate application of certain techniques and models used by the economist. A fairly early advance was the development of a simple conceptual model to depict the economic reaction to the atmosphere, as shown in Fig. 1.1 (McQuigg and Thompson, 1966). This example shows the flow of actual events as solid lines and the flow of weather information as broken lines, and it should be emphasized that the information can be of value only if it is effectively employed in the decision-making process. In other words, many decisions are made either with no information available or by managers who choose to ignore data which are available. Sewell *et al.* (1968b) have reviewed the extent to which more sophisticated econometric techniques, such as input–output analysis and linear programming, are relevant in an economic assessment of climate and, more recently, McQuigg (1971) has outlined the application of simulation models to the study of the economic response to weather events.

Substantial progress has been made by economists in many of these fields but, apart from a general lack of economic data comparable in quantity and quality with weather observations, it would appear that one of the most important deficiencies is that of a model capable of measuring the regional impact of climate on various economic activities. This problem was isolated by Ackerman (1966), who proposed the concept of an ideal weather-pattern model which could be used to determine the effect of weather elements on the system of economic production and consumption in any area but, as in the case of some of the other techniques, further research will be required before such a model can become operational.

The present stress on economic appraisals of the atmosphere should not, of course, be allowed to obscure the fact that, even when considered as a resource, the full implications of weather and climate go beyond the purely financial level. Thus, there is a continuing need for further research on the physical aspects of the atmosphere, particularly insofar as it illuminates our understanding of topoclimates, air-pollution climatology, weather forecasting, or weather modification. In addition, the notion of atmospheric resources raises wider issues in ecological, social, legal, and political fields, since man's use of the atmosphere is no longer necessarily individual or localized. As stated by Ostrom (1968), the atmosphere is a common-pool, flow resource but, unlike water-resource systems which are formed into natural watersheds suitable for recognition as definite boundaries for organizing management schemes, the atmosphere has no distinct boundaries which can be used to delineate sub-systems such as topoclimates or airsheds. It is more difficult in the atmosphere, therefore, to trace the adverse consequences as well as the possible benefits, which may well extend beyond those who make even a rational economic decision to adopt a certain course of action. This type of effect is most likely to result from human adjustment to the atmosphere involving either air pollution or weather modification.

The inadvertent modification of the atmosphere has been examined by a special study group (Massachusetts

Institute of Technology, 1971). More information is still required, however. For example, Scorer (1971) has pointed out that, compared with water, the residence time of pollution in the atmosphere is short, varying from a few days in temperate areas to a few weeks in the tropics, but until more is known about pollution in the stratosphere and the interaction between the stratosphere and the troposphere, it might be premature to think of climate as a renewable, flow resource. Similarly, Cooper (1968) has discussed the possible impact of both pollution and modification on plants and animals. In some cases, it was felt that amenity may well be damaged although, in the absence of further research, it would be difficult either to determine whether the effects would be reversible or to attach a specific monetary value to the consequences.

Meteorologists such as Stagg (1961), Mason (1966), and White (1967) have begun to accept the wider implications of the atmosphere and it appears both necessary and inevitable that, in the long term, the atmosphere should be comprehensively managed for the general good of mankind. The obstacles will be formidable, but the case for atmospheric management has been persuasively presented by Sewell (1968b), who believed that much more research effort was needed on the human dimensions of weather and climate. Sewell implied that, in the last resort, progress was tied up with the amount of financial support provided for research and claimed

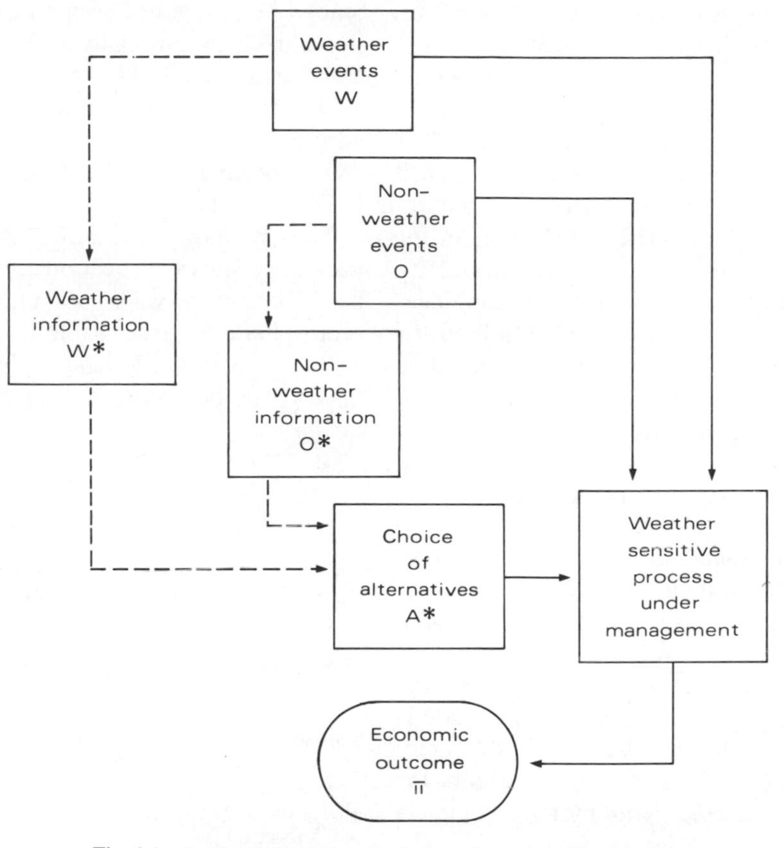


**Fig. 1.1 A schematic outline of relations between weather events, non-weather events, the choice of alternatives by management and the economic outcome of an enterprise. After McQuigg and Thompson (1966).**

that, while the US federal government allocated annually over $200 million for research and development in the physical sciences, the total research expenditure on the social science aspect was probably less than $100 000 per year. In the same way, although ESSA employed around 10 000 people, only 10 were considered to be professional economists. The problem of defining the priorities for research expenditure in the whole of the atmospheric sciences has been investigated by the US National Research Council (1971) who reported that, in the four years following 1960, the annual research expenditure in America rose from $37 million to $222 million. During the period 1970–79, the Research Council estimated that a total cost investment of $453 million annually will be required, comprising $119 million for weather-prediction research, $91 million for air-quality research, $60 million for weather- and climate-modification research, together with a further $183 million for major support facilities of which $140 million would be allocated to computer development. This level of atmospheric research expenditure was deemed to be necessary to ensure that the nation continued to obtain adequate benefits from its climatic resources.

### 1.3 The impact of atmospheric hazards

An important premise underlying all applied climatology is that man is in an essentially ecological relationship with the atmosphere and, although he possesses some limited ability to manipulate atmospheric processes, the most characteristic response to weather and climate is through adjustments which take place within the socio-economic framework. Even here, however, the scope for adjustment is often restricted both by an incomplete understanding of climatic variations and by the existence of strong, often conflicting, pressures from within society itself. These general principles are well illustrated by atmospheric hazards which arise from severe weather and are experienced largely as a result of the progressive occupation of traditional hazard zones such as the flood plain and the seashore (Burton and Kates, 1964). For example, the rapid upsurge in leisure activities in the USA over the last few decades has led to a demand for second homes located as close as possible to the shore along the eastern seaboard, and it has been estimated that approximately 125 000 structures have been built less than 3 m above mean sea level between Maine and North Carolina. This settlement has taken place notwithstanding the fact that the whole coast is subject to hurricanes and extra-tropical cyclones and Mather *et al.* (1965) have shown that storms of moderate to severe intensity may be expected on average once every 1·4 years along the coasts of New York and New Jersey. Since 1935, there has been a marked increase in coastal storm damage, especially in the New England and New York sectors, and, though this may be partially attributed to the occurrence of lower central pressures in closed barometric lows, the main cause was undoubtedly the greater occupancy of the coastal margins (Mather *et al.*, 1967).

Thus, perhaps because of, rather than in spite of, man's present social sophistication and technological strength, he is in many ways becoming more vulnerable to atmospheric hazards. Table 1.1 demonstrates the rising economic cost of selected geophysical hazards in the US and it can be seen that, even excluding the drought

| *Hazard* | *Loss of life* No. | *Loss of life* Period | *Annual or average annual property damage* Amount ($ million) | *Annual or average annual property damage* Period |
|---|---|---|---|---|
| Floods | 70 | 1955–64 | 1000 | 1966 |
| | | | 350–1000 | 1964 |
| | | | 290 | 1955–64 |
| Hurricanes | 110 | 1915–64 | 250–500* | 1966 |
| | | | 100 | 1964 |
| | | | 89 | 1915–64 |
| Tornadoes | 194 | 1916–64 | 100–200* | 1966 |
| | | | 40 | 1944–64 |
| | | | 300 | 1967 |
| Hail, wind and thunderstorms | | | 125–250* | 1966 |
| | | | 53 | 1944–53 |
| Lightning strikes and fire | 160 | 1953–63 | 100 | 1965 |
| Earthquakes | 3 | 1945–64 | 15 | 1945–64 |
| Tsunamis | 18 | 1945–64 | 9 | 1945–64 |
| Heat and insolation | 238 | 1955–64 | | |
| Cold | 313 | 1955–64 | | |
| Totals | 1106 | | 621–2174 | |

* Insured losses only.

**Table 1.1** Estimates of average annual losses from selected geophysical hazards in the US. Single year estimates are the level of average losses current to year cited. Property damage figures are in millions of dollars unadjusted unless otherwise noted. (After Burton, Kates, and White, 1968.)

hazard, atmospheric events account for the overwhelming majority of fatalities and property losses.

All of these losses do not derive from the wilful settlement of recognized hazard zones and we are only just beginning to unravel the complex, interactive web of human adjustment to severe weather events. Kates (1971) has proposed a general systems concept, reproduced in Fig. 1.2, whereby a *natural hazard* is depicted as the joint outcome of interaction between the *human use system*, which is defined as the smallest managerial unit capable of independent adjustment to the hazard, and the *natural event system*, which is seen in terms of statistical parameters such as magnitude, frequency, duration, and temporal spacing of the event. The resulting natural hazard produces *hazard effects*, which in turn lead to a complex *adjustment process control* subsystem. It is then through the adjustment process control system that adjustments to the hazard take place involving both *natural event modification adjustments* and *human use modification adjustments*. A detailed appraisal of many of these adjustments in relation to specific activities is contained in later chapters and, at this stage, it will be sufficient to mention just a few of the more general climatic hazards and indicate something of their effects.

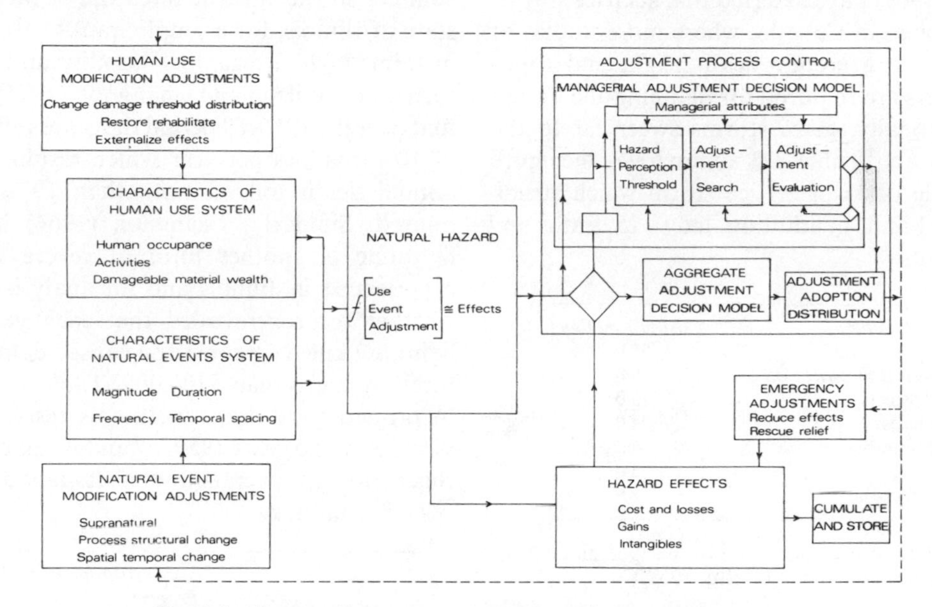


**Fig. 1.2 Human adjustment to natural hazards: a general systems model. After Kates (1971).**

River floods are one of the most widespread short-term hydrometeorological hazards, which arise from the concentration of population in well defined danger areas, and they produce a highly distinctive human response in the construction of storage dams and river training works. Individual events can cause considerable losses. Thus, a flood in the South Platte river basin in the western US during June 1965 claimed six lives and caused damage estimated at $5000 million (Rostvedt, 1970). Approximately 75 per cent of the damage occurred in the Denver metropolitan area which emphasizes the point made by Brater (1968) and others that, as the population density and land values continue to rise in urban areas, floods become a greater problem both economically and as a threat to safety. According to Burton (1970), the average annual cost of allowing uncontrolled development on floodplains in Canada is over $300 million while, in a multi-purpose control project designed for Toronto, the benefits for public recreation were considered to be double those deriving from flood control. Many areas suffer an alternation of water excess and water deficiency which has serious implications for agricultural communities, as detailed by Burley (1965) for part of south-eastern Australia and by Coulter (1966) for northern New Zealand.

A rather less publicized hazard, common to most mid-latitude areas in the winter, is that of ice accretion due to raindrops freezing after striking the ground. This causes a sheet of clear, glazed ice to cover all exposed surfaces and leads to widespread damage and disruption. McKay and Thompson (1969) have asserted that such ice storms are a major problem in Canada, where more studies of the variation in ice accretion with height and topographic parameters are required for planning and design purposes. Occasionally, glaze storms sweep far to the south in the US and Table 1.2 summarizes the more easily quantifiable costs of an ice storm which struck Tennessee in 1951 and, in addition, led to 25 deaths and 500 other accidents.

| *Type of damage* | *Cost in $ million* |
|---|---|
| Forests | 56·0 |
| Communications and power lines | 10·0 |
| Highways and streets | 15·0 |
| Fruit and nut trees | 4·0 |
| Buildings and plumbing | 4·3 |
| Livestock | 3·0 |
| Truck and grain crops | 1·6 |

**Table 1.2** Economic losses due to the 1951 glaze storm in Tennessee. (After Harlin, 1952.)

World-wide, however, the greatest hazard is presented by the severe convective storm which produces a number of specific threats including heavy rain, hail, strong winds, and lightning. Hail is a recurrent hazard during summer in several mid-continental areas, and the US is no exception. Changnon (1972) has stated that the total average annual national loss in crops and property for the US is around $315 million, of which $284 million is attributed to crop losses and represents about 1 per cent of the national crop production. Although many hail suppression experiments have been undertaken, the traditional human response to this hazard is crop insurance. At the present time, about 15 per cent of the national crop value in the United States is insured in this way and the available records provide a more reliable source of data for evaluating the economic impact of severe weather than exists for any other climatic hazard. As might be expected, the area of greatest hail losses is the Great Plains, where damage amounts to about $86 million each year, with the Corn Belt ranked second.

The major problem for human adjustment through either insurance or weather modification arises from the catastrophic storms which can create between $1 to $5 million in losses, comprising 15 to 75 per cent of the total annual loss in a State, during a single day. Severe summer storms are also a feature of the Mid-West. The state of Illinois, for example, ranks first in the US both in terms of hail insurance liability and in deaths from tornadoes. Wilson and Changnon (1971) have indicated that over the 1916–69 period the State suffered an average of 10 tornadoes per year which resulted in an average annual death toll of more than 19 with 110 people injured. Similarly, Changnon (1964) has shown that lightning is another form of severe weather hazard experienced in Illinois and an analysis of 34 years of weather records revealed that each year, on average, lightning killed 6 people and caused damage to property totalling more than $100 000. Table 1.3 lists the losses to property and crops in Illinois resulting from severe weather in the years 1950–57 and it can be seen that, over this period, the average annual damage amounted to well over $16 million.

| | *Average loss ($ 000s)* | | |
|---|---|---|---|
| | *Property* | *Crops* | *Total* |
| Hail | 460·9 | 3 680·0 | 4 140·9 |
| Heavy rains | 2 635·0 | 1 503·8 | 4 138·8 |
| Winds | 3 352·0 | 386·8 | 3 738·8 |
| Tornadoes | 2 453·6 | 10·2 | 2 463·8 |
| Winter storms | 1 780·3 | 0·0 | 1 780·3 |
| Lightning | 105·0 | 3·0 | 108·0 |
| Total | 10 786·8 | 5 583·8 | 16 370·6 |

**Table 1.3** Average annual property and crop losses in Illinois due to severe local storms, 1950–57. (After Changnon, 1972.)

The most spectacular weather hazard, however, is the hurricane, which poses such a recurrent threat to the Gulf coast and eastern seaboard of the US that, for some time, a review of the Atlantic hurricane season has been drawn up annually (Simpson and Pelissier, 1971). For example, Hurricane Camille was the most destructive of the thirteen Atlantic hurricanes recorded during the 1969 season by Simpson *et al.* (1970) and led to damage in excess of \$1·4 billion. Hurricanes are also the major category in so-called killer storms, defined by Cressman (1969) as severe weather events claiming more than 100 fatalities, and, without the benefit of the existing forecasting and warning services, it was estimated that the 350–400 deaths attributed to Camille in 1969 could easily have been into tens of thousands.

Thus, in an average season, hurricane damage amounted to \$300 million compared with a cost for aerial reconnaissance and communications of \$2·7 million and a cost of \$6·8 million for taking protective measures and evacuating inhabitants. The total cost was, therefore, around \$310 million of which about \$25 million was saved as a result of the forecasting and warning services. In the years when no hurricanes came ashore, Sugg estimated that the forecasting costs were still \$2·4

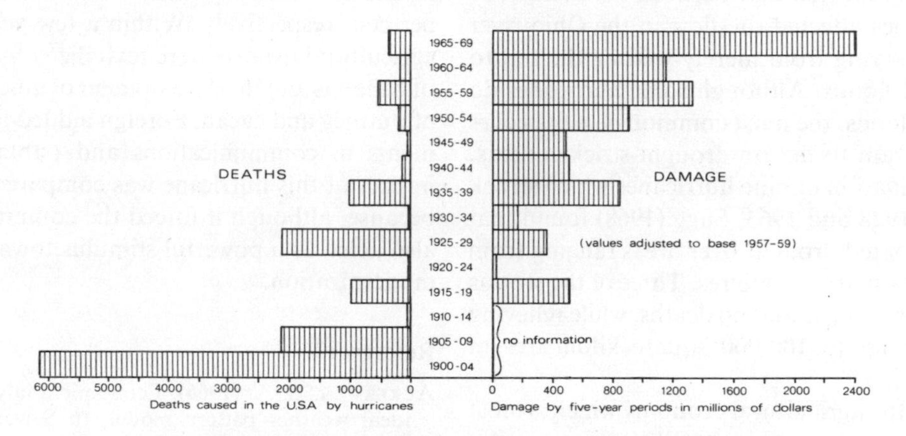


**Fig. 1.3 Trends in losses from hurricanes in the US summarized by 5-year periods. After Gentry (1970). Copyright 1970 by the American Association for the Advancement of Science.**

Until the recent advent of modification experiments, the development of improved forecasting techniques, combined with evacuation and other emergency actions, represented the only feasible human response to hurricanes. The long-term results of this policy in the US can be seen in Fig. 1.3, which illustrates that better advanced warnings of hurricanes have produced a progressive decline in hurricane-caused deaths since the beginning of this century although, at the same time, there has been an equally marked increase in the economic impact of these storms (Gentry, 1970). Despite the rising cost of hurricane damage, much of which would appear to be unpreventable by the application of any technology available at the present day, Sugg (1967) has argued that the existing warning systems do create appreciable financial savings even though, because of deficiencies in forecasting, the area warned is normally about three times as large as the area eventually damaged.

million, mainly for air survey work, but when hurricanes invaded the land it was considered that the warning system would result in a net benefit between three to four times the expenditure incurred.

In some other parts of the world, hurricane forecasting is less advanced and the loss of life remains exceptionally high. The coast of Bangladesh is a notoriously vulnerable area, largely because the low-lying coastal margins at the head of the Bay of Bengal are easily submerged by the attendant storm surge, and in November 1970 at least 300 000 people died in a hurricane. Koteswaram (1971) has pointed out that the sparse network of meteorological observations in the area has always hindered the early detection of tropical cyclones but information from aircraft, radar, and satellite sources is now helping to fill some of the gaps. In addition, mathematical models have been employed to calculate sea level changes in response to hurricanes moving up the Bay of Bengal and the results of these investigations could form the basis for an improved warning programme (Das, 1972 and Flierl and Robinson, 1972).

Although hurricanes may be regarded as the most severe type of weather hazard, they may also be used to demonstrate the principle that the net impact of atmospheric hazards is not always wholly disadvantageous. One of the most variable features of hurricane systems is the amount of precipitation which they release, and Relyea (1969) has reported that between 1871 and 1967 about 40 hurricanes affected the flow in the Ohio river with the effects varying from merely wetting the soil to the production of floods. Although some hurricanes do bring disastrous floods, the most common benefit occurs when they cause rain to fall on drought-stricken areas. Thus, in an examination of nine hurricanes which struck the US between 1928 and 1963, Sugg (1968) found that each storm terminated drought over areas ranging from 26 000 to 155 000 square kilometres. Three of the storms caused only minor damage and no deaths, while relieving drought in areas up to 100 000 square kilometres in extent.

Normally, it is the agricultural economy which is most sensitive to hurricane rainfall, and really intense precipitation combined with high winds can cause widespread damage to standing crops. This happened in September 1967 when Hurricane Beulah struck the Rio Grande valley in Texas and stripped early oranges and grapefruit from the trees resulting in a loss to the fruit industry alone of around $25 million (Flitters, 1970). To try to determine the balance between agricultural benefits and losses from hurricane-induced rainfall in the southeastern US, Hartman *et al.* (1969) devised a series of multiple regression equations for various crop yields expressed in terms of rainfall over 10-day time periods. It was concluded that, although the net effect depended a lot on the actual timing and track of individual hurricanes, storms experienced in the most active part of the hurricane season between August and early October were likely to be detrimental to yields, whereas July hurricanes could produce some benefit. Even hurricanes which cause considerable devastation can bring an economic boom to disaster areas as a result of insurance compensation, and fishermen have found significant increases in the lobster population in the Florida Keys after rough seas and high tides associated with the passage of hurricanes.

In some cases, longer-term benefits have been recognized, and Weaver (1968) has cited the consequences of Hurricane Janet, which swept over Grenada in the Windward Islands in September 1955 killing over 100 people and rendering a further 20 000 homeless. The short-term economic effects were disastrous, as the island relied heavily on the export crops of nutmeg and cacao, the cultivation areas of which were destroyed by 75 and 45 per cent respectively. Within a few years, however, the agricultural exports were revitalized by the introduction of bananas and the development of quick-growing species of nutmeg and cacao. Foreign aid led to major improvements in communications and public works and the impact of this hurricane was compared to that of a war because, although it forced the community into debt, it also acted as a powerful stimulus towards much-needed modernization.

## References


Ackerman, E. A. (1966). Economic analysis of weather: An ideal weather pattern model. In Sewell, W. R. D. (ed.), *Human Dimensions of Weather Modification*, Dept. of Geography, University of Chicago, Research Paper No. 105:61–75.

Brater, E. F. (1968). Steps toward a better understanding of urban runoff processes. *Water Resour. Res.*, **4**:335–48.

Burley, T. M. (1965). Flood and drought in the Hunter valley of New South Wales and their impact upon the agricultural community. *Tijd. voor Econ. Soc. Geogr.*, **56**:193–9.

Burton, I. (1970). Flood-damage reduction in Canada. In Nelson, J. G. and Chambers, M. J. (eds.), *Water*. Methuen, 77–108.

Burton, I. and Kates, R. W. (1964). The floodplain and the seashore: A comparative analysis of hazard-zone occupance. *Geogrl. Rev.* **54**:366–85.

Burton, I., Kates, R. W., and White, G. F. (1968). The human ecology of extreme geophysical events. *Natural Hazard Research Working Paper*, **1**:33. Quoted by Chorley, R. J. and Kennedy, B. A. (1971) *Physical Geography—A Systems Approach*. Prentice-Hall, London, 370 pp.

Chandler, T. J. (1970). *The Management of Climatic Resources*. An inaugural lecture delivered at University College, London, 17 Feb. 1970. H. K. Lewis and Co. Ltd, 30 pp.

Chang, J.-H. (1968). Progress in agricultural climatology. *Prof. Geogr.*, **20**:317–20.

Changnon, S. A. (1964). Climatology of damaging lightning in Illinois. *Mon. Weath. Rev.*, **22**:115–20.

CHANGNON, S. A. (1972). Examples of economic losses from hail in the United States. *J. Appl. Met.*, **11**:1128–37.

COOPER, C. F. (1968). Needs for research on ecological aspects of human uses of the atmosphere. In US National Science Foundation, *Human Dimensions of the Atmosphere*, NSF 68—18, Washington DC, 43–51.

COULTER, J. D. (1966). Flood and drought in northern New Zealand. *New Zealand Geographer*, **22**:22–34.

CRESSMAN, G. P. (1969). Killer storms. *Bull. Amer. Met. Soc.*, **50**:850–5.

CRUTCHFIELD, J. A. and SEWELL, W. R. D. (1968). Economic research aspects of human adjustment to weather and climate. In US National Science Foundation, *Human Dimensions of the Atmosphere*, NSF 68—18, Washington DC, 59–69.

CURRY, L. (1952). Climate and economic life: a new approach. *Geogrl. Rev.*, **42**:367–83.

DAS, P. K. (1972). Prediction model for storm surges in the Bay of Bengal. *Nature*, **239**:211–13.

FLIERL, G. R. and ROBINSON, A. R. (1972). Deadly surges in the Bay of Bengal: dynamics and storm-tide tables. *Nature*, **239**:213–15.

FLITTERS, N. E. (1970). Hurricane Beulah. A report in retrospect on the hurricane and its effect on biological processes in the Rio Grande Valley, Texas. *Int. J. Biometeor*, **14**:219–26.

GEIGER, R. (1965). *The Climate near the Ground*. Harvard University Press, 611 pp.

GENTRY, R. C. (1970). Hurricane Debbie modification experiments. *Science*, **168**:473–5.

HARE, F. K. (1966). The concept of climate. *Geography*, **51**:99–110.

HARLIN, B. W. (1952). The great southern glaze storm of 1951. *Weatherwise*, **5**:10–13.

HARTMAN, L. M., HOLLAND, D., and GIDDINGS, M. (1969). Effect of hurricane storms on agriculture. *Water Resour. Res.*, **5**:555–62.

HERBERTSON, A. J. (1905). The major natural regions of the world. *Geogrl. J.*, **25**:300–10.

HUNTINGTON, E. (1907). *The Pulse of Asia*. Houghton Mifflin, Boston.

HUNTINGTON, E. (1945). *Mainsprings of Civilization*. J. Wiley and Sons, New York.

KATES, R. W. (1971). Natural hazard in human ecological perspective: hypotheses and models. *Econ. Geogr.*, **47**:438–51.

KOTESWARAM, P. (1971). Cyclone distress mitigation measures in India. *W. M. O. Bulletin*, **20**:89–92.

LAMB, H. H. (1959). Our changing climate, past and present. *Weather*, **14**:299–318.

LAMB, H. H. (1969). The new look of climatology. *Nature*, **223**:1209–15.

LEIGHLY, J. (1949). Climatology since the year 1800. *Trans. Amer. Geophys. Un.*, **30**:658–72.

MARSH, G. P. (1864). *Man and Nature*. Sampson, Low and Son, London, 560 pp.

MARSH, G. P. (1874). *The Earth as Modified by Human Action*, Scribner, Armstrong and Co., New York, 656 pp.

MASON, B. J. (1966). The role of meteorology in the national economy. *Weather*, **21**:382–93.

MASSACHUSETTS INSTITUTE OF TECHNOLOGY (1971). *Inadvertent Climate Modification*. Report on the Study of Man's Impact on Climate. MIT Press, Cambridge, Mass., and London, 308 pp.

MATHER, J. R., ADAMS, H., and YOSHIOKA, G. A. (1965). Coastal storms of the eastern United States. *J. Appl. Met.*, **3**:693–706.

MATHER, J. R., FIELD, R. T., and YOSHIOKA, G. A. (1967). Storm damage hazard along the east coast of the United States. *J. Appl. Met.*, **6**:20–30.

MAUNDER, W. J. (1970). *The Value of the Weather*. Methuen, London, 388 pp.

MCKAY, G. A. and THOMPSON, H. A. (1969). Estimating the hazard of ice-accretion in Canada from climatological data. *J. Appl. Met.*, **8**:927–35.

MCQUIGG, J. D. (1971). Some attempts to estimate the economic response of weather information. *Weather*, **26**:60–8.

MCQUIGG, J. D. and THOMPSON, R. (1966). Economic value of improved methods of translating weather information into operational terms. *Mon. Weath. Rev.*, **94**:83–7.

MILLER, A. A. (1956). The use and mis-use of climatic resources. *Adv. of Science*, **13**:56–66.

MILLER, D. H. (1957). What climatologists need from other geographers. *Prof. Geogr.*, **9**:8–10.

MILLER, D. H. (1965a). Geography, physical and unified. *Prof. Geogr.*, **17**:1–4.

MILLER, D. H. (1965b). The heat and moisture budget of the earth's surface. *Advan. Geophys.* **11**:175–302.

OSTROM, V. (1968). Needs for research on the political aspects of the human use of the atmosphere. In US National Science Foundation, *Human Dimensions of the Atmosphere*, NSF 68–18, Washington DC, 71–9.

PERRY, A. H. (1971). Econoclimate—a new direction for climatology. *Area*, **3**:178–9.

RELYEA, C. M. (1969). Hurricanes and the Ohio River. *Trans. N.Y. Acad. Sci.*, **31**:42–55.

ROSTVEDT, J. O. *et al.* (1970). Summary of floods in the United States during 1965. *U.S. Geological Survey. Water Supply Paper*, 1850–E., 110 pp.

SCORER, R. S. (1971). New attitudes to air pollution—the technical basis of control. *Atmos. Envir.*, **5**:903–34.

SEWELL, W. R. D. (1968a). The problem in perspective. In US National Science Foundation, *Human Dimensions of the Atmosphere*, NSF 68–18, Washington DC, 1–17.

SEWELL, W. R. D. (1968b). Emerging problems in the management of atmospheric resources: the role of social science research. *Bull. Amer. Met. Soc.*, **49**:326–36.

SEWELL, W. R. D., KATES, R. W., and PHILLIPS, L. E. (1968a). Human response to weather and climate: geographical contributions. *Geogrl. Rev.*, **58**:262–280.

SEWELL, W. R. D., KATES, R. W., and MAUNDER, W. J. (1968b). Measuring the economic impact of weather and weather modification: A review of techniques of analysis. In US National Science Foundation, *Human Dimensions of the Atmosphere*, NSF 68–18, Washington DC, 103–12.

SIMPSON, R. H. *et al.* (1970). The Atlantic hurricane season of 1969. *Mon. Weath. Rev.*, **98**:293–306.

SIMPSON, R. H. and PELISSIER, J. M. (1971). Atlantic hurricane season of 1970. *Mon. Weath. Rev.*, **99**:269–77.

STAGG, J. M. (1961). Meteorology and the community. *Q. Jnl. Roy. Met. Soc.*, **87**:465–71.

SUGG, A. L. (1967). Economic aspects of hurricanes. *Mon. Weath. Rev.*, **95**:143–6.

SUGG, A. L. (1968). Beneficial aspects of the tropical cyclone. *J. Appl. Met.*, **7**:39–45.

THORNTHWAITE, C. W. (1953). Topoclimatology. *Proc. Toronto Met. Conference*, Amer. Met. Soc. and Roy. Met. Soc., 227–32.

THORNTHWAITE, C. W. (1956). Modification of rural microclimates. In Thomas, W. L. (ed.) *Man's Role in Changing the Face of the Earth*, University of Chicago Press, 567–83.

THORNTHWAITE, C. W. (1961). The task ahead. *Ann. Ass. Amer. Geogrs.*, **51**:345–56.

THORNTHWAITE, C. W. (1962). The geographer's role in climatology. *Sonderdruck aus der Hermann von Wissman—Festschrift*, Tubingen, 81–8.

TWEEDIE, A. D. (1967). Challenges in climatology. *Australian J. Sci.*, **29**:273–8.

U.S. NATIONAL RESEARCH COUNCIL (1971). *The Atmospheric Sciences and Man's Needs: Priorities for the Future*. National Academy of Sciences, Washington DC, 88 pp.

WEAVER, D. C. (1968). The hurricane as an economic catalyst. *J. Trop. Geog.*, **27**:66–71.

WHITE, R. M. (1967). Meteorology on a new threshold. *Bull. Amer. Met. Soc.*, **48**:250–7.

WILSON, J. W. and CHANGNON, S. A. (1971). *Illinois Tornadoes*. Illinois State Water Survey, Circ. 103, Urbana, 60 pp.

# 2. The mosaic of rural topoclimates

## 2.1 Introduction

The purpose of this chapter is to draw attention to the variety of small-scale climates which exist near the earth's surface in dominantly rural areas. These climates often show characteristics which are in sharp contrast to the broader generalizations of so-called regional climates mainly because of the interaction of the atmosphere with the underlying surface. Thus, although all climate is to some extent a function of the relationships at the earth–air interface, the importance of surface features increases markedly as the scale of climatic reference diminishes and it is only at the very lowest levels of the atmospheric boundary layer that surface influences become strong enough to create really special phenomena.

A contraction of the horizontal as well as the vertical scale may also disclose local modifications within the broadly uniform regional climate, which are obscured when a wider view is taken. Similarly, the time-scale of atmospheric events is significant, since any attempt to synthesize macroclimate usually relies on the analysis of steady-state conditions achieved over a period of many years, and the resulting averages smooth out the more intermittent local contrasts.

Although local climates are necessarily restricted in space and time, they are of great practical significance for the community because virtually all of man's social and economic activities are conducted near the bottom of the atmospheric boundary layer. In the rural context, therefore, local climatic differences are of particular importance for agriculture and forestry, but are by no means entirely limited to these two fields. In addition, it is at the topoclimatic scale that man is most likely to modify natural atmospheric processes, deliberately or not, by land-use changes and other means, and there is a growing need for a better understanding of the consequences of such actions.

Most rural topoclimates owe their origin to variations in either topography or the nature of the vegetation cover, and this chapter places an emphasis on the mechanisms and range of atmospheric behaviour of a group of local climates subjectively chosen on the basis of relief or surface characteristics. The group of climates selected varies from entirely natural topographic climates, such as those of valleys or coasts, to include the increasingly artificial creations due to the spread of agricultural crops and afforestation. On the other hand, a discussion of the entirely man-made climate within the urban environment is more conveniently deferred until the following chapter.

## 2.2 Climate and scale

The study of climate at the smallest scale has traditionally been carried out under the term *microclimatology*, which has described a whole range of effects from, say, the possible influence of an extensive forest area on nearby precipitation to humidity variations around individual leaves. Much of the early work in microclimatology was undertaken by meteorologists investigating the nature of vertical atmospheric gradients near the ground but it has proved difficult to place vertical limitations on these studies. For example, Geiger (1965) has specifically mentioned the lowest 2 m of air while Munn (1966) has dealt with the atmosphere up to a height of 50 to 100 m. In the US, experimental work on the Great Plains, Nebraska, reported by Lettau and Davidson (1957), confirmed the concentrated gradient of atmospheric phenomena in the lower layers as shown by the profiles of air temperature and wind speed in Fig. 2.1. It can be seen that, above this level prairie site around noon on a clear summer day, the surface temperature of the ground was some 10 degC warmer than the air at a height of 1 m compared with a gradient of only 2–3 degC in the next 100 m. Windspeed decreases towards ground level as a result of surface friction, and within the lowest 0·5 m there was as much difference in velocity as existed from 0·5 m up to 100 m. An even steeper gradient of temperature was found in the upper soil layers and the combined air and soil temperature curve in Fig. 2.1 serves to illustrate the magnitude of the fluxes taking place above and below the ground surface.

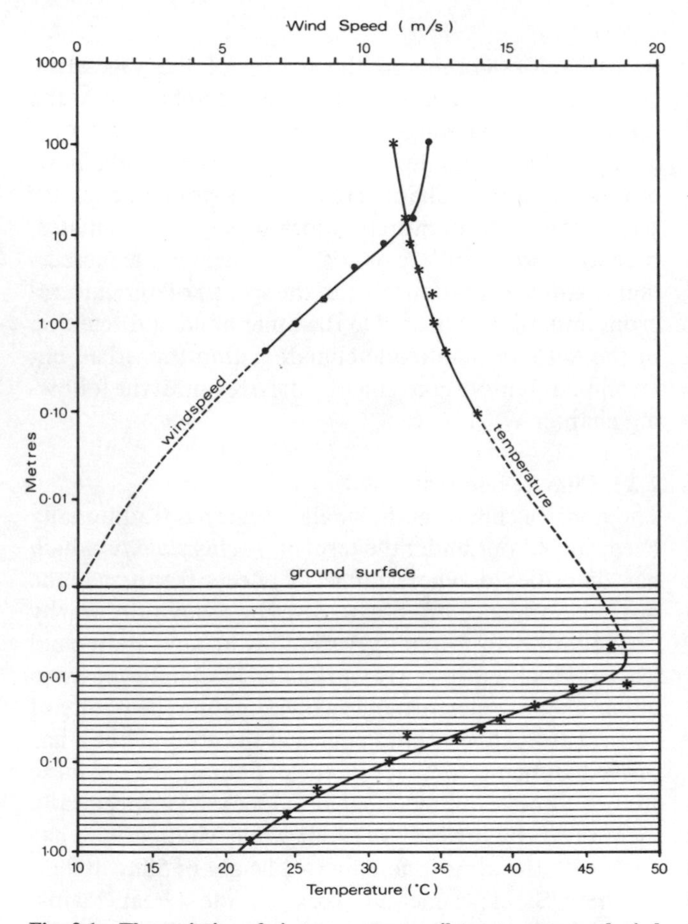


**Fig. 2.1 The variation of air temperature, soil temperature, and wind speed with height at a level site in prairie country near O'Neill, Nebraska. Measurements taken at 1435 CST on 31 August 1953, during anticyclonic weather. Compiled from data tabulated in Lettau and Davidson (1957).**

Much earlier observations by Johnson (1929) on the diurnal variation of air temperature with height indicated that, over downland in southern England with clear conditions in June, the range at 1·2 m was some 3 degC greater than at 7 m. The same effect, much reduced, was still visible with cloudy skies in December; similarly, early work by Heywood (1931) showed the tendency for the diurnal range of wind speed to increase towards ground level. As emphasized by Deacon (1969), these features owe their existence to the transfer mechanisms of heat, moisture, and momentum at the earth–air interface, and it is these same physical processes which play such an important role in applied climatology.

A new impetus in climatology was advocated by Thornthwaite (1953), who stressed the horizontal rather than the vertical gradient in the atmosphere and proposed the term *topoclimatology*. This suggested a more geographical approach with an appreciation of the influence of relief, aspect, vegetation, and land-use contrasts, plus the stated necessity for fieldwork and mapping techniques as an aid to understanding the resulting local climatic patterns. Again, it has been known for many years that horizontal gradients exist, even over apparently uniform surfaces. For example, Best (1931) established during the summer of 1929 that, over a distance of 15 m within a grass-covered meteorological enclosure, the air was not thermally homogenous at the standard height of 1·2 m and that temperature differences up to about 1 degC persisted for up to 30 min, and occasionally for longer. The work of Rider *et al.* (1963) is an example of a small-scale study of the horizontal transfer of heat and moisture.

Whatever the merits of particular terminology, all workers would agree that the unity of small-scale climates stems from the essential fact that they lie within the restricted zone of primary earth–air interchange where localized variations in surface parameters such as altitude, albedo, moisture content, or surface roughness modify the regional fluxes to produce distinctive local climatic differences. These differences may be created either *in situ* or they may arise from the modified properties of airmasses advected over the place concerned. In the first case, the influence of the earth's surface can be regarded as *active* as, for example, when a mountain range generates a largely internal circulation of thermally driven winds, as opposed to the *passive* effects of such topography in introducing mechanical disturbances in an airstream flowing across the mountains.

In the present state of knowledge, it is difficult to place precise areal limitations on such characteristics. There is probably a general feeling that *microclimate* should now be restricted to smaller-scale events than *topoclimate*, and Barry (1970) has proposed the general categories listed in Table 2.1. Equally, Atkinson (1970) has put the case for *mesoclimatology*, a term which did not come into the literature until the early 1950s. According to Atkinson, most authorities would regard 15–150 km as an appropriate spatial scale for mesoclimatology, although the concept has been used to cover a range from 1·5 to 800 km.

| System | Approximate characteristic dimensions: Horizontal scale (km) | Vertical scale (km) | Time scale |
|---|---|---|---|
| Global wind belts | $2 \times 10^3$ | 3 to 10 | 1 to 6 months |
| Regional macroclimate | $5 \times 10^2$ to $10^3$ | 1 to 10 | 1 to 6 months |
| Local (topo) climate | 1 to 10 | $10^{-2}$ to $10^{-1}$ | 1 to 24 h |
| Microclimate | $10^{-1}$ | $10^{-2}$ | 24 h |

**Table 2.1** Spatial systems of climate. (After Barry, 1970.)

It will be apparent that both discontinuity and overlap are present in these areal definitions, which tend to represent the extreme upper limits of activity and, to avoid any possible confusion, the term topoclimate will be used in this book to include both mesoclimate and microclimate. The time-scale restriction of up to 24 h is important because most topoclimates show their best development under calm conditions accompanied by either a strong positive or a strong negative radiation balance. In other words, a clear-sky anticyclone is the most favourable synoptic type and, although such situations may well last for several days, the small-scale climatic response is frequently best observed on a diurnal basis. These irregular, local events represent real challenges and opportunities in the atmospheric environment which must be recognized and understood before full use can be made of our climatic resources. Most topoclimates depend on the natural and largely unchanging physical characteristics of an area but, with the increasing modification of the rural environment by man, particularly for agricultural purposes, it is necessary to include some reference to these effects.

## 2.3 Mountain climates

The systematic deterioration of climate which takes place with increasing altitude is a well-known characteristic of all upland areas (Manley, 1945a), but within the average lapse-rate relationship lie more detailed responses of the atmosphere to topography. Indeed, in its broadest interpretation, relief permits the development of the strongest topoclimatic contrasts. One of the most obvious effects of a mountain barrier is the added turbulence which it sets up in a passing air flow which, in turn, may alter local precipitation and temperature patterns. Although mean windspeeds increase with altitude on windward slopes, data obtained by Lawrence (1960) from 11 sites between 230 m and 350 m above sea level in the Lancashire Pennines showed a complex pattern. At 2 m above ground, three distinct vertical zones with different height/velocity gradients were recognized and the sheltering effect of minor contour irregularities was found to be significant even at the highest levels.

In high mountain areas, it is not unusual for the strongest surface winds to occur just to leeward of the hill crest, and this feature is especially evident when, instead of simply increasing turbulence, topography introduces a more organized oscillatory disturbance into the air flow. Where long ridges lie at right-angles to the path of air, there is a possibility of lee waves forming downwind of the obstacle with a favourable combination of meteorological and topographic conditions. Meteorologically such standing waves are most likely to occur in a stable airstream sandwiched between layers of lesser stability, thus allowing some return back to the surface of air displaced upwards, as shown in Fig. 2.2 (Wallington, 1960).

Although wind direction should be consistent with

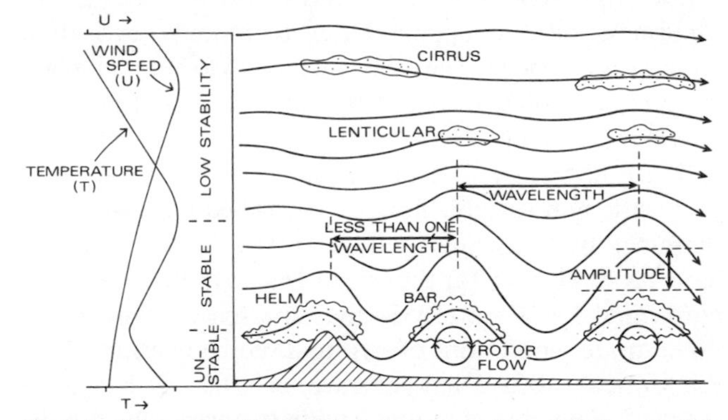


**Fig. 2.2 Lee waves and associated cloud forms downwind of a mountain barrier. After Wallington (1960).**

height, there should also be a marked increase of velocity with altitude. From a topographic viewpoint, the greatest number of waves is produced by the longest ridges, as shown by Scorer (1951), and a large expanse of level ground to leeward helps to maintain a simple oscillatory system compared to the more complex, out-of-phase flows which sometimes occur over dissected terrain. Since the vertical displacement of the air is proportional to the height of the obstacle, the amplitude of the waves depends on the shape of the barrier (steep, high ridges producing the largest waves). On the other hand, the visual evidence for lee waves usually comes from cloud forms and Ludlam (1952) has indicated the possibility of orographic cirrus clouds forming in central England downwind of a ridge rising to only 224 m above sea level and some 122 m above the surrounding country. Once established, such wave clouds may persist for many hours (George, 1959). Under certain conditions, the disturbed airflow may extend up to the stratosphere, with consequent implications for aviation. Corby (1954) has described examples of standing waves associated with several large mountain ranges, including the spectacular 'Bishop Wave' initiated in winter airstreams over the Californian Sierra Nevada and similar oscillations over the Southern Alps of New Zealand.

Lee waves produce strong surface winds down the lee slope of relief obstacles and in Britain the most detailed investigation of such a *helm* wind was conducted in the Northern Pennines by Manley (1945b). Here, the downslope wind is accelerated by the steep topography of the Crossfell escarpment overlooking the Eden valley, and Manley has reported a NE wind of great strength and steadiness on the upper slopes commonly exceeding 20 m/s above 450 m. The other surface wind characteristic is severe turbulence caused by rotor flow, the position of which is usually shown by bar clouds or ragged cloud formation. Such low-level turbulence is often indicated by erratic and variable flow, as described for east of the Pennines by Gray and Stewart (1965). Similar features have been reported for the Isle of Man, where it has been confirmed by Ward (1953) that the local helm breaks down under conditions of strong daytime heating. Once the critical surface temperature of 13·3 °C has been reached, detached cumulus begins to form, which indicates the existence of convection currents powerful enough to overcome the descending flow in the standing wave system.

Another type of downslope lee wind is the warm dry air flow known as *föhn* from extensive pioneer work in the Alps, although the North American term *chinook* is sometimes used synonymously. In a review paper, Brinkmann (1971) has drawn attention to the problem of defining föhn mechanisms and characteristics. The so-called thermodynamic explanation advanced over a century ago—which traditionally requires a loss of moisture by precipitation on the windward slopes with air cooling at the saturated adiabatic lapse rate above the condensation level, followed by the descent of air warming at the dry adiabatic rate to leeward with a consequent increase in temperature plus a reduction in both relative and absolute humidity—is not completely tenable. Föhn can occur without windward precipitation and, although it is widely accepted that subsidence of air from higher levels is responsible for the warmth and dryness of the winds, the exact synoptic processes which cause the warm air to descend and replace the colder, denser air at ground level are not yet fully understood, but could be linked to a lee-wave effect.

Some of the problems of interpretation are no doubt due to the existence of several types of föhn (Riehl, 1971). Thus Beran (1967) has amplified the earlier work of Glenn (1961) on the North American chinook, which extends along a zone between 320 and 480 km wide from Alberta to New Mexico, and recognized the five types of chinook situation shown in Fig. 2.3. Type A represents the original thermodynamic concept involving precipitation, while type B occurs when air is brought from higher levels and warmed by compression at the dry rather than the saturated adiabatic lapse rate. Type C is associated with a shallow mass of cold polar or arctic air lying against the eastern slope of the Rockies with perturbations forming on the interface between the cold air and warmer air aloft. Type D is a case of warm air advection preventing nocturnal radiation cooling and Type E depicts the situation when subsidence occurs as a result of the flow of a highly stable air layer becoming blocked by mountains.

A study of the chinook in Alberta by Brinkmann and Ashwell (1968) suggested that the amplitude of the chinook wave system depends largely on the strength of the upper westerly winds and that with light winds aloft the

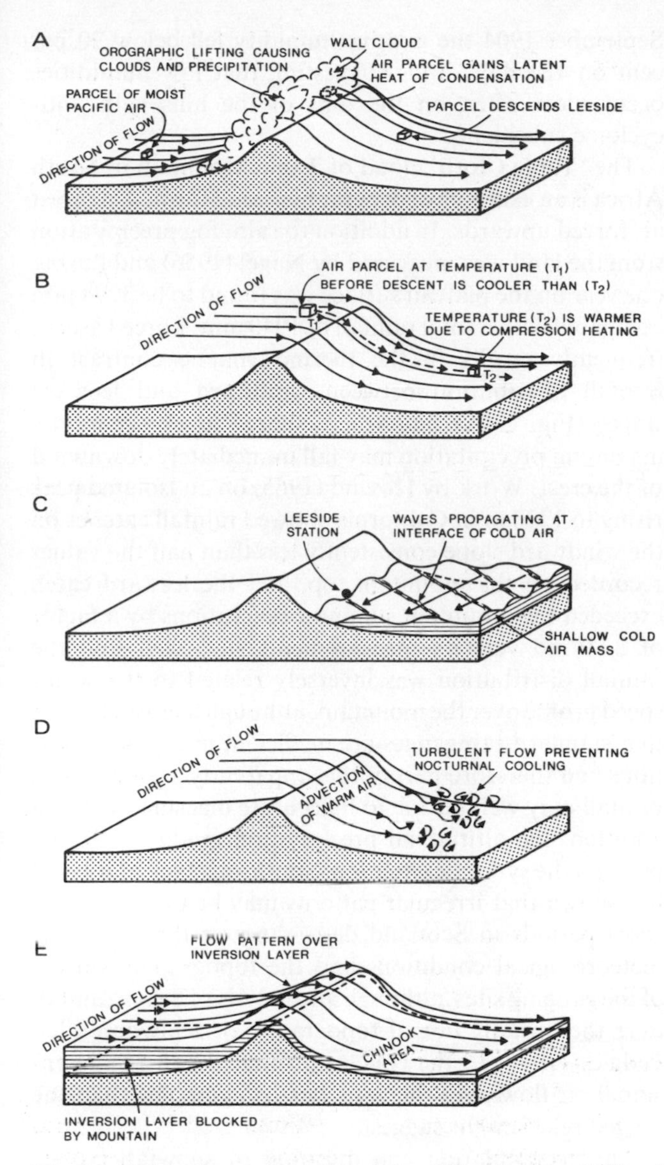


**Fig. 2.3 Schematic representation of five different types of chinook wind. After Beran (1967).**

chinook current may not reach the ground. There was also some evidence that the linear band of cloud known as the 'chinook arch' (Thomas, 1963) had its western edge close to the mountains in the morning but moved eastwards towards Calgary during the day. This led to the conclusion that the preferred area for chinook development between Calgary and the mountains would also be a favoured area for ranching, as it would have a lighter snow cover and so be more suitable for winter cattle grazing.

Most föhn-like winds produce abrupt and rapid changes in temperature and humidity accompanied by faster winds down the mountain slopes. These fluctuations are especially severe east of the Rocky Mountains, where violent chinook winds commonly exceed hurricane force, and Turner (1966) has reported a rise in temperature of 25·5 degC in one hour in Canada. In Britain, föhn conditions are neither as frequent nor as spectacular, although Lockwood (1962) has claimed that the detection of examples may be hampered by the small number of observing stations in the preferred areas north of the Cairngorms and in the lee of Snowdonia, the Lake District, and the Pennines. Nevertheless, only six föhn examples were found between 1944 and 1958. All occurred in stable airstreams associated with anticyclonic conditions and produced negligible rainfall to windward, while the available evidence suggested some subsidence due to blocking of air by the mountain range (as shown in Fig. 2.3E), probably aided by descending air caused by large-amplitude lee waves aloft.

Typical British föhn conditions occurred to the lee of Snowdon on 24 March 1945, under a deep southerly airstream. At 0600 GMT, before the arrival of the föhn, it was calm with an air temperature of 11·7 °C and relative humidity 49 per cent. At 0700, however, the windspeed was 11·6 m/s, with air temperature 17·2 °C and relative humidity 23 per cent. In some areas, the highest winter temperatures are due to föhn influences and Lawrence (1953) has quoted a night temperature of at least 15·6 °C in north-east Scotland in mid-February 1945.

The nature of airflow around mountains has an important influence on the local distribution of precipitation and cloud forms, and some of these relationships have been illustrated diagrammatically by Pedgley (1967a). Figure 2.4A shows the typical situation of a moist airstream with extensive layered cloud rising over western Britain and indicates how the increase in water content of the upcurrents on the windward slope gives rise to heavier rainfall around the summits, which is subsequently replaced by a further decrease in intensity with a raising of the cloud base on the leeward side. Figures 2.4B and C indicate the effect of sharp breaks in topography in producing leeward eddies which may be either cloud-free or cloudy respectively, and Fig. 2.4D shows something of the complex eddying which may take place around isolated summits often giving banner-type clouds.

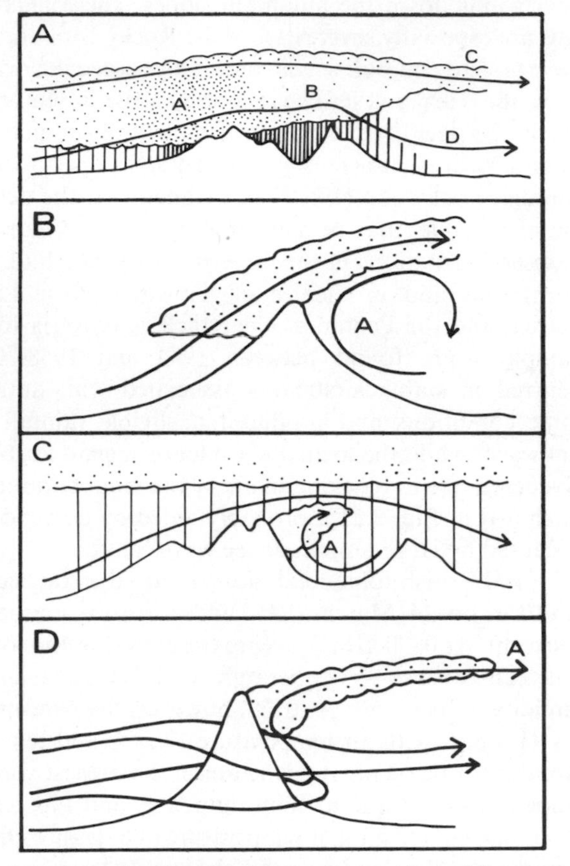


**Fig. 2.4 Some characteristic airflow patterns and cloud forms around mountains. After Pedgley (1967a).**

Not surprisingly there is a marked increase in cloud duration with altitude and Glasspoole (1953) has reported that the summit of Great Dun Fell at 848 m in Cumberland is in cloud throughout the year with an average frequency varying from 37 per cent at 1500 h to 62 per cent at 2400 h. On the other hand, Green (1967) has claimed that, though the summit of Ben Nevis at 1343 m may have a relative humidity of almost 100 per cent associated with a complete cloud cover for about 70 per cent of the time, between October 1884 and September 1904 the relative humidity fell below 20 per cent on 181 days, thus suggesting that low humidities occur most often on the tops of the hills with anti-cyclonic subsidence.

The 'Table Cloth' cloud of Table Mountain in South Africa is an example of orographic cloud formed in moist air forced upwards. In addition to rain, fog precipitation from the land was measured by Nagel (1956) and during one year on the plateau surface was found to be 3294 mm compared with a total rainfall of 1940 mm. Forced ascent frequently results in the thermodynamic contrast in rainfall distribution between windward and leeward slopes (Fig. 2.4A) but, where sharp ridges occur, the maximum precipitation may fall immediately downwind of the crest. Work by Hovind (1965) on an isolated peak rising to 1220 m in California showed rainfall catches on the windward slope consistently less than half the values recorded on the mountain top, and the leeward catch exceeded the mountain summit observations by a factor of 2. Wind velocity measurements suggested that the rainfall distribution was inversely related to the wind-speed profile over the mountain, although it is well known that standard raingauges are inefficient in exposed locations and therefore part of this apparently asymmetrical rainfall may well be due to inaccurate measurements. In addition, the altitudinal precipitation gradient will depend on the synoptic situation. Smithson (1969 and 1970) has shown that irregular patterns may be expected over short periods in Scotland depending on the prevailing meteorological conditions and the topographic variety of the gauging sites, although an analysis of heavy rainfall over the broadly domal topography of Snowdonia led Pedgley (1970) to the conclusion that the three-dimensional air flow over the area was less complex than the ragged relief might suggest.

The accumulation and duration of snowfall is very dependent on altitude, although Manley (1939) was able to show that observed occurrences in Britain departed as much as 10 per cent from expected values owing to local topographic factors. More recently, Manley (1971), has stressed the problem of making reliable snow-cover observations: the smooth, south-facing slope of Skiddaw in the English Lakes may be bare only two to three days after a snowfall, whereas deep drifts may still linger on the shaded north-facing hillsides around the head of

Borrowdale. Similarly, the sunny south-west slopes of the northern Pennine escarpment have a frequency of snow cover about 10 per cent less than that at Moor House which is at the same altitude, but about 8 km distant on the leeward side of the watershed. Such differences are partly dependent on aspect but also reflect the higher initial accumulations of snow which occur just to the lee of the mountain crest. Green (1968) has demonstrated how the location of the most persistent snowbeds shows the combined influence of both aspect and microtopography, and Pedgley (1967b) has emphasized how the shape of individual snowdrifts reveals the effect of highly local patterns of air currents, which is an important factor in view of the significance of snow redistribution over the exposed, treeless uplands of Britain.

## 2.4 Valley climates

Valley climates owe their distinctiveness primarily to two separate but often related factors. These are, first, the degree of slope together with the aspect of the valley sides and, second, the differences in relative relief which inevitably arise in such areas. It will be convenient to treat each of these principal factors separately but it should also be noted that the climatic characteristics of a particular valley will also depend on the orientation to the prevailing wind and the specific geometry of the relief features involved.

*Slope factors*. The energy balance of sloping ground is determined both by the angle of the slope and its orientation, and these relationships have been examined in detail by Geiger (1969). In general, the combined effects are greatest where there is the largest angle of incidence of solar radiation to the surface, and in steeply dissected areas the north-facing slope may be largely in shadow while the south-facing valley side is experiencing almost perpendicular rays. Data for Arosa at 1890 m above sea level in the Swiss Alps quoted by Kendrew (1938) show that in mid-winter a steep south-facing slope may receive 238 times as much energy as a comparably inclined slope to the north, and, even with the higher altitude of the midday sun in summer, the south-facing slope still enjoys an appreciable advantage. Such radiation inequality leads inevitably to different heat fluxes between slopes, as shown by data gathered in western Oregon, US, by Lowry (1959) and shown in Table 2.3.

| | *S75°* | *S15°* | *0°* | *N15°* | *N75°* |
|---|---|---|---|---|---|
| 15 June | 154 | 314 | 306 | 280 | 38 |
| 15 December | 238 | 130 | 73 | 13 | 0 |

**Table 2.2** Mean heat (g/cal cm$^2$) received in 24 h on south-facing (S) and north-facing (N) slopes at Arosa. (After Kendrew, 1938.)

| | *5°S slope* | *10°N slope* |
|---|---|---|
| Altitude | 375 m | 418 m |
| Whole-spectrum radiation surplus | +77 | +49 |
| Evapotranspiration | −5 | −5 |
| Sensible-heat flux | −47 | −27 |
| Heat flow into ground | −25 | −16 |

**Table 2.3** Heat fluxes (langleys/h) on south- and north-facing slopes averaged over 1100–1400 h on a clear day in August. (After Lowry, 1959.)

It can be seen that around midday on a summer day the south-facing slope receives about 50 per cent more radiation than the north-facing hillside, and that the heat taken up by the air and soil is increased.

In an attempt to relate variations in the radiant energy flux to differences in both the heat and moisture balances, Rouse and Wilson (1969) undertook a two-year study of NNW- and SSE-facing forested hillslopes in Quebec. The greatest topographically determined differences in both total and direct-beam radiation were observed between the steepest portions of the two slopes, and differential snowmelt led to a complete removal of snow some three weeks earlier on the S-facing slope with the result that the north slope soils had stored 50 per cent more water than their southern counterparts by early spring. A specific investigation of evapotranspiration rates in sloping terrain by Rouse (1970) suggested that, when soil moisture is non-limiting, the rate of water loss is simply a direct function of the net radiation level, although work at a similar mid-latitude site in New Zealand by Jackson (1967) indicated that slope and aspect are influential, especially at the equinox. So far, there have been few published studies of the heat balance of a valley and the work of Greenland (1973) in the southern Alps of New Zealand is unusual in this respect.

Although a good deal is now known about the radiation climatology of different slopes, experimental observations to determine resultant contrasts in ground and air temperatures have been less successful. This is partly due to the elimination of small thermal differences by wind turbulence on exposed hillslopes but may also be attributed to the complicating effects of other factors. For example, although the total amount of daily solar radiation is received equally around true noon, the temperature does not follow the same symmetrical regime because of the thermal inertia of the earth–air system and the changes in the utilization of energy which take place during the day. In temperate latitudes, incoming radiation in the morning may be largely used in evaporation and drying the soil compared with the larger proportion available in the afternoon for heating the soil and the air. Because of this, the warmest slopes in the northern hemisphere are often those orientated towards south-west rather than due south. On the other hand, the recurrent development of convective cloud during the afternoon, as reported for the Innsbruck area by Geiger (1969), may lead to the shielding of SW slopes from direct radiation, thus producing higher values to the SE orientations.

Soil temperature differences between sunny and shady sites can be quite marked, especially during spells of settled weather. Work by Soons and Rainer (1968) in New Zealand showed that in early September the temperature at 4 mm depth on a slope facing NE reached a maximum of only about 13 °C around true noon compared with 18 °C attained on the slope inclined to the NW some 2½ h after the peak of incoming radiation. Observations by Howe (1955) on a south-facing coastal slope in Wales throughout a fine week in April indicated considerable contrasts in the thermal range of the soil between sunny and shaded sites at the 50 mm depth, and on two consecutive June days the exposed site experienced maxima 9·1 and 10·0 degC respectively higher than the shaded position.

*Mean daily temperature ranges (°C) during an April week in 1954 at Aberystwyth*

| *Soil (sunny)* | *Soil (shady)* | *Air* |
|---|---|---|
| 6·7–17·8 | 4·4–7·8 | 5·8–12·8 |

As might be anticipated, air temperature contrasts due to aspect are much smaller than observed earth temperature, but measurements at 1·5 m in the Italian Appennines reported by Smith (1967a) for an April day were about 3·3 degC higher on the south-facing slope than at equivalent altitudes orientated towards the north.

Such differences in the radiation and heat budget of slopes are often enough to generate a system of thermally driven valley winds. The intensity and frequency of these local circulations depend on a complex of factors, as shown in the comprehensive reviews by Geiger (1965) and Flohn (1969), but the clearest examples usually occur in steep topography with very light gradient winds. In all cases, the valley winds have a definite diurnal regime with upslope *anabatic* winds during the daylight hours being replaced by downslope *katabatic* flows at night. For example, Mendonca (1969) has described the diurnal reversal of air flow on the slopes of a volcano, where the local wind was controlled by the radiation balance of the lava surface, the strength of the free air flow, and the level of the trade wind inversion. In this case, balloon measurements revealed that the upslope wind was some 600 m deep compared with only 55 m for the downslope flow and it was deduced that six times more air was transported by the upslope wind than by the nocturnal downslope flow.

In some areas, such as the Rhone valley, the upvalley wind can deform trees and Yoshino (1964) has used such evidence to illustrate changes in the mean velocity of the wind. Such winds are strongly influenced by the topographic detail of the valley bottom and, in the same area, Barsch (1965) has used similar evidence to investigate the funnelling of the mistral wind. Upslope winds owe their origin to the development of thermals over upland areas early in the day, as shown by MacCready (1955), and the diurnal reversal of valley circulations may take place within one hour after local sunrise or sunset, according to observations by Urfer-Hennenberger (1967) for a valley in the Swiss Alps.

Some of the early attempts to model valley wind systems, such as that by Defant (1951) illustrated in Fig. 2.5, have shown a symmetrical flow pattern which is largely independent of slope aspect, although it is known that local air flows are sensitive to changing thermal conditions. Thus, Gleeson (1951) put forward a theory of cross-valley winds arising from the differential heating of slopes and this was confirmed by experiments in Canada by MacHattie (1968), who found that winds blow to the more intensely insolated east-facing slope during the morning and move to the west-facing slope by the late afternoon.

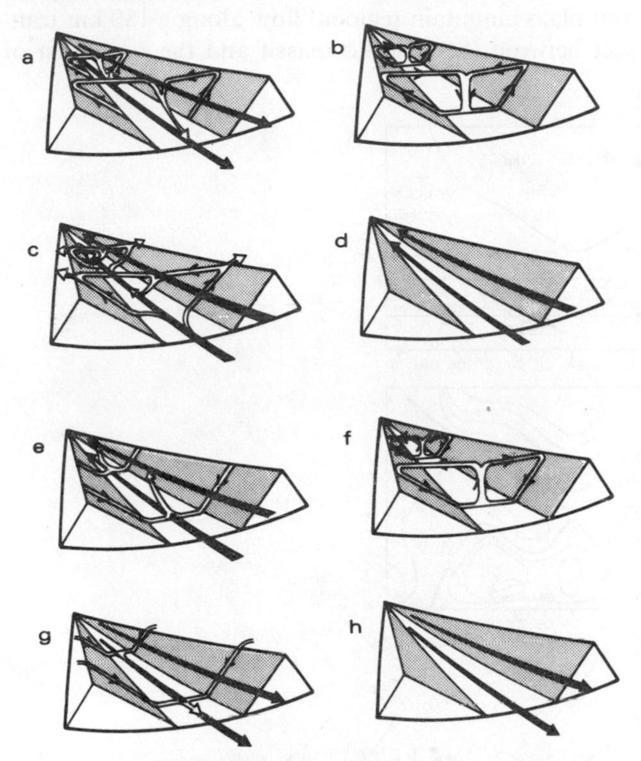


**Fig. 2.5 A schematic representation of the diurnal variations of slope and valley winds.**
**(a) Sunrise—valley cold, plains warm.**
**(b) Mid-morning—valley and plains have similar temperatures.**
**(c) Noon and early afternoon—valley warmer than plains.**
**(d) Late afternoon—valley still warmer than plains.**
**(e) Evening—valley only slightly warmer than plains.**
**(f) Early night—valley and plains have similar temperatures.**
**(g) Middle of night—valley colder than plains.**
**(h) Pre-dawn—valley much colder than plains.**
**After Defant (1951).**

*Elevation factors*. At night, valley climates depend much less on slope aspect, since all slopes experience the same amount of back radiation to the atmosphere, and it is the influence of height differences which becomes significant. Even the inclination of the slope is unimportant, although there is some evidence to suggest that angles of less than 2° are incapable of supporting the katabatic drainage of cold air which largely determines the distribution of minimum temperatures on the lower slopes. Under calm nocturnal conditions, a temperature inversion develops near the ground and, since the air cooled by contact with the upper slopes is environmentally cooler and denser than the air at lower levels, it begins to flow downslope under the influence of gravity. Over open ground on valley interfluves the initial flow is weak and shallow, but as the circulation develops, and especially where it is channelled by topography from large and very cold surface catchment areas, the winds may achieve much greater strength and depth. Thus, katabatic winds are best developed along the edges of the continental ice-caps, and Loewe (1950) has reported a virtually constant circulation which flows off the Antarctic ice-sheet with a mean annual velocity of 19 m/s and a depth of 300 m.

Katabatic winds have been reported from many areas including the equatorial Andes (Lopez and Howell, 1967) and the Gulf of Carpentaria in northern Australia, where a shallow layer of cold air accumulates at the foot of a highland zone to create a hydraulic jump in the lower atmosphere (Clarke, 1972), but one of the most detailed investigations was that by Tyson (1968) into the nocturnal flow within the Bushman's valley, South Africa. This valley has a range of local relief of 1585 m between the valley floor and the top of the Drakensberg escarpment and Fig. 2.6A shows a section through the nocturnal *mountain wind* with the anabatic *valley wind* above and the *gradient wind* aloft. It can be seen that, as in other areas, the katabatic flow had a characteristic onset phase with rapid initiation and deepening. This can cause significant temperature fluctuations and Thompson (1967) reported a temperature drop of 15 degC/h associated with the onset of a nocturnal wind at the mouth of a small canyon near Salt Lake City. Figure 2.6A shows that a similar surging effect continued throughout the night with a pre-reversal invigoration occurring immediately before the breakdown of the wind. This type of surging is not well understood but may be due to the intermittent release of cold, dense air temporarily ponded up behind obstacles further up the valley.

Figure 2.6B illustrates the dissipation of the mountain wind which, from being at its deepest over the centre of the valley at 0700 h, was at its shallowest at this point by 0830, and by 0900 h the down-valley flow had weakened sufficiently to allow a cross-valley circulation to develop. This cellular circulation was strongest and deepest over the western slope, which was 5 degC hotter than the eastern slope. It can be seen from Fig. 2.6C that the down-slope drainage of cool air off the eastern side of the valley created a shallow inversion which is evident in the 0900 lapse rate but by 0930 the lapse rates near the ground were unstable and the anabatic valley wind filled the lower part of the valley.

More recently, Tyson and Preston-Whyte (1972) have noted that a larger-scale, diurnally reversing wind occurs above the valley level in the form of a mountain-plain and plain-mountain regional flow along a 180 km transect between the Lesotho massif and the east coast of

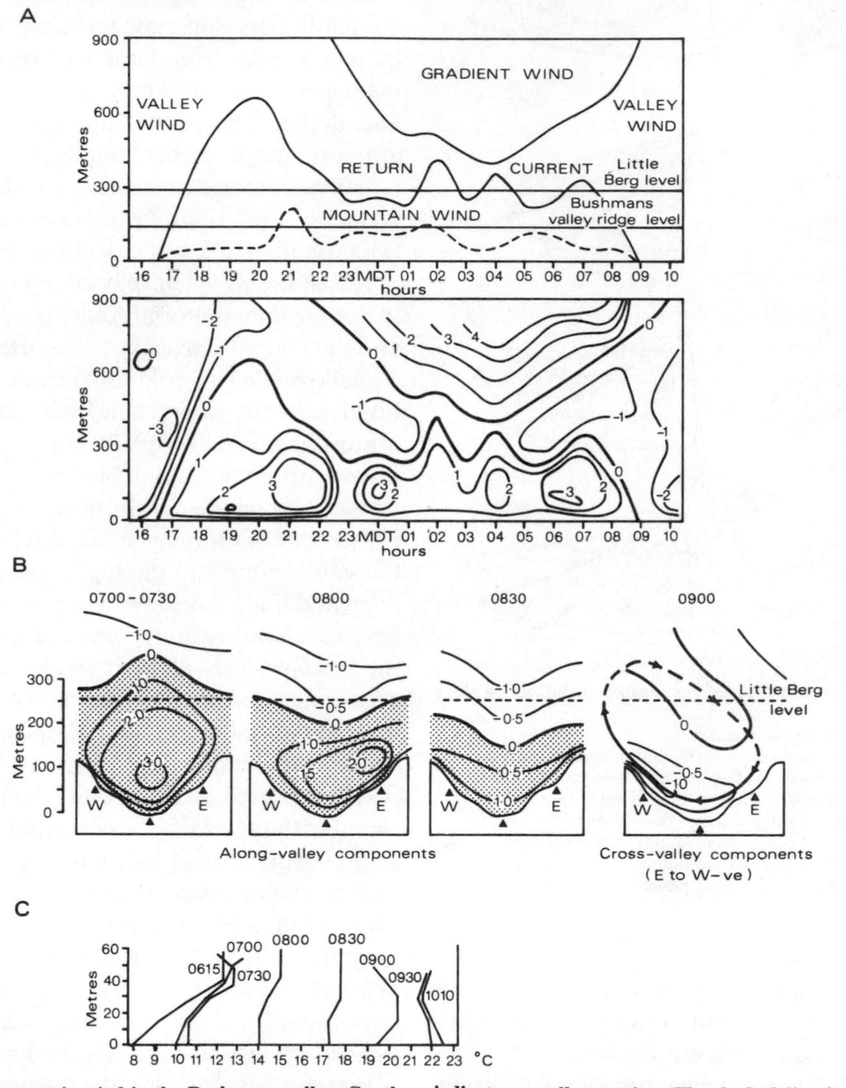


Fig. 2.6 Features of the mountain wind in the Bushmans valley, South Africa, on 1–2 May 1965.
(A) Time sections of the mountain wind showing wind speeds in m/s as components of motion parallel to the axis of the valley. Positive components indicate down-valley motion; negative components indicate up-valley motion. The dashed line in the upper section shows the height of maximum flow.
(B) Time sections to show the decay of the mountain wind on 2 May 1965.
(C) Lapse rate conditions associated with B.
After Tyson (1968).

South Africa. Other investigators have attempted to relate the volume and velocity of the cold air drainage to the net radiation balance of the slope, and Bergen (1969), working on a forested slope 3000 m above sea level in Colorado, found that some prediction of the downslope variation in wind speed was possible. In Britain, katabatic flows are probably confined to some 5 m above ground level and classic experiments in a Cotswold valley by Heywood (1933) revealed that, over a 21-month period, the velocity of the wind rarely exceeded 1 m/s and that the gravitational energy of the circulation was so low as to prevent a continuing wind down the main valley. Stronger winds have been recorded under more favourable conditions and McGinnigle (1963) observed speeds of 2·5–5 m/s for katabatic flows from the snow-covered western slopes of the northern Pennines associated with a winter anticyclone. A comprehensive account of nocturnal winds in Britain was presented by Lawrence (1954).

Relative elevation exerts a primary control on valley temperatures. This is especially true of the nocturnal thermal properties of the lower slopes and valley bottoms which, because of the ecological and economic significance of low minima in agricultural areas, have attracted so much attention in the literature, but it also applies to the daytime values. In detail, individual valleys reveal highly specific characteristics, but most of the larger valleys show a recognizable threefold altitudinal division. The lowest zone near to the valley floor is representative of the so-called 'valley climate', where the outstanding temperature extremes are recorded. With gentle slopes and fairly open valleys, it appears that unusually low night minima may be mainly attributed to the katabatic drainage of cold air into the lowest 'frost hollows' but, in areas of more dissected relief, the temperatures also depend on the partial stagnation of air in the enclosed valley bottoms together with the strongly negative nocturnal radiation balance due to the shielding of insolation by the horizon.

On the slopes above the 'valley climate' lies the 'thermal zone', which experiences a much reduced diurnal range of temperature with night-time minima above those in the lower valley. Dunbar (1966) has claimed that the concept of the thermal belt was discovered in the fruit-growing area of North Carolina in 1858, and it is known that this slope zone corresponds to the layer of air at several tens of metres above level ground, which is cool by day because of its distance from the surface and is subject to a temperature inversion at night. This warm layer migrates up the valley slopes as the valley progressively fills with cold air below and the development of this feature has been traced in parts of the western US by Young (1921) and, more recently, in South Africa by Keen (1968). The final zone, which may be distinguished in deep valleys, represents that part of the atmosphere lying above the level of nocturnal temperature inversions, where there is a regular positive lapse rate of temperature with height.

Several workers, such as Davies (1952), regard diurnal range rather than the minimum temperature as the best index of a valley climate. This is an acceptable criterion as long as it is appreciated that it is the lowering of minima, rather than the raising of maxima, which is chiefly responsible for the large daily ranges observed. The general progressive increase in range from summit to valley bottom has been well illustrated for a Californian valley by Waco (1968), where the lowest site recorded a diurnal variation over 33 degC on occasion compared with sites further up the valley as shown in Fig. 2.7. Much earlier observations in Britain by Hawke (1933) indicated that, for a suitable topographic location, the daily range in south-east England might exceed 28 degC. More recent investigations in Scotland by Dight (1967) have shown not only that this range is reached in more northerly latitudes in June, but also that diurnal fluctuations in excess of 19·5 degC were achieved in all months except September and October. Thus, in sheltered glens, at least, large daily ranges are not purely a function of the summer months and Dight has stressed the significance of slightly higher maxima in this context. On the other hand, data collected by MacHattie (1970) for a valley on the eastern margin of the Canadian Rockies showed an inversion of 1·7 degC in the daily maximum temperature in the first 90 m above the valley floor. This relative depression of daytime valley temperatures was attributed to increased evapotranspiration in the moist valley bottom compared to the drier slopes above, with a consequently higher fraction of the incoming energy being used up as latent rather than sensible heat.

The lowest minimum temperatures occur in frost pockets which are often a cumulative assemblage of three important physical factors—relief, sub-soil characteristics, and the prevailing meteorological conditions. Broadly speaking, the most severe effects are found in sharply-incised depressions at a free-draining site, which precludes the retention of surface moisture, under the synoptic influence of a continental anticyclone. One of the most famous frost pockets in the world, which combines these factors, is the Gstettner Alm sink-hole, which

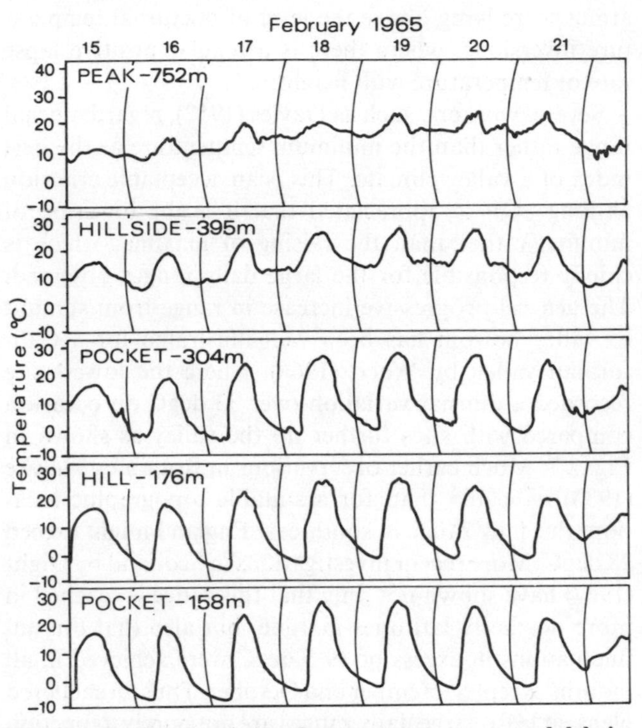


**Fig. 2.7 The diurnal variation of screen temperatures recorded during 15–22 February 1965, at five stations in Cold Canyon, Santa Monica Mountains, California. After Waco (1968).**

is a funnel-shaped doline formed in limestone at 1270 m above sea level near Lunz in Austria. Although this depression is only about 150 m deep, on a single night the temperature has decreased by 27 degC from top to bottom, and the temperature at the lowest point has fallen below − 50 °C on several occasions.

Nowhere in Britain is capable of producing such extreme values and it is normally accepted that a dry valley in the Chilterns near Rickmansworth is the most continental lowland site in the country. In a study lasting for 156 calendar months from 1938 to 1942, Hawke (1944) recorded only two months without ground frost and severe frosts occurred even in the warmest months of July and August. These features were due mainly to the combined effects of a fairly large topographic catchment source for katabatic drainage, a porous layer of sandy gravel overlying chalk bedrock and the general inland situation well removed from the possibility of coastal amelioration. Other areas show the same basic controls working even in relative isolation. For example, Lawrence (1956) has shown that minor contour inflections are enough to cause some canalization of katabatic flows in Herefordshire. Similarly, Oliver (1966) used the Santon Downham record to demonstrate the low minima which may be expected over sandy hollows, while Richardson (1956) has emphasized thc importance of short-term airmass relationships when assessing valley minima.

The distribution of low minimum temperatures near the valley floor may well be quite localized. King (1952) revealed that probable variations in wind turbulence were responsible for substantial thermal discrepancies within the Rickmansworth frost hollow and some sites have failed to give a true picture of the local conditions. For example, Catchpole (1963) documented the Houghall frost hollow in north-east England on the basis of records obtained between 1925 and 1945 for a site on the lower valley flank at 49 m above sea level. Later, the meteorological station was re-sited only some 11 m lower down on the river floodplain, but Smith (1967b) was able to show that the new site produced appreciably lower minima.

The thermal gradients associated with valley bottom inversions in southern England were investigated by instrumented traverses by Lawrence (1958) and Harrison (1967 and 1971). The general conclusions which emerged were that, on radiation nights, the total temperature increase with height was about 1·5 degC per 30 m but over the lower valley slopes the average was rather more than 3 degC per 30 m. Over short height ranges, the increase exceeded 5 degC per 30 m and it was claimed that height differences as small as 3 m could lead to significant variations in the liability to frost.

Similar vehicle traverses were undertaken across a broad valley with a height range of some 275 m in Pennsylvania by Hocevar and Martsolf (1971). Figure 2.8 illustrates that a consistently intimate relationship was found between altitude and air temperature just before dawn during radiation frost conditions on three spring mornings. In fact, most of the temperature variation could be explained in terms of relative elevation above the valley bottom and there was an overall mean temperature change of 6·2 degC per 100m. It was suggested, therefore, that this relationship could be used as a means of approximating the relative frost danger from contour maps of similar valleys.

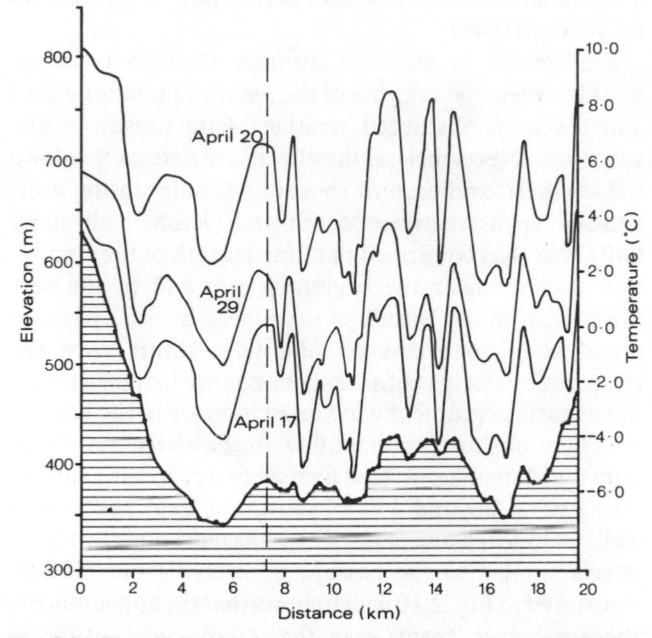


**Fig. 2.8 The relationship between minimum temperature and relief as indicated by three traverses across the Nittany valley, Pennsylvania, in the spring of 1968. The broken vertical line indicates the location of just one of the 'thermal belts' identified. After Hocevar and Martsolf (1971).**

## 2.5 Coastal climates

The juxtaposition of large land and water surfaces inevitably gives rise to modifications of regional climate, and the concept of oceanicity and continentality is recurrent in considerations of climatic behaviour at scales rather greater than those discussed in this book. Whatever the implications may be at the continental level, it must be stressed that the mechanisms are basically similar to any other edge effect. Thus, not only is the major impact of oceanicity usually confined to a narrow coastal strip, but the detailed pattern of climate near a coast or lake shore is also controlled by local topography in much the same way as further inland.

This theme may perhaps be illustrated best by reference to the well-known, but often exaggerated, amelioration of minimum temperatures which is expected along coastlines. Although this can be a highly significant factor, it cannot be guaranteed for all areas, especially where rising ground inland facilitates the drainage of cold air at night. There is ample, longstanding evidence of such local variations in Britain: Spence (1936) compared temperatures on an offshore island with those at an identical height in a small valley 6 km inland and found that in the valley the average maximum temperature was some 2·2 degC higher in summer with mean minima 1·7 degC lower in autumn. Similarly, Manley (1944) drew attention to the Bridlington site on the Yorkshire coast, where only 1·5 km from the sea the frost liability is virtually equal to that at an inland station because it is near the foot of a long slope stretching down the flank of the Wolds. Other temperature variations near the British coast have been described by Howe (1953) and Reynolds (1956), while several writers such as Thomas (1956) and Douglas (1960) have shown the influence of coastal topography on other climatic phenomena such as airflow. For example, in the exposed western peninsula of Pembrokeshire, Oliver (1960) was able to demonstrate the localized steering of the prevailing wind by small-scale topographic features.

This in no way denies the fact that genuine coastal climates do exist as a result of the marked variations in heat budgets between land and water. Although differences occur in net radiation between land and sea due to contrasts in albedo and other factors, these are much less important than the differences in the utilization of the incoming energy. Sellers (1965) has shown that, while 90 per cent of the net radiation over the oceans is used up in evaporation with only 10 per cent going to warm the air, these two methods of heat loss are more or less equally important on land. Above the oceans, the sensible heat flux remains small and almost constant during day and night, but over land it varies considerably. Other things being equal, the sensible heat flux from ground to air increases as the surface becomes drier and as the temperature difference between surface and atmosphere becomes larger.

In the sub-tropical deserts, nearly all of the incoming energy is used to heat the air and it is hardly surprising that the low-latitude arid coastlands exhibit some of the best developed features of coastal climates. These energy-balance differences are also important in high latitudes, especially on a seasonal basis when the latent heat flux from the nearby ocean in winter may well be the only positive element in the energy balance. This dependence of the coastal climate on the sea has been well illustrated by Miller (1965) for Copenhagen, where advected heat transported in the marine air contributes a downward flux of sensible heat equivalent to about 50 1g/day from September to February that helps to counter a large radiation deficit. This means that winter temperatures are more equable, and that evapotranspiration and the growing season are extended well into the winter.

Horizontal heat-budget contrasts are often capable of producing a diurnally reversing sea or lake breeze when there is only a light geostrophic wind. Essentially, the sea-breeze is a simple thermally driven wind, and several of the basic physical assumptions have been used by Estoque (1961) to formulate a theoretical model of this coastal airflow. A cross-section through a typical sea breeze is presented in Fig. 2.9, where it can be seen that convection currents over the heated land surface initiate a continuous circulation whereby the ascending air over the land is replaced by a flow of denser, colder air moving off the sea or lake surface onto the land. When the land surface becomes cooler than the water at night, a reverse circulation will be set up.

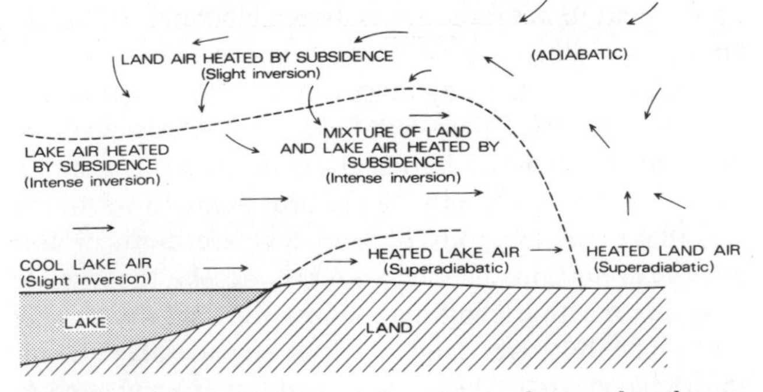


**Fig. 2.9 A schematic representation of a sea breeze when the geostrophic wind is light. After Munn (1966).**

As shown by Flohn (1969), the strength and frequency of the circulation depends largely on the thermal gradient between land and sea. Therefore, because the horizontal surface contrasts are stronger during daylight hours, the sea-breeze blowing inland during the day is generally better developed than the nocturnal land-breeze. On many low-latitude coasts, the sea-breeze has almost clockwork regularity and controls the diurnal pattern of coastal weather. Pedgley (1958) has illustrated the abrupt arrival of the Mediterranean sea-breeze by observations taken at Ismailia some 70 km inland on the Suez Canal and similar observations have been reported for Batavia by Wexler (1946).

Sea-breezes in the mid-latitudes tend to be more weakly developed because of the reduced radiation input and the more disturbed weather characteristic of the westerlies. Nevertheless, the effects are detectable along the southern and eastern shores of Britain during anti-cyclonic spells between March and October, although Gill (1968) has confirmed that the strength of the circulation declines near the beginning and end of the sea-breeze season. In Britain, a sea-breeze is usually recognized when an excess of land temperature over sea temperature occurs before either a change in the direction of the surface gradient wind or an increase in the velocity of a light onshore gradient flow in the afternoon. Findlater (1963) has shown that British sea-breeze circulations may extend beyond 150 m above sea level and spread well out to sea, while the arrival of the marine air produces effects similar to the passage of a cold front. This is illustrated in Fig. 2.10 which shows that the appearance of the sea-breeze front near the south coast causes a characteristic change in wind direction, accompanied by an increase in wind speed, plus a rapid fall in temperature, and a rise in relative humidity.

Simpson (1964 and 1967) concentrated on the inland travel of sea-breeze fronts from the south coast of England and concluded that the average time of passage of the fronts varied from 1100 h on the coast to 2100 h about 65 km to the north, which represented the approximate maximum penetration. The speed of inland progress of most of the fronts showed a gradual increase from 1·5 m/s or less near the coast to about 4·0 m/s in the later stages. The most vigorous sea breezes penetrate more than 60 km inland, and, in a comprehensive analysis over

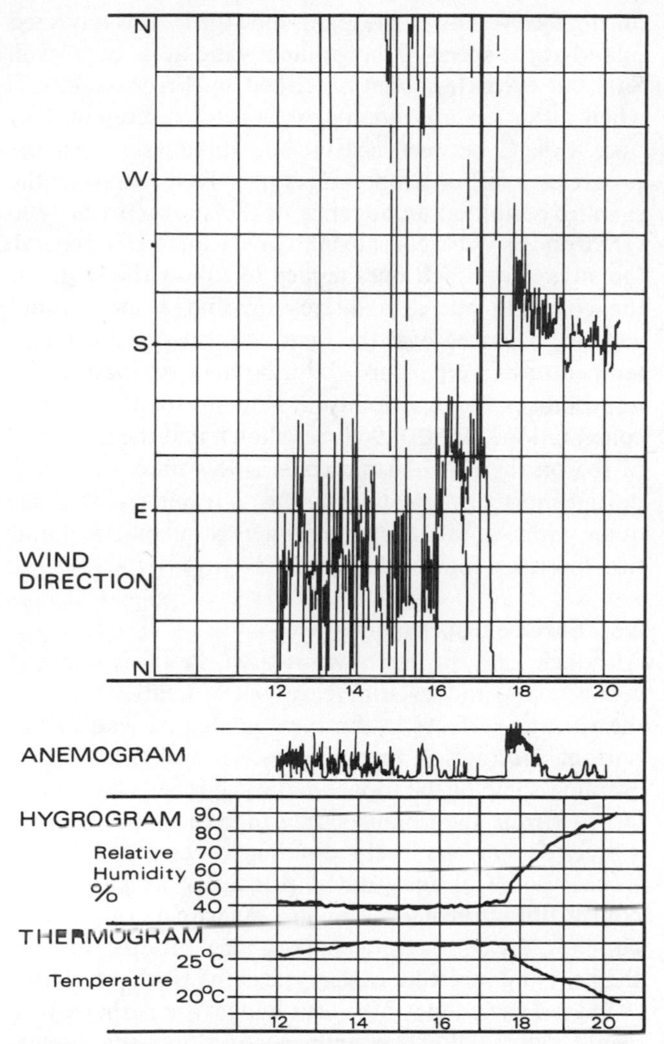


**Fig. 2.10 The arrival of the sea breeze front at Porton, southern England, on 24 July 1959. After Elliot (1964).**

6 years, Elliot (1964) found 47 well-defined occurrences at Porton Down about 40 km distant from the south coast. On almost all occasions the noon wind at 915 m was less than 5 m/s and the screen temperature rose to that of the mean monthly sea temperature around the Isle of Wight by 1000 GMT and continued to climb. However, Watts (1955) has claimed that a 5·5 degC temperature difference between land and sea is enough to initiate a sea-breeze, and has warned about discrepancies of this order which can occur between sea temperatures and inshore temperatures, especially where shallow water lying over tidal mud or sandflats can be warmed or cooled quickly, thus indicating the need for careful measurements.

On the other hand, it would be misleading to consider sea-breezes entirely in isolation. Thus, Findlater (1964) has shown that the sea-breeze may be related to inland convection and Moffitt (1956) has provided evidence of a katabatic offshore flow originating on the South Downs, which is frequently reinforced by the nocturnal land-breeze, whenever the air temperature near the coast falls by 8 or 9 degC from the daytime maximum. Sometimes sea-breezes are surprisingly infrequent, as when Stevenson (1961) made his investigation on the Yorkshire coast in the anticyclonic summer of 1959. On many days the land temperature exceeded the sea temperature by 8 to 16 degC, but during 5 months a definite diurnal windshift, as opposed to an increased velocity in the warmer hours, occurred on only 13 days.

Large lakes in the mid-latitudes may also be associated with strong local circulations, and Moroz (1967) has drawn attention to the breeze of Lake Michigan, which blows inland with a maximum velocity in excess of 7 m/s and carries the lake breeze front more than 16 km over the land. This lake breeze has also been investigated by Lyons (1972) who, in a mesoscale field study extending over 307 days during three summers, found that the breeze occurred on no less than 111 of the observed days and sometimes reduced temperatures on Chicago beaches by over 16 degC compared with values 10 km inland. Given the presence of a suitably weak, synoptic-scale pressure pattern, it was also found that the establishment of the lake breeze could be predicted with an accuracy of 90 per cent, whenever the inland maximum temperature was forecast to exceed the mean surface temperature of the lake and non-convective cloudiness was not expected to reduce the sunshine duration to less than 60 per cent of the clear-sky value.

Although the sea-breeze is most readily recognizable when steady pressure conditions support a light offshore gradient wind, Eddy (1966) has not only shown how the sea-breeze effect can be distinguished from the usual onshore airflow on the Texas coast, but also how the sea-breeze is related to a summer maximum of rainfall which runs parallel to the coast about 40 km inland. Similarly,

Lumb (1970) has demonstrated how the diurnal incidence of thunderstorms around Lake Victoria is dependent on land–lake circulations in the area. In a more detailed study, Harman and Hehr (1972) found that the lake breezes from Lake Superior and lakes Michigan–Huron converged on more than half of all the non-frontal days from late June through to August in 1970 and this convergence apparently gave rise to isolated diurnal rainshowers over eastern Upper Michigan. Rainfall from the cloud systems associated with the convergence accounted for all non-frontal precipitation in the study area and it was claimed that such rainfall contributed over 20 per cent of the total summer rainfall at some stations.

Some coasts experience regular on-shore advection unrelated to sea-breeze circulations, and this can also lead to characteristic variations in local climate. One such example is found on the east coast of Britain where, during the late spring and early summer, the development of high pressure over Scandinavia often promotes a light easterly airflow with considerable surface stability from off the North Sea. The base of this moist airstream is cooled by the low sea-surface temperatures at this time of year so that sea fog and low stratus cloud is advected inland across the coast. As outlined by Smith (1970a), this *haar* effect modifies both sunshine duration and maximum temperatures with significant differences occurring within a few kilometres according to the detailed advance and decay of the cloud sheet. Thus, Alexander (1964) has shown the effect of inshore water temperatures in the Eden estuary in eastern Scotland on the progress of the haar. Here the low stratus may recede with the cooler waters of the ebbing tide as the estuarine sands and muds are exposed and then effectively act as the land surface, while the poor visibility similarly re-advances with the next tide.

Sparks (1962) has stressed the importance of falling inland temperatures in controlling the spread of the haar across East Anglia. For a landward advance to take place, the coastal temperature must fall below the clearance temperature which, under the prevailing conditions of onshore airflow, is the surface value which is just sufficient to evaporate the leading edge of the cloud sheet as it is continuously formed offshore. If temperatures further inland fall below this clearance threshold during the evening or the night, then the haar is advected inland at the speed of the gradient wind at its own level. Such an event has been described by Freeman (1962) when, after an afternoon temperature discrepancy of over 8 degC between coastal and inland stations, the progressive fall of temperatures over East Anglia in the evening permitted an advance of the stratus from 1800 GMT on the Wash coast to 0001 GMT inland. In general, the advancing isochrones tended to follow the shape of the coastline but it is interesting that a more rapid advance occurred over the Fens where lower maximum temperatures were recorded during the day. Despite the importance of poor visibility and low stratus at some east coast stations, Smith (1967) has shown that the incidence of fog on the Lincolnshire coast is less than 1 per cent during most daylight hours. This is a much lower frequency than at inland stations where atmospheric pollution increases fog development.

## 2.6 Surface topoclimates

Although perhaps less obvious than the basic contrast between land and sea, differences in the detailed fabric of the various surfaces in the rural landscape lead to important contrasts in topoclimates. However, before we examine some of the topoclimatic contrasts which arise largely from the complex interdigitation of different surface covers over level ground, it will be convenient to discuss some of the general principles which lead to contrasting surface topoclimates. As with all other topoclimates, the contrasts depend on local modifications of the heat and moisture budgets near the earth's surface.

Spatial variations of the heat budget are perhaps most significant, because they influence not only the thermal pattern of small-scale climates but also the water balance by effecting local control over evaporation rates. In turn, local contrasts in the heat budget are partly determined by differences in the reflection coefficient or *albedo* of land surfaces. This is the proportion of incoming solar radiation which is reflected back into the atmosphere without heating the earth's surface, and it varies from less than 10 per cent over dark fir trees to as much as 85 per cent over freshly fallen snow. The normal range of albedo over Britain is about 10–30 per cent and this has been confirmed by measurements at ground level (Cole and Green, 1963), from a comparison of ground and aircraft observations by Barry and Chambers (1966), and partially from the use of aerial photographs by Bendelow (1969).

In measurements in Israel, conducted by Stanhill *et al.* (1966), albedo varied from 12 per cent for a pine forest to 37 per cent for vegetation in a desert wadi. Sometimes, highly detailed differences can be detected, and in some airborne tropical measurements over Barbados by Chia (1967) it was found that the albedo for bare soil varied with the colour, the moisture of the surface layer, the presence of limestone concretions, and the roughness of the surface. On the other hand, Dirmhirn and Belt (1971) concluded that only small differences in albedo existed between individual soil and vegetation components for three sage-brush range sites in the inter-mountain region of south-eastern Idaho.

Where vegetation is involved, there may be seasonal changes in albedo. For example, Berglund and Mace (1972) have shown that sphagnum-sedge bog in Minnesota reached a maximum summer albedo of 16·1 per cent in June, owing to the increased reflection from new vegetation growth, compared to a minimum value of 11·6 per cent recorded for old vegetation in April–May. In contrast, the solar energy reflected from an adjacent stand of black spruce remained between 6 and 8 per cent throughout the growing season, probably because of the three-dimensional structure of the canopy which precluded the presence of a simple reflecting surface. Similarly, the trees showed little snow interception on the crowns, whereas, with a snow cover of 50–60 per cent, the winter albedo of the bog surface increased from 13·5 to 35·1 per cent, and a fresh, continuous snow cover produced an albedo of 81·7 per cent. Few measurements of albedo are available for low latitudes but average data for Nigerian grassland obtained by Oguntoyinbo (1970) were appreciably lower than those for similar vegetation outside the tropics, possibly because of a smaller leaf area index, less ground cover, and also, perhaps, greater solar elevations.

The way in which albedo and other factors influence the radiation balance and the surface temperatures of various natural ground covers in temperature latitudes has been investigated at the Rothamsted experimental station in Hertfordshire. Monteith and Szeicz (1961) measured the diurnal fluctuation in the radiative fluxes over a grass cover 400 mm high with an albedo of 26 per cent. The results for a cloudless August day are shown in Fig. 2.11. The net radiation, *RN*, is made up of the balance of incoming and outgoing energy at the earth's surface according to the equation

$$RN = RS\downarrow - RS\uparrow + RL\downarrow - RL\uparrow$$

where $RS\downarrow$ is the downward shortwave radiation (insolation),
$RS\uparrow$ is the upward shortwave radiation reflected from the surface.
$RL\downarrow$ is the downward longwave radiation,
$RL\uparrow$ is the upward longwave radiation.

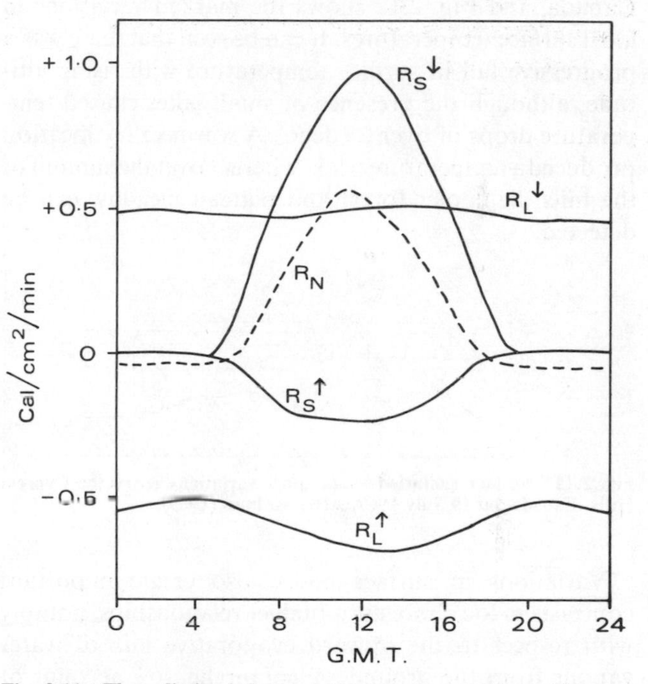


**Fig. 2.11 The radiation balance over a grass cover 400 mm high on a clear August day at Rothamsted. After Monteith and Szeicz (1961).**

It can be seen from Fig. 2.11 that the small negative balance of net radiation during the night gave way to a large positive balance around noon and that this diurnal pattern was principally dependent on the downward flux of shortwave radiation, which was very large compared to either the upward flux of shortwave radiation (determined by albedo) or the upward flux of longwave radiation (determined by the temperature of the grass surface).

In further experiments Monteith and Szeicz (1962) used a radiometer to measure the effective surface temperature of various natural ground covers on cloudless summer days and found that, although the maximum temperature of tall crops and open water was close to the maximum air temperature, a bare soil surface exceeded the air temperature by 20 degC.

It will be apparent that such variations in surface-radiation balance will create real thermal contrasts in the rural landscape. Holmes (1969) has described a temperature transect taken with an airborne infra-red thermometer across an upland area in south-west Canada, and Fig. 2.12 shows the marked variations in local surface temperatures. It can be seen that there was a progressive fall in surface temperature with rising altitude, although the presence of small lakes caused temperature drops of over 25 degC. A warm valley location produced a temperature peak, whereas over the summit of the hills the cooler forest and plateau meadow can be detected.

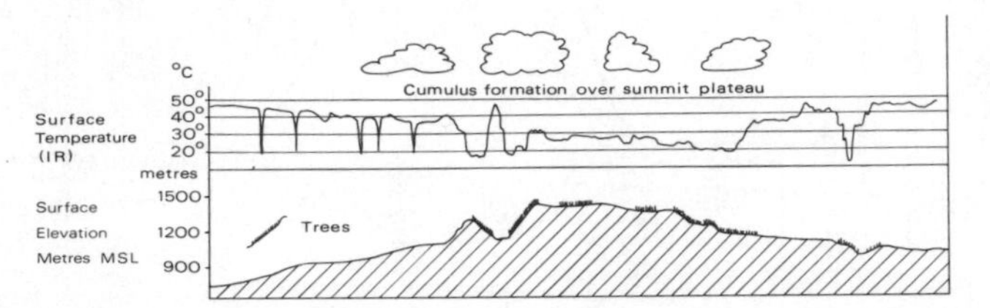


**Fig. 2.12 Surface radiation temperature variations across the Cypress Hills, Canada, on 19 July 1967. After Holmes (1969).**

Variations in surface cover also create important contrasts in local moisture-budget relationships, notably with respect to the upward evaporative flux of water vapour from the ground. A comprehensive account of evaporation problems is not appropriate here and reference may be made to general summaries like Penman (1963), Baier (1965), and Ward (1971). More briefly, the total evaporation loss is composed of two quite different processes and given the general term *evapotranspiration*. As the term implies, this is made up of direct evaporation from water, soil, and other surfaces and subject to purely meteorological factors, whereas transpiration from vegetated areas occurs mainly through leaf stomata and is a biological process partially dependent on plant physiology. Over open water and extensive uniform plant covers with unlimited water supplies, it has been customary to recognize a state of *potential evapotranspiration* which is the theoretical maximum rate of water output from an area of short, green crop under the prevailing meteorological conditions. However, even this simplifying assumption is difficult to observe in the field, largely because all evaporation processes require an understanding of both the heat–budget equation and the mechanism of turbulent transfer near the earth's surface, since these factors, respectively, determine the supply of heat energy available for evaporation and the efficiency with which water vapour is subsequently removed away from an evaporating surface. Not surprisingly, *actual evapotranspiration* from areas of mixed surface character is even more complex owing to the increasing importance of highly localized plant and soil factors as natural surfaces begin progressively to dry out.

Some of these factors, such as the colour and aspect of the surface, have been discussed earlier, and Seginer (1969) undertook a computer simulation experiment to assess the quantitative changes to evapotranspiration which could occur as a result of heat–budget modifications. By assuming that the plant albedo could be increased to 40 per cent, it was suggested that, under clear-sky conditions, the reductions in net radiation and evapotranspiration were both likely to be around 30 per cent compared with the values for an average albedo of 25 per cent for a field of natural vegetation.

Since transpiration is the most significant method of water loss for most vegetated surfaces, it follows that the ecological variables which determine the availability of soil water to the plant become more and more significant as the soil dries out. Such variables include the depth and efficiency of the plant-rooting system as well as the depth and texture of the soil. Similarly, variations in the shape and structure of the plant cover increase the local aerodynamic roughness of the surface which, in turn, raises the vapour transport. Thus, Rijtema (1968) has shown that the coefficient of turbulent exchange rises by a factor of 5 with an increase in vegetation height from a short-cut grass surface at about 20 or 30 mm to a plant height of 900 mm and, in some cases, such differences are capable of boosting local evaporative losses above the theoretical maximum rate according to the potential evapotranspiration concept.

Differences in the characteristics of natural surfaces, especially with regard to plant cover, produce spatial contrasts in the distribution of the incoming components of the water budget. The interception of precipitation by foliage and other plant surfaces is an important element in the water balance of vegetated areas, since this proportion of the precipitation is directly evaporated back to the atmosphere without entering into the land-based part of the hydrological cycle (Reynolds, 1967). Most experiments have been conducted on forested areas, which are described later, but Clark (1940), for example, showed that prairie grasses in Nebraska had a foliage area between 3 and 20 times greater than that of the underlying soil surface and that about two-thirds of the precipitation from heavy rainstorms and up to 97 per cent of rain from very light showers failed to reach the ground. Other vegetation may be effective in trapping snowfall, as outlined by Wilken (1967) for brush vegetation in an upland area of the western US.

In addition to precipitation, there may well be differences in condensation brought about either by hoarfrost or dewfall. Weight measurements of the accumulation of rime and hoarfrost on lodge-pole pine trees in the eastern US, by Berndt and Fowler (1969), suggest that such deposition adds some 75–100 mm of moisture to the total winter supply and comprises almost 10 per cent of the annual precipitation in the area. Dew is an equally difficult moisture element to measure and may be defined as the deposition of water droplets on radiatively chilled vegetation surfaces by the direct condensation of water vapour from the surrounding air.

Monteith (1957), working above a short grass cover in southern England, made a distinction between *distillation* and *dewfall*. Although both of these processes represent fluxes of water vapour and latent heat between the atmosphere, soil, and grass cover, distillation is the transfer of water vapour from soil to grass, which exceeded dewfall on windless nights because heat flux from the soil and net radiation loss were about equal, thus making transfers of sensible and latent heat from the atmosphere almost negligible. However, the measured rate of distillation was only 1–2 mg/cm$^2$ h compared with dewfalls of 3–4 mg/cm$^2$ h due to turbulent transfer of water vapour from the atmosphere when the wind velocity at 2 m became stronger than 0·5 m/s. Long (1958) conducted dewfall experiments at Rothamsted for various agricultural crops and found that dew did not usually form without the presence of an inversion in the vapour pressure gradient either within the crop or at the surface. For condensation to take place on dry foliage, the surface temperature should be at a lower temperature than the dewpoint of the air, although for wet surfaces the foliage temperature need only be at the dewpoint. For example, on clear nights the observed leaf temperature of potatoes was normally 0·5 to 1·0 degC lower than the air a few mm above the surface and sometimes up to 2 degC less. More recent observations of the vapour pressure gradients within a wheat crop by Burrage (1972) have indicated that condensation more than 60 cm above the soil surface tended to occur as dewfall whereas, below this level, distillation was more likely. In general, the rate of dewfall was normally about double that for distillation and dew duration periods lasted up to 14 h with depositions ranging up to 0·33 mm per night.

It will be clear, therefore, that different surface covers are capable of producing distinctive topoclimates and attention will now be drawn briefly to the effects of some rural surfaces on the temperature and humidity characteristics of the overlying air layers. A progression will be followed from bare soil, through low plant covers to forested areas, which represent the maximum floristic modification of local climates at the earth's surface.

## 2.7 Climatic variations near bare ground

The diurnal fluctuation in air temperature near ground level is largely dependent, in the absence of advective influences, on the temperature variations of the underlying surface which, in turn, are controlled by the diurnal pattern of net radiation receipt and utilization by the earth's surface. Since the air is heated from below, it follows that, with increasing distance from the ground surface, air temperatures are less closely related to surface conditions. Characteristically, this leads to both a progressive reduction with height of the range of air-temperature variation and a phase lag of air temperature behind ground temperature. This may be illustrated by Fig. 2.13, which shows the typical trend of bare ground and air temperatures at about screen level on an anticyclonic summer's day in the mid-latitudes.

It can be seen that the difference in the amplitude of

air and surface temperature variation is caused mainly by the difference in the maximum values. Assuming that there is a lack of surface moisture, which would otherwise use up a substantial proportion of the incoming heat energy in evaporation, the observed thermal contrasts are essentially due to the influence of different heat transfer mechanisms. Heat flux into the bare ground is achieved by conduction. This is a slow process which restricts the absorption layer near the surface, thereby creating a large surface diurnal range. In comparison,

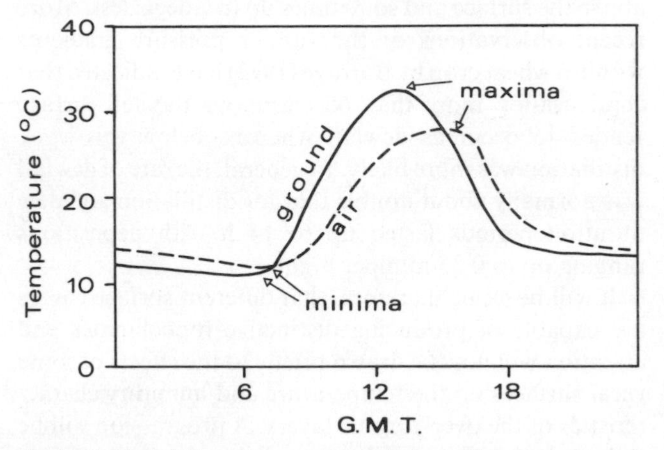


**Fig. 2.13 Typical diurnal temperature variations of bare ground and air near the ground on a clear summer day.**

heat transfer in the air occurs mainly by eddy diffusion and convection, which spread the surface temperature variations through a much greater vertical depth with a consequently smaller heat increase. The lag of air temperature behind the ground heating cycle, which lags in turn behind the radiation cycle, is due to the thermal inertia of the earth–air system. At noon, heat energy is arriving at the earth's surface faster than it can be dissipated and this excess continues to accumulate for a short time after noon. The temperature of both ground and air continues to climb until the maximum surface temperature is reached, shortly followed by the air, when there is equilibrium between heat arrival and dispersal. The phasing of minimum temperatures depends on the reverse process, so that maxima occur just after the solar noon and minima just after local sunrise.

Over a bare, dry surface, the daytime ground temperatures are determined by the restricted heat conduction into the earth. It is difficult to measure surface temperatures accurately, but Australian data obtained by West (1952) and illustrated in Fig. 2.14 show a reduction in the amplitude of the diurnal heating wave from about 30 degC at 25 mm depth to only 2–3 degC at 300 mm. Similarly, Fuchs and Tanner (1968) made measurements in a sandy soil around the incoming radiation peaks in summer and found that with a temperature of 43 °C, the surface was 2 degC hotter than the soil at 1 mm depth, 8 degC warmer than at 25 mm, and 23 degC warmer than at the 300 mm depth.

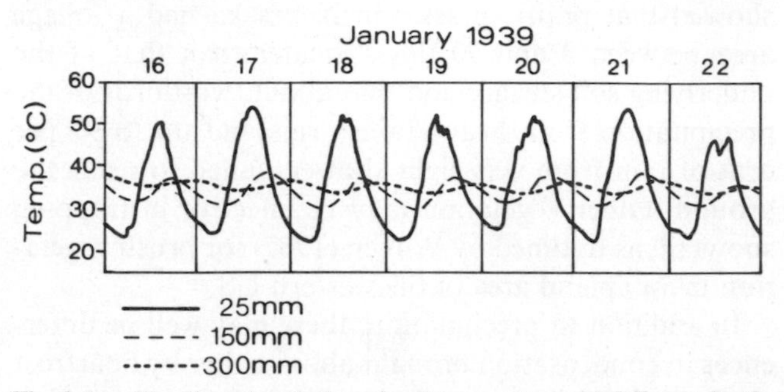


**Fig. 2.14 The diurnal variation of soil temperatures at different depths during clear summer weather at Griffith, Australia, between 16 and 22 January 1939. After West (1952).**

Even at screen level over the arid surfaces of the subtropical deserts, the mean maximum temperatures in the warmest month may lie between 45 and 50 °C, but Priestley (1966) has shown that with a wet surface, which is capable of removing much of the available heat energy by evaporation, the highest mean maxima are unlikely to exceed 33 °C anywhere in the world. This is a reminder that many soils contain water and air spaces as well as soil particles and that the climatic response of any particular surface depends on its detailed composition and nature as emphasized by Rider (1957). Thus, early experiments by Johnson and Davis (1927) were conducted on the near-surface thermal character of bare earth, clay, sand, rubble, tarmacadam, and short grass on Salisbury Plain, England, and the highest maximum (53 °C) was recorded in the tarmacadam and the lowest maximum (38 °C) occurred under grass. A similar discrepancy was noted with regard to the mean diurnal temperature range which in mid-summer varied between 33 °C in the tarmacadam to 16 °C below the grass surface compared with 14 °C in the air.

As far as bare soil surfaces are concerned, a sandy soil, with its high proportion of air spaces and low conductivity, will in general support a larger diurnal temperature range than occurs over a clay soil. These differences are most apparent with respect to night minima, and Brunt (1945) claimed that on cloudless nights the greatest fall in temperature between sunset and sunrise in Britain occurred at the sandy South Farnborough site. Similarly, Manley (1944) drew attention to the low minima recorded over the lighter soils of East Anglia and concluded that, over fairly flat country, air minima on the Breckland could be more than 3 degC lower than values recorded in other parts of eastern England on radiation nights.

More attention has probably been paid to night-time minima than day-time temperatures over bare surfaces. Although a temperature inversion often occurs in the lower atmosphere during anticyclonic nights, this does not necessarily imply that the lowest temperatures are recorded at the ground surface itself. Fairly continuous evidence from India since the 1930s has shown that the minimum temperature may be observed anywhere between 100 and 300 mm above the surface in the absence of the advection of colder air from the surrounding area, and subsequent experiments have confirmed the existence of this phenomenon in the mid-latitudes. Lake (1956) found that over bare soil in southern England the minimum temperature on radiation nights was commonly found between 40 and 150 mm above ground level and concluded that, under conditions of extreme stability, when eddy mixing is very low near the surface, longwave radiation from the air may become the dominant element in radiation exchange. More recently, Oke (1970) made measurements over grass, snow and, bare soil surfaces in Ontario, Canada, where the height and intensity of minima were strongly correlated with windspeed, surface roughness, and stability. Under calm conditions, air minima over flat, bare soil often occurred as high as 500 mm above the surface and the difference between air and surface soil minima was often more than 3 degC.

Despite the anomalies in lapse rate which can occur very near the ground surface at night, the normal temperature inversion usually results in appreciable differences between screen minima and so-called *grass minimum* temperatures. Thus, German evidence quoted by Geiger (1965) demonstrated that on cloudless anticyclonic nights the temperature at 50 mm could be up to 6·5 degC lower than the screen value. On the other hand, it is well known that a bare soil surface will cool more slowly than a grass surface, largely because the upper parts of the vegetation cover operate as a highly efficient radiation exchanger. At night, therefore, the minimum is lower over grass, because the relatively stagnant air layer trapped within the grass cover insulates the air above from the warmer soil surface, which is also heated by the upward conduction to ground level of heat energy stored from the day. The relationship between grass and bare soil minima has been investigated by Gloyne (1953).

Large diurnal temperature variations are often associated with large daily fluctuations in atmospheric humidity. Over wet surfaces, the major water-vapour flux is upwards, owing to the overriding influence of evaporation, and the downstroke of water vapour is small in both absolute and relative terms, although in clear summer weather there is often a net movement of water vapour towards the ground to form dew before sunrise. After sunrise, energy is available for evaporation and this produces a marked increase in vapour pressure near ground level in the early morning. This vapour remains near the surface until about midday and the vapour pressure may not begin to fall until convection and eddy diffusion increase the air turbulence in the early afternoon. The vapour-pressure lapse rate remains positive through the afternoon until, in the evening, the progressive fall in temperature and air movement lead to the re-establishment of the humidity inversion which continues through the night.

## 2.8 The effect of low plant covers

Traditionally, all standard meteorological observations are taken above level ground with a short grass cover. Although closely mown grass is the most uniform of all vegetated surfaces and provides a broadly representative guide to climatic conditions over all such areas, the convention has certain disadvantages. Apart from the problems of extrapolating such observations over 'non-standard' surfaces for frost prediction over roads or airfield runways, for example, it is known that it is impossible to achieve comparability between such surfaces. Thus, observations of night temperatures by Ritchie (1969) over concrete and grass-covered surfaces showed not only that the minima above concrete were slightly higher than the values over grass but also that the artificial surfaces produced a less erratic spatial distribution of temperatures. With an increase in grass length, such spatial variations become more marked. Monteith (1956) has demonstrated how the rate of snow melt may be accelerated over shorter grass by the superior thermal contact between the soil surface and the base of the snow layer compared with a longer cover, while Norman *et al.* (1957) have confirmed the insulating advantage of longer grass in terms of winter air minima. This thermal difference was found to be greatest during the early winter period and in two consecutive winters the accumulated temperatures below 0 °C were more than twice as great in herbage 25 mm high compared with grass between 300 and 450 mm long.

Grass this long is capable of supporting distinctive climatic gradients within the vegetation cover itself, as shown by the work of Waterhouse (1955) on a 3-year-old grass surface 600 mm high in eastern Scotland. On a clear summer day, typical profiles were shown as in Fig. 2.15. The rapid decrease of windspeed from the top of the grass surface produces relatively still air within the stand,

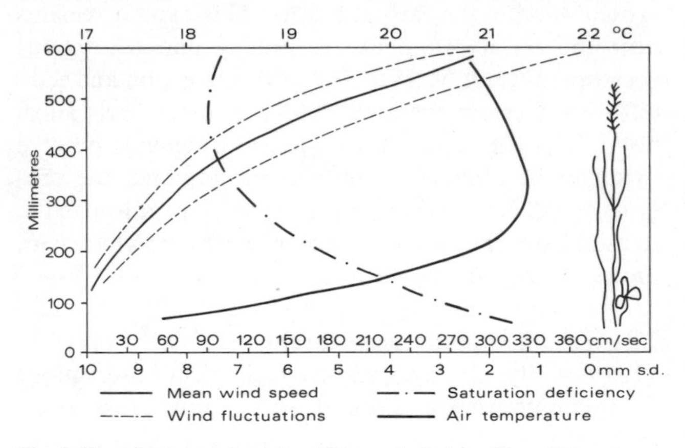


**Fig. 2.15 Air temperature, humidity, and wind profiles within a grass cover on a June afternoon in eastern Scotland. Sunny conditions after rain on the previous day. After Waterhouse (1955).**

thus leading to the heating of the air mainly by conduction from the warm leaves and stalks. The warmest layer is found where the cover density has not yet increased sufficiently to reduce the incident radiation too much and occurs about halfway within the stand. Humidity, on the other hand, characteristically increases towards ground level and is commonly high within dense meadow grass as a result of the vegetation's restricting evaporation from the soil and at the same time releasing water vapour via transpiration.

With increasing diversity of shape and structure, the climatic response of a low vegetation cover becomes more complex and this complexity can be seen in various types of agricultural crop. Several general reviews of crop environment have been published, such as those of Van Wijk (1963) and Long *et al.* (1964), but, in addition, a number of specific investigations have been conducted. For example, Brown and Covey (1966) examined the transfer of energy within a mature corn crop and found that 46 per cent of the net radiation was used for transpiration compared with 32 per cent in the sensible heat flux, 13 per cent for soil evaporation, and 6 per cent in the soil heat flux. On average, 16 per cent of the net radiation penetrated to the ground beneath the dense crop.

Paton (1948) studied the temperature profile within a Scottish wheatfield during a settled summer spell and found a day-time structure similar to that reported for grass by Waterhouse (1955). By 1400 h the temperature at middle levels within the crop was as much as 28 degC higher than at the wheatheads 1070 mm above ground level and up to nearly 4 degC higher than the temperature at the same height above bare soil. The lowest night minima were recorded at the same crop level and the diurnal range in this layer reached 19·5 °C compared to only 11·1 °C nearer the ground. This was attributed by Paton to the extreme stagnation of air within the crop.

A more comprehensive series of experiments on wheat was conducted by Penman and Long (1960) and representative profiles taken throughout a summer day within a wheat crop 700 mm high are depicted in Fig. 2.16. It was found that spatial variations occurred within the wheatfields at the same level, so that an average temperature difference of 0·8 degC and a vapour pressure difference of 1·0 mb persisted for several hours at positions only 8 m apart. The location and persistence of the anomalously warm spots appeared to depend on the chance penetration of more direct incident radiation through gaps in the crop canopy and the spots moved around as the angle of solar radiation varied through the course of the day.

Similar records were gathered for a potato crop by Broadbent (1950), who compared temperatures and humidities in a standard screen with measurements made at 150 mm within the crop. In clear summer weather, the maximum crop temperature was up to 7·2 degC higher than in the screen and the minimum usually just over 1 degC lower. By day, crop humidity was appreciably higher than in the screen with a difference of nearly 4 degC in dew-point temperature reducing at night to only about 1 degC corresponding to the measured difference in mean minimum air temperatures. Many tropical field crops show comparable characteristics as illustrated in a study of the radiative and aerodynamic features of irrigated cotton in a semi-arid area of Israel by Stanhill and Fuchs (1968).

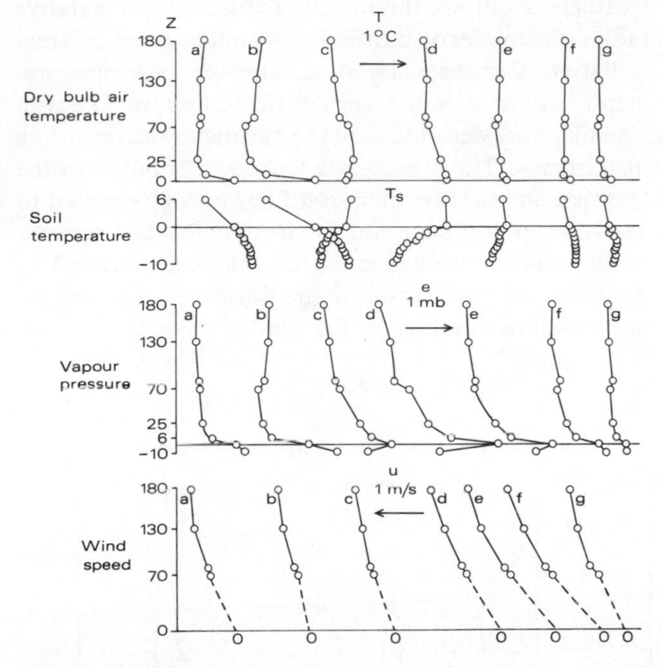


**Fig. 2.16 Profiles of air temperature $T$, soil temperature $T_s$ (with Z-scale enlarged 5 times) vapour pressure $e$ and wind speed $u$ at (a) 00, (b) 04, (c) 08, (d) 12, (e) 16, (f) 20, and (g) 24 h. Observations taken at Rothamsted on 18 June 1956 with wheat crop 700 mm high. After Penman and Long (1960).**

## 2.9 Forest climates

Since the modifying effect of vegetation on the local climatic environment increases with the vertical extent and horizontal spread of the plant cover concerned, it follows that forest influences are dominant in many areas. These influences arise from the relative obstruction offered by the tree canopies to vertical fluxes of heat and moisture between the atmosphere and the ground surface and from the barrier effects of forest stands to similar horizontal transfers. The complexity of the resulting modifications is due to interrelationships between both the biological and the physical factors involved. On the botanical side, considerations such as the type of species, the size, age, and density of the stand, together with the presence of an understorey, may all be important, while the continuing development of a forest through time and the varying seasonal conditions of deciduous trees in particular means that one is dealing with a highly dynamic variable. On the physical side, the same type of forest association may show contrasting behaviour in different climatic regimes and under different weather sequences, and variations in local topography may even obscure the modifying effects of a tree cover.

One of the basic problems in forest meteorology is to ensure genuine comparability between a forest site and one in the open, and work by Fritts (1961) on ravine forest in Illinois has shown that there was often more difference in daily maximum temperatures between north- and south-facing sides of a valley than between observations made within the forest and in an adjacent clearing. It should be emphasized that it is the cumulative effect of all the physical and biological factors which determines the distinctive character of the forest climate, but it will be convenient here to deal systematically with each of the main climatic parameters.

The modification of local energy budgets by the forest can be demonstrated in terms of both radiation and temperature conditions. Recent investigations have suggested that the albedo of forests tends to remain fairly constant. For example, Stewart (1971) concluded that the effect of canopy wetness on a pine forest in eastern England modified the mean summer albedo by less than 1 per cent except at large solar zenith angles. As might be expected, however, some of the most significant variations occur on a seasonal basis with deciduous species and Dewalle and McGuire (1973), working with an uneven-aged, mixed oak stand in Pennsylvania, found that the growing season albedo ranged from 16 to 18 per cent compared with a snow-free dormant season albedo of between 12 and 14 per cent.

The primary effect of a forest canopy is to intercept

and absorb most of the incoming solar radiation so that within dense tropical stands, built up of one or more storeys, the radiation intensity on the forest floor may average less than 1 per cent of that recorded in the open. The penetration of insolation is largely a function of stand density and Cheo (1946) has described thinning experiments on 25-year-old red pine in Minnesota, which produced a progressive increase in light intensity from 15·4 per cent to 60·2 per cent of that in the open with a reduction in tree density from 6510 to 1310 per hectare.

Figure 2.17 illustrates the vertical distribution of net radiation in a young spruce forest near Munich in Germany on a clear day in high summer and it can be seen that the maximum radiation exchange was located in the upper part of the canopy. On average, it was found that 88 per cent of the radiative energy was transferred in the upper half of the stand compared with only 12 per cent below that level. With deciduous trees, the relationship is less straightforward and it seems that, while even in dense stands as much as 70 per cent of the brightness outside the wood may be transmitted to the forest floor in winter, the fraction may drop to around 15 per cent in the period of maximum leaf development (Ovington and Madzurik, 1955).

In many ways, it is misleading to generalize in such a way about radiation conditions, because the interception efficiency of the forest depends largely on the proportions of diffuse and direct sunlight which are received in the open. Most canopy interception takes place with direct radiation under clear skies since, with overcast conditions, the incoming diffuse radiation has a greater chance of penetrating to within the trunk space. Thus, Anderson (1964) has shown that, in a mixed deciduous wood near Cambridge, the light received in the forest may vary during a single month from about 13 per cent to 38 per cent of that recorded outside and Fig. 2.18 demonstrates the irregularity of the daily transmission under British weather conditions. It is difficult to obtain representative radiation samples on the forest floor and, even with direct radiation, Clements (1966) has stressed how measurements within a white spruce plantation in Ontario, Canada, were dependent on the placing of the recording instruments. These problems may be attributed to the fact that any spot on the forest floor may be exposed to variations in radiation ranging from only a few per cent of the value in the open to the full solar beam depending on the actual position of the sun during the day and the movement of branches by the wind.

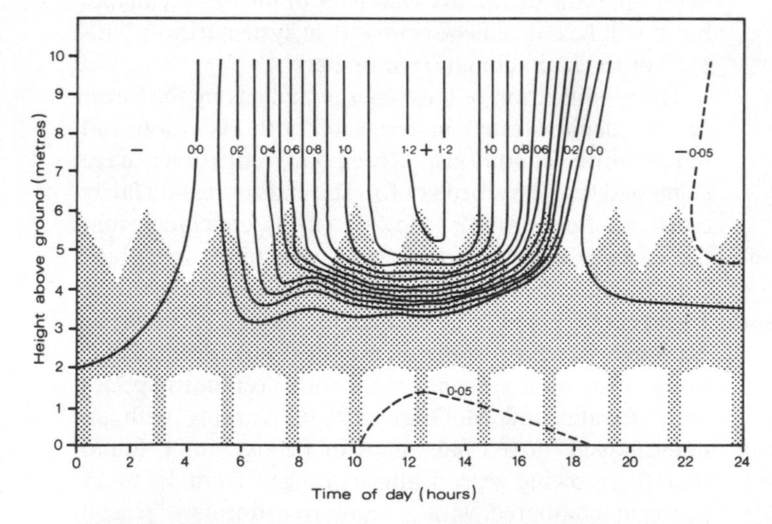


**Fig. 2.17 The vertical distribution of net radiation on a hot day in midsummer within a young spruce forest near Munich. After Baumgartner (1967).**

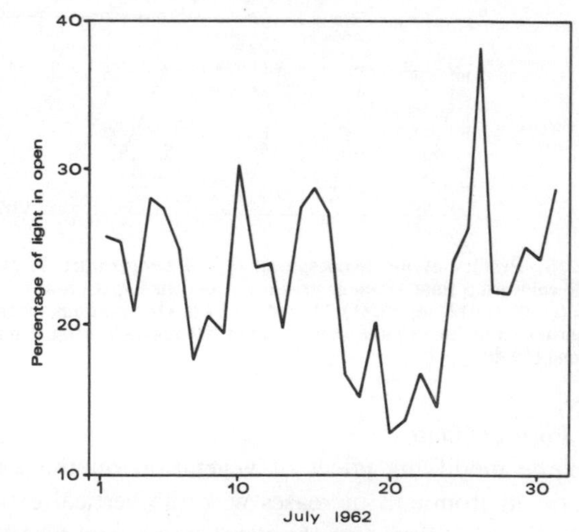


**Fig. 2.18 Daily totals of irradiance in a large clearing of a mixed deciduous wood near Cambridge, England, expressed as a percentage of daily totals in the open. After Anderson (1964).**

In general, it appears that pine canopies pose the greatest sampling difficulties and, on the basis of measurements in Connecticut, Reifsnyder *et al.* (1971) found that to sample the instantaneous radiation beneath a pine canopy with a standard error of 10 mly/min required 412 radiometers compared with only 18 for a hardwood canopy. To obtain a daily average needed 10 and 1 radiometers respectively. The effectiveness of several time- and space-sampling schemes for the solar radiation transmitted to the forest floor has been discussed by Gay *et al.* (1971). Vezina and Boulter (1966) have shown how the selective absorption and reflection of tree canopies influence the quality of light inside the forest. The canopy not only intercepts but also filters radiation by absorbing a high proportion of the ultraviolet radiation and this selectivity was found to be greatest with a deciduous rather than a coniferous canopy, especially when the incoming radiation was direct sunlight.

Air temperatures within forests are closely controlled by the factors governing radiation. Just as the tree canopy acts as a barrier to incoming radiation during the day, so it also tends to blanket the outgoing nocturnal radiation, thereby insulating the trunk space. A relative reduction in the range of thermal variation is a general characteristic of most mid-latitude forests and this feature operates over a variety of time-scales. The seasonal effect is demonstrated in Fig. 2.19, which indicates differences in average monthly temperatures between forests composed of Norway spruce, Scots pine, beech, and *forteto* oak maquis compared with temperature conditions in the open as represented by the horizontal

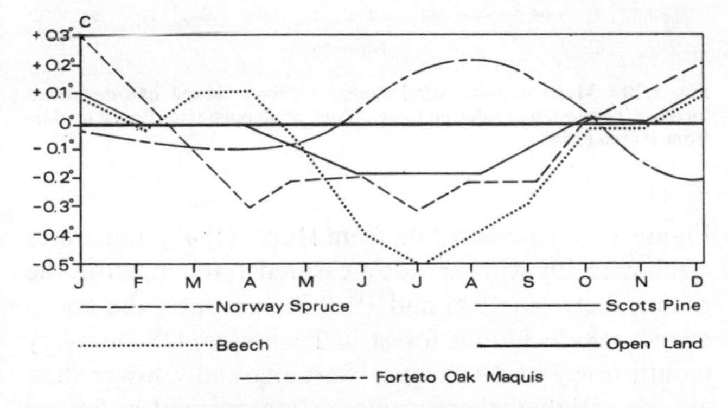


**Fig. 2.19 Mean monthly forest temperatures compared with thermal conditions in the open. After Food and Agriculture Organisation (1962).**

line. It can be seen that, with the exception of the maquis, the effect of all three forest types is to raise winter and lower summer values relative to unforested conditions, although there are important differences between the species. The smallest influence is exerted by the Scots pine which depresses summer temperatures by only 2 degC and has little effect in winter compared with Norway spruce which modifies temperatures at both seasons by about 3 degC. The major effect, however, is created by beech which depresses mean July temperatures by 5 degC. According to Food and Agriculture Organization (1962), this influence cannot be explained solely in physical terms and is attributed to the fact that beech transpires more water during the growing season than any other species. Since transpiration causes a loss of heat, it follows that a larger reduction of temperatures will occur during the summer months with the more hygrophilous species such as beech, although during the winter any temperature influences are almost entirely dependent on physical controls with the cessation of transpiration during long periods.

A similar reduction in the range of thermal behaviour within the forest is often evident on a diurnal basis. In most cases, the diurnal difference in range between forest and open, even under favourable anticyclonic weather, will rarely exceed 5 or 10 degC but continentality and altitude can combine to produce more exceptional results. For example, Kittredge (1948) has drawn attention to a site at Fort Valley, Arizona, where at an altitude of 2210 m the diurnal range in winter within a coniferous forest was 16·6 degC less than in the open.

On the other hand, Fig. 2.19 shows that the evergreen xerophilous woodland found in Mediterranean areas and known as 'maquis' results in opposite temperature effects from those described so far. As with other forest influences, it appears that this apparent anomaly, which is usually most marked during summer, is due to a particular combination of physical and physiological factors. In the first place, the low, dense cover causes extreme stagnation of air which is then trapped close to both the primary heat source of the forest canopy and the secondary source of heat radiating from the woodland floor. Secondly, and probably more important, is the fact that this vegetation adjusts to the long summer drought by drastically reducing the transpiration rate.

This means that no physiological cooling is available and the result is very high maximum temperatures within the still air.

The thermal effects of some tropical forests appear to be dissimilar to those associated with the mid-latitudes, largely because the greater variety of species and the strongly stratified structure of the vegetation tend to produce a more complex response. Evidence provided by Hales (1949) for the Panama jungle suggests that the dense upper canopy acts as the principal heat exchange surface and therefore shows a greater range of diurnal temperature than the undergrowth, and Kunkel (1966) found mean air temperatures in the Liberian rainforest to be from 1 to 3 degC lower than over open land. Even more complex relationships exist where an intermediate storey occurs between the upper canopy and the undergrowth. Data quoted by Richards (1952) for a rainforest at an altitude of 300 m on the Philippine Islands show that, although daily maxima recorded in the middle storey are intermediate between those of the upper canopy and the undergrowth, the night-time minima are higher than either at canopy or near ground level since the middle storey is insulated by stagnant air both above and below.

Even discontinuous forest canopies can create special topoclimates and Miller (1956) has described the persistent warmth found under sparse stands of lodgepole pine in the Sierra Nevada during winter and spring when the air temperature may often reach 15 °C over a deep snow cover. It was found that more than half the incoming radiation was absorbed by the foliage, and in calm conditions the surface temperature of the pine needles exceeded the air temperature by 10 degC, thus providing a heat source sufficient to promote such a warming in the trunk space that the normal nocturnal temperature inversion actually strengthened during the course of the morning from about 3 °C to over 16 °C. Similarly, Hurst (1967) has claimed that minimum screen temperatures within a coniferous forest in Suffolk are higher than over bare soil in the open, even with only two-thirds canopy cover.

Soil temperatures also reflect the temperature modification associated with forests and Smith (1970b), working on a small Pennine plantation thinned by windblow, has shown that differences between mean daily soil temperatures at the 100 mm depth at adjacent open and wooded sites are statistically significant.

One of the most obvious changes brought about by woodland is the modification of airflow both around and within the forest cover. Again, the forest geometry is a dominant factor in the detailed response, but in most cases there is a substantial reduction in wind movement, so that mean annual speeds are usually lowered to between 20 and 50 per cent of those in the open and vary from about 0·1 to 1 m/s within mature woodland (Food and Agriculture Organisation, 1962). The density of the forest is important and thinning experiments by Jemison (1934) on closely spaced coniferous forests in Idaho showed that velocities in uncut, half-cut, and clear-cut zones were respectively 0·1, 0·4, and 0·6 m/s, and some of the largest proportional reductions averaging around 11 per cent have been recorded within dense maquis woodland.

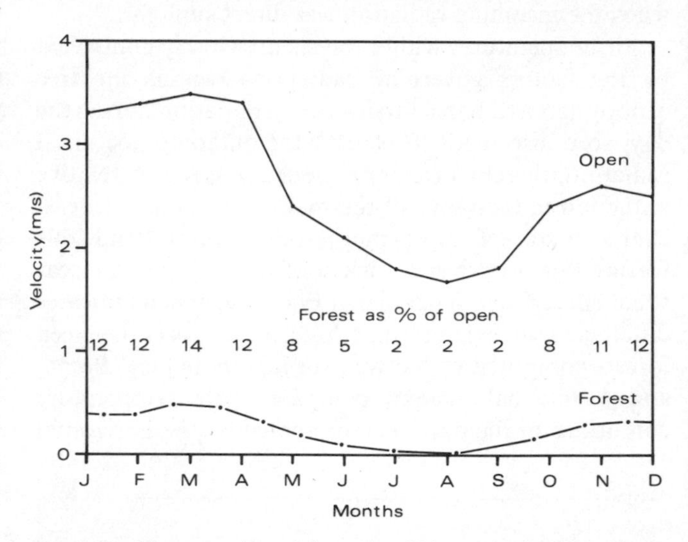


**Fig. 2.20 Mean monthly wind speeds within a mixed oak-deciduous forest in Tennessee in relation to velocities at an open site. Based on data from Hursh (1948).**

Figure 2.20, based on data from Hursh (1948), illustrates mean monthly wind speeds measured at 0·6 m above the ground between 1936 and 1939 for an open site and a mixed oak-deciduous forest in Tennessee, US. In every month, the forest velocities were markedly lower than outside, although there are interesting seasonal variations with the greatest absolute reductions taking place with the higher windspeeds of winter and the largest proportional reductions, which are somewhat exceptional, occurring in summer. This latter effect may be attributed partly to the increased foliage developed during summer and Fig. 2.21 indicates the seasonal contrast in wind profile for similar speeds around a 115-year-old German oak stand with a greater penetration of air movement before the leaves open.

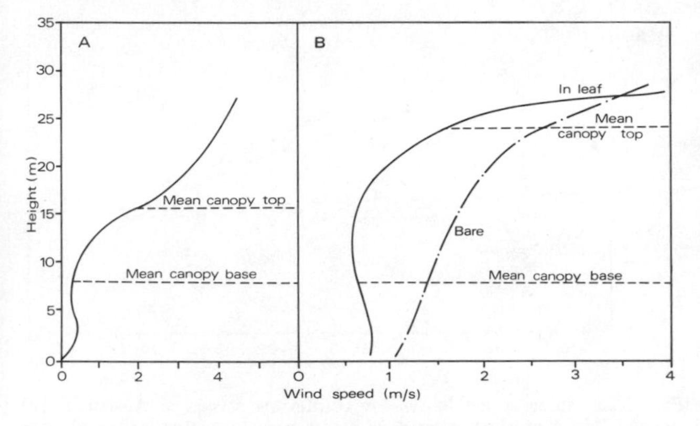


**Fig. 2.21 Wind-speed profiles associated with different forest covers. (A) A dense stand of mixed pine. After Oliver (1971). (B) An oak grove showing the influence of seasonal variations in foliage. After Geiger (1965).**

However, it is known that large variations in wind velocity outside have little effect within the forest and this is shown in Fig. 2.21, which depicts the characteristic wind profile found inside both deciduous and coniferous stands. Over open areas, the ground surface acts as the friction layer creating a steeply decreasing windspeed gradient within the lowest few metres. For a forest, the canopy operates as the effective friction surface and the sharp reduction in velocity is maintained through the air trapped inside the crowns. Measurements for a pine stand of Ponderosa pine at 1220 m in northern California have shown that the greatest relative reduction takes place within the crowns, and reductions of 15 per cent and 24 per cent were recorded respectively for outside winds of 4·5 and 2·2 m/s (Fons, 1940). Below the crowns, there is a tendency for a slight increase in windspeed to occur, as shown in Fig. 2.21, especially if there is only a sparse herbaceous cover, before the velocity decreases rapidly to its lowest level near the forest floor.

The moisture balance is one of the most controversial aspects of forest climates. In general, the reduction in direct radiation, maximum temperature, and windspeed all tend to produce high relative humidities within the forest especially during the summer months of maximum transpiration. From data collected in Germany and Switzerland, it is evident that the proportional increase in relative humidity over the open during this season is largely dependent on this physiological activity with percentage values of 9·4 per cent for beech, 8·5 for Norway spruce, 7·9 for larch, and 3·9 for Scots pine (Food and Agriculture Organisation, 1962). Less conclusive results are obtained when vapour pressure is considered, but there is little doubt that a forest environment is very effective in reducing evaporation. A survey of four separate experiments with evaporation pans in a variety of forests in the US and Australia indicated that open-water evaporation from the forest floor is rarely more than 50 per cent of that in the open and that the greatest percentage reductions occurred in summer when evaporation opportunity was at a maximum. Similar results have been achieved with piché evaporimeters and, in a comprehensive comparison between a forested and a clear-cut ridge in Ontario, MacHattie and McCormack (1961) concluded that the forest cover reduced evaporation to one-half to two-thirds of that in the cleared area.

Externally, forests may well have different evapotranspiration regimes from areas of adjacent vegetation and Tajchman (1971) made a comparative study of water losses from a 70-year-old stand of Norway spruce, a two-year-old crop of alfalfa, and a potato field near Munich during one summer season. It was found that, although the total net radiation over the forest amounted, respectively, to 20 per cent and 16 per cent more than that recorded for the alfalfa and the potatoes, the evapotranspiration from the forest varied from 14 per cent greater than that for the potatoes to 4 per cent less than the water loss from the alfalfa. Since soil water was unlimited at all sites, it was concluded that the well developed rooting system of the alfalfa crop was probably a more important factor than the greater net radiation or the higher rates of ventilation experienced in the tree canopy in promoting evapotranspiration.

A great deal has been written about the influence of forests on local precipitation. In one of the most penetrating reviews, Penman (1963) has re-emphasized that, irrespective of any modification of air humidity, forests are incapable of increasing the total precipitation received by any area. This is because the major source of precipitable moisture in virtually all climatic regimes is the vapour advected over land areas by the global wind system. The release of this atmospheric vapour as rain is dependent on the presence of favourable precipitation-forming mechanisms, normally involving the widespread uplift of air as at frontal surfaces or orographic barriers, rather than on local evaporation or atmospheric humidity relationships. Indeed, the only way in which precipitation could possibly be increased by trees would be where a forest exists along the crest of a ridge lying athwart the passage of damp, oceanic air and so forces the air to rise to marginally higher levels in the atmosphere than it would otherwise achieve.

On the other hand, there is ample evidence that a woodland cover reduces the amount of precipitation recorded within the forest. This is due to the process of *interception*, whereby rain or snow may be collected and suspended within the canopy or on other surfaces, especially during the initial part of any storm, before subsequent direct evaporation back to the atmosphere. For almost every storm, therefore, the precipitation penetrating to the forest floor either by throughfall or stem flow will be less than that received at a comparable open site, although the actual reduction involved is highly variable in both absolute and relative terms. For example, interception will be most effective if a given quantity of precipitation falls in a series of isolated showers rather than in one heavy rainstorm, since, at the end of each shower, there will be an increased opportunity for evaporation which will help to make room for more interception from the following shower. If all the precipitation came in one storm the finite interception capacity of the forest cover would soon be reached and all the excess water would penetrate to the forest floor. Figure 2.22, which summarizes data quoted by FAO (1962) for coniferous stands in Australia and Japan, indicates that the proportion of interception loss decreased from over 80 per cent to less than 20 per cent with an increase in total rainfall on either a monthly or a storm basis. Seasonal variations may also occur and Ståfelt (1963) has shown that, over a four-year period, crown interception by a spruce stand in southern Sweden averaged 57 per cent of the May–August precipitation compared with only 52 per cent between September and April, the difference presumably being attributable to the lighter rainfall and higher evaporation rates of summer.

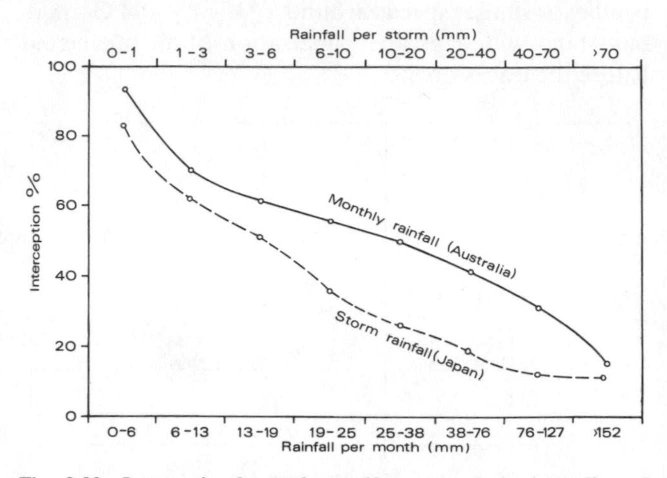


**Fig. 2.22 Interception losses for coniferous stands in Australia and Japan. Based on data quoted in Food and Agriculture Organisation (1962).**

The biological characteristics of the forest are also important and interception generally increases with the age, height, basal areas, and density of the stand. For example, Cheo (1946) reported that experimental thinning of *pinus resinosa* in Minnesota from 6500 to 1300 trees per hectare reduced the interception loss from 40 per cent to 11 per cent and more recent work, such as that of Czarnowski and Olzewski (1968) on a deciduous stand in Poland, has sought statistical relationships between interception and forest geometry. It should be emphasized that coniferous species are normally more efficient interceptors than deciduous trees, even when the latter are in full leaf. Ovington (1954) experimented for two years with 12 different species in English plantations and found that, whereas interception for all four deciduous species averaged less than 35 per cent, the 8 conifers recorded higher values up to 54 per cent for Douglas fir (*pseudotsuga taxifolia*). Helvey (1967) has claimed that interception by white pine stands in North Carolina may be up to 100 per cent more than that associated with native hardwoods.

The main reason for this widely accepted difference

appears to be the contrasting nature of deciduous and coniferous leaves, since the needle-like structure of the latter offers a greater total surface area to the atmosphere which, in turn, means more water storage through surface tension forces and also more opportunity for evaporation from the leaf. Most throughfall to the forest floor takes place by drip from overhead foliage. Thus, Rogerson and Byrnes (1968) have shown that, although gross throughfall averaged 80 per cent of the summer rainfall under a natural hardwood stand and a red pine plantation in Pennsylvania, stem flow was only 2 per cent of overall precipitation for the hardwoods and 1 per cent for the pines. In areas of light rainfall, stem flow may be well under 1 per cent.

A wide range of figures is given in the literature for average interception loss. Kittredge (1948) quoted values from 3 to 48 per cent for various species in the US, and Reynolds and Leyton (1963) have claimed about 30 per cent for coniferous woodland in England. In one of the more comprehensive reviews, Zinke (1967) has suggested an average annual loss in the US between 20 and 40 per cent for conifers and between 10 and 20 per cent for hardwoods. However, in view of the complexity of interception and the difficulty of replicating measurements, as stressed by Jackson (1971) for a tropical area in northern Tanzania, values of mean interception loss should be treated with caution.

## 2.10 Inadvertent modification of rural climates

This chapter would be incomplete without a brief comment on the unconscious changes which man sometimes brings about in rural topoclimates. Such modifications may be considered quite separately from the planned results of manipulating the rural environment through irrigation or the erection of shelter-belts, and the agricultural impact of these and similar activities will be discussed in a later chapter. This is not to imply, however, that unconscious modifications do not arise as byproducts from such actions. For example, a farmer irrigating his fields is concerned only with managing the water economy of the area to which water is applied, although, in fact, the changed evapotranspiration conditions will be reflected in the heat budget of the area too, and advection is also likely to promote the horizontal transfer of some of the modified air downwind from the irrigated site.

From what has already been said, it will be clear that any change in the environmental conditions at the earth's surface will have some potential for modifying local climatic behaviour, and with the spread of certain agricultural practices such as ploughing, drainage, irrigation, and artificial cropping patterns—including afforestation —sometimes on a regional scale, it follows that resulting changes must be taking place in the lowest atmospheric layers. For the most part, the available evidence suggests that these changes, insofar as they affect heat budget and airflow relationships, are small, but there is growing concern about the influence of rural land-use conversion on the water balance. A consideration of some of these hydrological effects is deferred until a later chapter and here we shall be mainly occupied with the topoclimatic gradients which sometimes arise with sharp discontinuities in land-use.

Some climatic modifications are quite isolated and temporary. For example, Rider (1965) has shown that the burning of corn stubble on Salisbury Plain may easily produce local fire-formed cumulus cloud up to the condensation level when winds are light. On the other hand, Warner (1968) has reported changes in rainfall over the last 60 years in a sugar-producing area of Queensland, Australia. During the three months of the cane-harvesting season, there was a reduction of rainfall at inland stations of about 25 per cent compared with areas upwind of the smoke from the cane fires. This rainfall reduction, which was broadly correlated with increasing cane production, was tentatively attributed to the effect of smoke particles hindering the coalescence process of rain formation.

Marked thermal zones may exist in the lower atmosphere as a result of changes in surface albedo and Paltridge (1968) was able to trace thermals, with temperatures up to 6 degC above ambient temperature conditions, to two large sheds with black-painted iron roofs. Airflow patterns near ground level are readily disturbed even by small obstructions and Rider (1952) showed that a hedge 1·68 m high was capable of reducing windspeed at 1 m at distances up to over 18 m upwind and 76 m downwind of the obstacle, with a maximum reduction down to only 12 per cent of the observed speed in the open. Even differences in the height of a cropped grass surface may be influential, as shown by Taylor (1962), and Fig. 2.23 depicts the definite increase in wind velocity at a height of 1 m as air moves from a longer to a shorter grass cover.

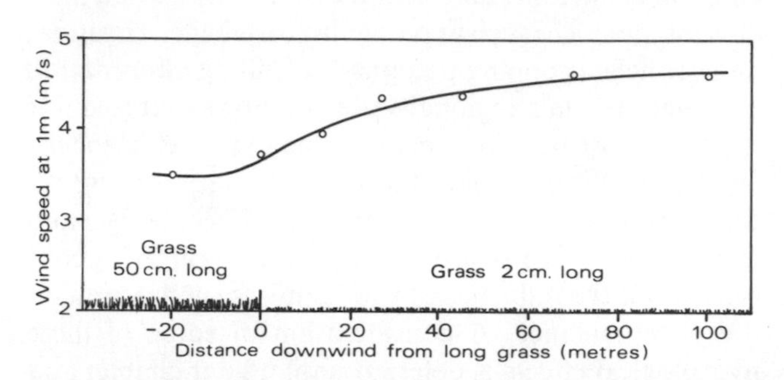


**Fig. 2.23 Acceleration of the wind speed near ground level as a result of moving from a rough surface to a smoother one. After Taylor (1962).**

A more obvious intrusion into the rural landscape occurs with reservoir construction. Gregory and Smith (1967) investigated the effect of a small reservoir in the northern Pennines, and, over a two-month summer period, found a regular diurnal cycle in windward-leeward temperature differences, although at no time did the reservoir alter downwind temperatures by more than 3 degC. In northern Italy Cambié (1969) has reported that the inflow of cold water from hydroelectric installations into the upper part of Lake Garda has lowered the water temperature sufficiently to reduce tourism during the summer and increase the incidence of lake-side fogs in winter. Similarly, Holmes (1970) has drawn attention to the thermal consequences of certain agricultural land-use patterns on the atmospheric boundary layer in Canada. The temperature contrasts shown in Fig. 2.24 depend largely on differences in surface moisture conditions, with dry farmland and uncultivated prairie producing warming effects compared with the cooling associated with irrigated land. The lower part of the diagram indicates that these thermal influences are detectable up to and beyond 60 m above ground level. Not surprisingly, the edge effects associated with irrigated land tend to be most marked in relatively arid environments, as illustrated by the work of Dyer and Crawford (1965) in California and by the observations made by Davenport and Hudson (1967) in the Sudan Gezira.

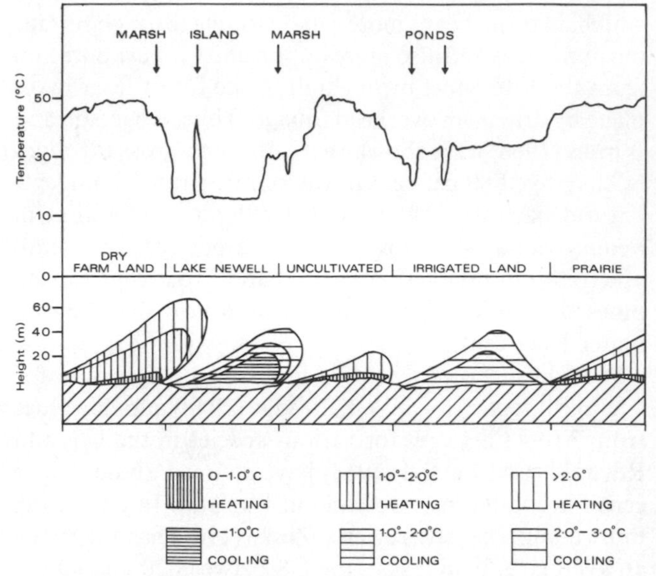


**Fig. 2.24 Surface radiation temperature variations and average profiles of air temperature recorded during August afternoons in 1968 over rural land in Alberta. After Holmes (1970).**

## References


ALEXANDER, L. L. (1964). Tidal effects on the dissipation of haar. *Met. Mag.*, **93**:379–80.

ANDERSON, M. C. (1964). Studies of the woodland light climate: Seasonal variation in the light climate. *J. Ecol.*, **52**:643–63.

ATKINSON, B. W. (1970). Meso-systems in the atmosphere. *Can. Geogr.*, **14**:286–308.

BAIER, W. (1965). The interrelationships of meteorological factors, soil moisture and plant growth. *Int. J. Biometeorol.*, **9**:5–20.

BARRY, R. G. (1970). A framework for climatological research with particular reference to scale concepts. *Trans. Inst. Brit. Geogr.*, **49**:61–70.

BARRY, R. G. and CHAMBERS, R. (1966). Albedo variations in southern Hampshire and Dorset. *Weather*, **21**:60–5.

BARSCH, D. (1965). Les arbres et le vent dans la vallée méridionale du Rhône. *Rev. de Géographie de Lyon*, **40**:35–45.

BAUMGARTNER, A. (1967). The balance of radiation in the forest and its biological function. In Tromp, S. W. and Weihe, W. H. (eds) *Biometeorology Vol. 2 Pt. 2,* Pergamon Press., Oxford, 743-54.

BENDELOW, V. C. (1969). A determination of the albedo of Morecambe Bay, north of a line from Aldingham to Bare. *Met. Mag.*, **98**:305–9.

Beran, D. W. (1967). Large amplitude lee waves and chinook winds. *J. Appl. Met.*, **6**:865–77.

Bergen, J. D. (1969). Cold air drainage on a forested mountain slope. *J. Appl. Met.*, **8**:884–95.

Berglund, E. R. and Mace, A. C. (1972). Seasonal albedo variations of black spruce and sphagnum-sedge bog cover types. *J. Appl. Met.*, **11**:806–12.

Berndt, H. W. and Fowler, W. B. (1969). Rime and hoarfrost in upper-slope forests of eastern Washington. *J. Forestry*, **67**:92–5.

Best, A. C. (1931). Horizontal temperature differences over small distances. *Q. Jl. R. Met. Soc.*, **57**:169–75.

Brinkmann, W. A. R. (1971). What is a foehn? *Weather*, **26**:230–9.

Brinkmann, W. A. R. and Ashwell, I. Y. (1968). The structure and movement of the chinook in Alberta. *Atmosphere*, **6**:1–10.

Broadbent, L. (1950). The microclimate of the potato crop. *Q. Jl. R. Met. Soc.*, **76**:339–454.

Brown, K. W. and Covey, W. (1966). The energy-budget evaluation of the micro-meteorological transfer processes within a cornfield. *Agric. Met.*, **3**:73–96.

Brunt, D. (1945). Some factors in micro-climatology. *Q. Jl. R. Met. Soc.*, **71**:1–10.

Burrage, S. W. (1972). Dew on wheat. *Agric. Met.*, **10**:3–12.

Cambié, G. M. (1969). Il fiume che ha sete. *Vie d'Italia e del mondo*, **2**:263–5.

Catchpole, A. J. W. (1963). The Houghall frost hollow. *Met. Mag.*, **92**:121–9.

Cheo, K. H. (1946). Ecological changes due to thinning red pine. *J. Forestry*, **44**:369–71.

Chia, L.-S. (1967). Albedo of natural surfaces in Barbados. *Q. Jl. R. Met. Soc.*, **93**:117–20.

Clark, O. R. (1940). Interception of rainfall by prairie grasses, weeds and certain crop plants. *Ecol. Monographs*, **10**:243–77.

Clarke, R. H. (1972). The morning glory: an atmospheric hydraulic jump. *J. Appl. Met.*, **11**:304–11.

Clements, J. R. (1966). Solar radiation in a forest clearing. *Weather*, **21**:316–17.

Cole, J. A. and Green, M. J. (1963). Measurements of net radiation over vegetation and of other climatic factors affecting transpiration losses in water catchments. *Int. Ass. Sci. Hydrol.*, Pub. No. 62:190–202.

Corby, G. A. (1954). The airflow over mountains. *Q. Jl. R. Met. Soc.*, **80**:491–521.

Czarnowski, M. S. and Olszewski, J. L. (1968). Rainfall interception by a forest canopy. *Oikos*, **19**:345–50.

Davenport, D. C. and Hudson, J. P. (1967). Changes in evaporation rates along a 17-km transect in the Sudan Gezira. *Agric. Met.*, **4**:339–52.

Davies, J. L. (1952). Some effects of aspect upon valley temperatures in south Cardiganshire. *Geography*, **37**:19–23.

Deacon, E. L. (1969). Physical processes near the surface of the earth. In Landsberg, H. E. (ed.) *World Survey of Climatology. Vol. 2, General Climatology 2*, Elsevier Publishing Co., Amsterdam, 266 pp.

Defant, F. (1951). Local winds. In *Compendium of Meteorology*, Amer. Met. Soc., Boston, 655–72.

Dewalle, D. R. and McGuire, S. G. (1973). Albedo variations of an oak forest in Pennsylvania. *Agric. Met.*, **11**:107–13.

Dight, F. H. (1967). The diurnal range of temperature in Scottish glens. *Met. Mag.*, **96**:327–34.

Dirmhirn, I. and Belt, G. H. (1971). Variation of albedo of selected sagebrush range in the inter-mountain region. *Agric. Met.*, **9**:51–61.

Douglas, C. K. M. (1960). Some features of local weather in south-east Devon. *Weather*, **15**:14–17.

Dunbar, G. S. (1966). Thermal belts in North Carolina. *Geogrl. Rev.*, **56**:516–26.

Dyer, A. J. and Crawford, T. V. (1965). Observations of the modification of the microclimate at the leading edge. *Q. Jl. R. Met. Soc.*, **91**:345–8.

Eddy, A. (1966). The Texas coast sea-breeze: a pilot study. *Weather*, **21**:162–70.

Elliot, A. (1964). Sea breezes at Porton Down. *Weather*, **19**:147–50.

Estoque, M. A. (1961). A theoretical investigation of the sea-breeze. *Q. Jl. R. Met. Soc.*, **87**:136–46.

Findlater, J. (1963). Some aerial explorations of coastal airflow. *Met. Mag.*, **92**:231–43.

Findlater, J. (1964). The sea breeze and inland convection—an example of their interrelation. *Met. Mag.*, **93**:82–9.

Flohn, H. (1969). Local wind systems. In Landsberg, H. E. (ed.) *World Survey of Climatology. Vol. 2, General Climatology 2*, Elsevier Publishing Co., Amsterdam, 266 pp.

Fons, W. L. (1940). Influence of forest cover on wind velocity. *J. Forestry*, **38**:481–6.

Food and Agriculture Organisation (1962). *Forest Influences*. Forestry and Forest Products Studies, No. 15, United Nations, Rome, 307 pp.

Freeman, M. H. (1962). North Sea stratus over the Fens. *Met. Mag.*, **91**:357–60.

Fritts, H. C. (1961). An analysis of maximum summer temperatures inside and outside a forest. *Ecology*, **42**:436–40.

Fuchs, M. and Tanner, C. B. (1968). Surface temperature measurements of bare soil. *J. Appl. Met.*, **7**:303–5.

Gay, L. W., Knoerr, K. R. and Braaten, M. O. (1971). Solar radiation variability on the floor of a pine plantation. *Agric. Met.*, **8**:39–50.

Geiger, R. (1965). *The climate near the ground.* (Trans. by Scripta Technica Inc.) Harvard University Press, 611 pp.

Geiger, R. (1969). Topoclimates. In Landsberg, H. E. (ed.) *World Survey of Climatology. Vol. 2, General Climatology 2*, Elsevier Publishing Co., Amsterdam, 266 pp.

GEORGE, D. J. (1959). A persistent wave cloud over north Breconshire, 13 April 1958. *Weather*, **14**:233–6.
GILL, D. S. (1968). The diurnal variation of the sea-breeze at three stations in north-east Scotland. *Met. Mag.*, **97**:19–24.
GLASSPOOLE, J. (1953). Frequency of cloud at mountain summits. *Met. Mag.*, **82**:156–7.
GLEESON, T. A. (1951). On the theory of cross-valley winds arising from differential heating of the slopes. *J. Met.*, **8**:398–405.
GLENN, C. L. (1961). The chinook. *Weatherwise*, **14**:175–82.
GLOYNE, R. W. (1953). Radiation minimum temperatures over grass surfaces and bare soil surface. *Met. Mag.*, **80**:263–7.
GRAY, A. and STEWART, W. J. (1965). Orographic effects at Acklington. *Met. Mag.*, **94**:8–11.
GREEN, F. H. W. (1967). Air humidity on Ben Nevis. *Weather*, **22**:174–84.
GREEN, F. H. W. (1968). Persistent snowbeds in the W. Cairngorms. *Weather*, **23**:206–9.
GREENLAND, D. (1973). An estimate of the heat balance in an Alpine valley in the New Zealand southern Alps. *Agric. Met.*, **11**:293–302.
GREGORY, S. and SMITH, K. (1967). Local temperature and humidity contrasts around small lakes and reservoirs. *Weather*, **22**:497–505.
HALES, W. B. (1949). Micrometeorology in the tropics. *Bull. Amer. Met. Soc.*, **30**:124–37.
HARMAN, J. R. and HEHR, J. G. (1972). Lake breezes and summer rainfall. *Ann. Ass. Amer. Geogrs.*, **62**:375–87.
HARRISON, A. A. (1967). Variations in night minimum temperatures peculiar to a valley in mid-Kent. *Met. Mag.*, **96**:257–65.
HARRISON, A. A. (1971). A discussion of the temperatures of inland Kent with particular reference to night minima in the lowlands. *Met. Mag.*, **100**:97–111.
HAWKE, E. L. (1933). Extreme diurnal ranges of air temperature in the British Isles. *Q. Jl. R. Met. Soc.*, **59**:261–5.
HAWKE, E. L. (1944). Thermal characteristics of a Hertfordshire frost-hollow. *Q. Jl. R. Met. Soc.*, **70**:23–40.
HELVEY, J. D. (1967). Interception by Eastern White Pine. *Wat. Resour. Res.*, **3**:723–9.
HEYWOOD, G. S. P. (1931). Wind structure near the ground and its relation to temperature gradient. *Q. Jl. R. Met. Soc.*, **57**:433–52.
HEYWOOD, G. S. P. (1933). Katabatic winds in a valley. *Q. Jl. R. Met. Soc.*, **59**:47–58.
HOCEVAR, A. and MARTSOLF, J. D. (1971). Temperature distribution under radiation frost conditions in a central Pennsylvania valley. *Agric. Met.*, **8**:371–83.
HOLMES, R. M. (1969). A study of the climate of the Cypress Hills. *Weather*, **24**:324–30.
HOLMES, R. M. (1970). Meso-scale effects of agriculture and a large prairie lake on the atmospheric boundary layer. *Agron. J.*, **62**:546–9.
HOVIND, E. L. (1965). Precipitation distribution around a windy mountain peak. *J. Geophys. Res.*, **70**:3271–8.
HOWE, G. M. (1953). Observations on local climatic conditions in the Aberystwyth area. *Met. Mag.*, **82**:270–4.
HOWE, G. M. (1955). Soil-warmth in sunny and shaded situations. *Weather*, **10**:49–52.
HURSH, C. R. (1948). *Local climate in the Copper Basin of Tennessee as modified by the removal of vegetation.* US Dep. Agric. Circ. No. 774, Washington, DC.
HURST, G. W. (1967). Further studies of minimum temperature in the forest of Thetford Chase. *Met. Mag.*, **96**:135–142.
JACKSON, I. J. (1971). Problems of throughfall and interception assessment under tropical forest. *J. Hydrol.*, **12**:234–54.
JACKSON, R. J. (1967). The effect of slope, aspect and albedo on potential evapotranspiration from hillslopes and catchments. *J. Hydrol.* (NZ), **6**:60–8.
JEMISON, G. M. (1934). The significance of the effect of stand density upon the weather beneath the canopy. *J. Forestry*, **32**:446–51.
JOHNSON, N. K. (1929). A study of the vertical gradient of temperature in the atmosphere near the ground. *Meteorol. Office, London, Geophys. Mem.*, **46**:32 pp.
JOHNSON, N. K. and DAVIS, E. L. (1927). Some measurements of temperature near the surface in various kinds of soil. *Q. Jl. R. Met. Soc.*, **53**:45–59.
KEEN, C. S. (1968). The development of a nocturnal warm air layer within a shallow valley. *S. African Geogrl. J.*, **50**:135–8.
KENDREW, W. G. (1938). *Climatology*. Oxford University Press, 2nd Ed., 383 pp.
KING, M. B. (1952). A comparison of the daily extremes of temperature at two stations in a Chiltern dry valley. *Q. Jl. R. Met. Soc.*, **78**:102–3.
KITTREDGE, J. (1948). *Forest influences*. McGraw-Hill, New York.
KUNKEL, G. (1966). Einige Temperaturmessungen im liberianischen Regenwald. *Zeitschrift für Meteorologie*, **19**:51–9.
LAKE, J. V. (1956). The temperature profile above bare soil on clear nights. *Q. Jl. R. Met. Soc.*, **82**:187–97.
LAWRENCE, E. N. (1953). Föhn temperature in Scotland. *Met. Mag.*, **82**:74–9.
LAWRENCE, E. N. (1954). Nocturnal winds. *Meteorological Office, Professional Notes* 7, No. 111, HMSO, 13 pp.
LAWRENCE, E. N. (1956). Minimum temperatures and topography in a Herefordshire valley. *Met. Mag.*, **85**:79–83.
LAWRENCE, E. N. (1958). Temperature and topography on radiation nights. *Met. Mag.*, **87**:71–5.
LAWRENCE, E. N. (1960). Variation of surface wind velocity with height in hilly terrain. *Met. Mag.*, **89**:287–92.
LETTAU, H. H. and DAVIDSON, B. (eds) (1957). *Exploring the Atmosphere's First Mile*. Vol. 2, Site Description and Data Tabulation. Pergamon Press, New York, 578 pp.
LOCKWOOD, J. G. (1962). Occurrence of föhn winds in the British Isles. *Met. Mag.*, **91**:57–65.

LOEWE, F. (1950). A note on katabatic winds at the coasts of Adelieland and King George V Land. *Geofis. Pura. Appl.*, **16**: 159–62.

LONG, I. F. (1958). Some observations on dew. *Met. Mag.*, **87**: 161–8.

LONG, I. F., MONTEITH, J. L., PENMAN, H. L., and SZEICZ, G. (1964). The plant and its environment. *Meteorologische Rundschau*, **17**: 97–101.

LOPEZ, M. E. and HOWELL, W. E. (1967). Katabatic winds in the equatorial Andes. *J. Atmos. Sci.*, **24**: 29–35.

LOWRY, W. P. (1959). Energy budgets of several environments under sea-breeze advection in western Oregon. *J. Meteorol.*, **16**: 299–311.

LUDLAM, F. H. (1952). Hill-wave cirrus. *Weather*, **7**: 300–6.

LUMB, F. E. (1970). Topographic influences on thunderstorm activity near Lake Victoria. *Weather*, **25**: 404–10.

LYONS, W. (1972). The climatology and prediction of the Chicago lake breeze. *J. Appl. Met.*, **11**: 1259–70.

MACCREADY, P. B. (1955). High and low elevations as thermal source regions. *Weather*, **10**: 35–40.

MCGINNIGLE, J. B. (1963). Katabatic winds at Acklington during a very cold spell. *Met. Mag.*, **92**: 367–71.

MACHATTIE, L. B. (1968). Kananaskis Valley winds in summer. *J. Appl. Met.*, **7**: 348–52.

MACHATTIE, L. B. (1970). Kananaskis Valley temperatures in summer. *J. Appl. Met.*, **9**: 574–82.

MACHATTIE, L. B. and MCCORMACK, R. J. (1961). Forest microclimate: a topographic study in Ontario. *J. Ecol.*, **49**: 301–23.

MANLEY, G. (1939). On the occurrence of snow-cover in Great Britain. *Q. Jl. R. Met. Soc.*, **65**: 2–27.

MANLEY, G. (1944). Topographical features and the climate of Britain. *Geogrl. J.*, **103**: 241–58.

MANLEY, G. (1945a). The effective rate of altitude change in temperate Atlantic climates. *Geogrl. Rev.*, **35**: 408–17.

MANLEY, G. (1945b). The Helm wind over Cross Fell. *Q. Jl. R. Met. Soc.*, **71**: 197–215.

MANLEY, G. (1971). The mountain snows of Britain. *Weather*, **26**: 192–200.

MENDONCA, B. G. (1969). Local wind circulation on the slopes of Mauna Loa. *J. Appl. Met.*, **8**: 533–41.

MILLER, D. H. (1956). The influence of open pine forest on daytime temperature in the Sierra Nevada. *Geogrl. Rev.*, **46**: 209–18.

MILLER, D. H. (1965). The heat and water budget of the earth's surface. *Advan. Geophys.*, **11**: 175–302.

MOFFITT, B. J. (1956). Nocturnal wind at Thorney Island. *Met. Mag.*, **85**: 268–71.

MONTEITH, J. L. (1956). The effect of grass-length on snow melting. *Weather*, **11**: 8–9.

MONTEITH, J. L. (1957). Dew. *Q. Jl. R. Met. Soc.*, **83**: 322–41.

MONTEITH, J. L. and SZEICZ, G. (1961). The radiation balance of bare soil and vegetation. *Q. Jl. R. Met. Soc.*, **87**: 159–70.

MONTEITH, J. L. and SZEICZ, G. (1962). Radiative temperature in the heat balance of natural surfaces. *Q. Jl. R. Met. Soc.*, **88**: 496–507.

MOROZ, W. J. (1967). A lake breeze on the eastern shore of Lake Michigan. Observations and model. *J. Atmos. Sci.*, **24**: 337–55.

MUNN, R. E. (1966). *Descriptive micrometeorology*. Academic Press, New York and London, 245 pp.

NAGEL, J. F. (1956). Fog precipitation on Table Mountain. *Q. Jl. R. Met. Soc.*, **82**: 452–60.

NORMAN, M. J. T., KEMP, A. W., and TAYLOR, J. E. (1957). Winter temperatures in long and short grass. *Met. Mag.*, **86**: 148–52.

OGUNTOYINBO, J. S. (1970). Reflection coefficient of natural vegetation, crops and urban surfaces in Nigeria. *Q. Jl. R. Met. Soc.*, **96**: 430–41.

OKE, T. R. (1970). The temperature profile near the ground on calm clear nights. *Q. Jl. R. Met. Soc.*, **96**: 14–23.

OLIVER, H. R. (1971). Wind profiles in and above a forest canopy. *Q. Jl. R. Met. Soc.*, **97**: 548–53.

OLIVER, J. (1960). Wind and vegetation in the Dale Peninsula. *Field Studies*, **1**: 2, 12 pp.

OLIVER, J. (1966). Low minimum temperatures at Santon Downham, Norfolk. *Met. Mag.*, **95**: 13–17.

OVINGTON, J. D. (1954). A comparison of rainfall in different woodlands. *Forestry*, **27**: 41–53.

OVINGTON, J. D. and MADZURIK, H. A. I. (1955). A comparison of light in different woodlands. *Forestry*, **28**: 141–6.

PALTRIDGE, G. W. (1968). A note on thermals observed between 300 and 900 metres at Cardington, Bedfordshire. *Met. Mag.*, **97**: 56–8.

PATON, J. (1948). Temperatures and airflow within a wheatfield. *Weather*, **3**: 22–6.

PENMAN, H. L. and LONG, I. F. (1960). Weather in wheat: an essay in micro-meteorology. *Q. Jl. R. Met. Soc.*, **86**: 16–50.

PENMAN, H. L. (1963). *Vegetation and hydrology*. Tech. Comm. No. 53. Commonwealth Agricultural Bureaux, 124 pp.

PEDGLEY, D. E. (1958). The summer sea-breeze at Ismailia. *Met. Office*, Rep. No. 19, HMSO, London, 18 pp.

PEDGLEY, D. E. (1967a). Weather in the mountains. *Weather*, **22**: 266–75.

PEDGLEY, D. E. (1967b). The shapes of snowdrifts. *Weather*, **22**: 42–48.

PEDGLEY, D. E. (1970). Heavy rainfalls over Snowdonia. *Weather*, **25**: 340–50.

PRIESTLEY, C. H. B. (1966). The limitation of temperature by evaporation in hot climates. *Agric. Met.*, **3**: 241–6.

REIFSNYDER, W. E., FURNIVAL, G. M., and HOROWITZ, J. L. (1971). Spatial and temporal distribution of solar radiation beneath forest canopies. *Agric. Met.*, **9**: 21–37.

REYNOLDS, E. R. C. (1967). The hydrological cycle as affected by vegetation differences. *J. Inst. Wat. Engrs.*, **21**: 322–30.

REYNOLDS, E. R. C. and LEYTON, L. (1963). Measurement and significance of throughfall in forest stands. In Rutter, A. J. and Whitehead, F. H. (eds). *The water relations of plants.* Blackwell Scientific Publications, Oxford, 125–41.

REYNOLDS, G. (1956). Local temperature variations in Wirral. *Weather,* **11**:15–16.

RICHARDS, P. W. (1952). *The tropical rain forest.* Cambridge University Press, Cambridge, 450 pp.

RICHARDSON, W. E. (1956). Temperature differences in the South Tyne Valley, with special reference to the effects of air mass. *Q. Jl. R. Met. Soc.,* **82**:342–8.

RIDER, D. J. (1965). Convection from burning cornfields. *Weather,* **20**:238–41.

RIDER, N. E. (1952). The effect of a hedge on the flow of air. *Q. Jl. R. Met. Soc.,* **78**:97–101.

RIDER, N. E. (1957). A note on the physics of soil temperature. *Weather,* **12**:241–6.

RIDER, N. E., PHILIP, J. R., and BRADLEY, E. F. (1963). The horizontal transport of heat and moisture—a micro-meteorological study. *Q. Jl. R. Met. Soc.,* **89**:507–31.

RIEHL, H. (1971). An unusual chinook case. *Weather,* **26**:241–6.

RIJTEMA, P. E. (1968). On the relation between transpiration, soil physical properties and crop production as a basis for water supply plans. *Inst. Land and Water Management Res.,* Tech. Bull. 58, Wageningen.

RITCHIE, W. G. (1969). Night minimum temperatures at or near various surfaces. *Met. Mag.,* **98**:297–304.

ROGERSON, T. L. and BYRNES, W. R. (1968). Net rainfall under hardwoods and red pine in Central Pennsylvania. *Water Resour. Res.,* **4**:55–7.

ROUSE, W. R. (1970). Relation between radiant energy supply and evapotranspiration from sloping terrain: an example. *Can. Geogr.,* **14**:27–37.

ROUSE, W. R. and Wilson, R. G. (1969). Time and space variations in the radiant energy fluxes over sloping forested terrain and their influence on seasonal heat and water balances at a middle latitude site. *Geograf. Ann.,* **51A**:160–75.

SCORER, R. S. (1951). Forecasting the occurrence of lee waves. *Weather,* **6**:99–103.

SEGINER, I. (1969). The effects of albedo on the evapotranspiration rate. *Agric. Met.,* **6**:5–31.

SELLERS, W. D. (1965). *Physical climatology.* University of Chicago Press, Chicago and London, 272 pp.

SIMPSON, J. E. (1964). Sea-breeze fronts in Hampshire. *Weather,* **19**:208–20.

SIMPSON, J. E. (1967). Aerial and radar observations of some sea-breeze fronts. *Weather,* **22**:306–16.

SMITH, F. J. (1967). A comparison of the incidence of fog at a coastal station with that at an inland station. *Met. Mag.,* **96**:77–81.

SMITH, K. (1967a). A temperature transect across the Lima valley, north Italy. *Weather,* **22**:363–6.

SMITH, K. (1967b). A note on minimum screen temperatures in the Houghall frost hollow. *Met. Mag.,* **96**:300–2.

SMITH, K. (1970a). Climate and weather. In Dewdney, J. C. (ed.) *Durham County and City with Teesside.* British Association Local Executive Committee, Durham, 58–74.

SMITH, K. (1970b). The effect of a small upland plantation on air and soil temperatures. *Met. Mag.,* **99**:45–9.

SMITHSON, P. A. (1969). Effects of altitude on rainfall in Scotland. *Weather,* **24**:370–76.

SMITHSON, P. A. (1970). Influence of topography and exposure on airstream rainfall in Scotland. *Weather,* **25**:379–86.

SOONS, J. M. and RAINER, J. N. (1968). Microclimate and erosion processes in the Southern Alps, New Zealand. *Geograf. Ann.* **50A**:1–15.

SPARKS, W. R. (1962). The spread of low stratus from the North Sea across East Anglia. *Met. Mag.,* **91**:361–5.

SPENCE, M. T. (1936). Temperature changes over short distances in the Edinburgh district. *Q. Jl. R. Met. Soc.,* **62**:25–31.

STÅFELT, M. G. (1963). On the distribution of the precipitation in a spruce stand. In Rutter, A. J. and Whitehead, F. H. (eds) *The water relations of plants,* Blackwell Scientific Publications, Oxford.

STANHILL, G. and FUCHS, M. (1968). The climate of the cotton crop: Physical characteristics and microclimate relationships. *Agric. Met.,* **5**:183–202.

STANHILL, G., HOFSTEDE, G. J., and KALMA, J. D. (1966). Radiation balance of natural and agricultural vegetation. *Q. Jl. R. Met. Soc.,* **92**:128–40.

STEVENSON, R. E. (1961). Sea-breezes along the Yorkshire coast in the summer of 1959. *Met. Mag.,* **90**: 153–62.

STEWART, J. B. (1971). The albedo of a pine forest. *Q. Jl. R. Met. Soc.,* **97**:561–4.

TAJCHMAN, S. J. (1971). Evapotranspiration and energy balances of forest and field. *Water Resour. Res.,* **7**:511–23.

TAYLOR, R. J. (1962). Small-scale advection and the neutral wind profile. *J. Fluid Mech.,* **13**:529–39.

THOMAS, T. M. (1956). Topography and weather in S.W. Pembrokeshire. *Weather,* **11**:183–6.

THOMAS, T. M. (1963). Some observations of the chinook arch. *Weather,* **18**:166–70.

THOMPSON, A. H. (1967). Surface temperature inversions in a canyon. *J. Appl. Met.,* **6**:287–96.

THORNTHWAITE, C. W. (1953). Topoclimatology. *Proc. Toronto Meteorol. Conf. R. Met. Soc. (London)*, 227–32.

TURNER, R. W. (1966). Pincher Creek, *Weather,* **21**:412–13.

TYSON, P. D. (1968). Nocturnal local winds in a Drakensberg valley. *S. African Geogrl. Jl.,* **50**:15–32.

TYSON, P. D., and PRESTON-WHYTE, R. A. (1972). Observations of regional topographically-induced wind systems in Natal. *J. Appl. Met.,* **11**:643–50.

URFER-HENNENBERGER, C. (1967). Zeitliche Gesetzmassigkeiten des Berg und Talwindes. *Veroffentl. Schweiz. Meteorol. Z. Anst.,* **4**:245–52.

VAN WIJK, W. R. (ed.) (1963). *Physics of plant environment*. North-Holland Publishing Co., Amsterdam, 382 pp.

VÉZINA, P. E. and BOULTER, D. W. K. (1966). The spectral composition of near ultraviolet and visible radiation beneath forest canopies. *Can. J. Bot.*, **44**: 1267–84.

WACO, D. E. (1968). Frost pockets in the Santa Monica mountains of southern California. *Weather*, **23**: 456–61.

WALLINGTON, C. E. (1960). An introduction to lee waves in the atmosphere. *Weather*, **15**: 269–76.

WARD, F. W. (1953). Helm wind effect at Ronaldsway, Isle of Man. *Met. Mag.*, **82**: 234–7.

WARD, R. C. (1971). Measuring evapotranspiration: a review. *J. Hydrol.*, **13**: 1–21.

WARNER, J. (1968). A reduction in rainfall associated with smoke from sugar-cane fires—an inadvertent weather modification? *J. Appl. Met.*, **7**: 247–51.

WATERHOUSE, F. L. (1955). Microclimatological profiles in grass cover in relation to biological problems. *Q. Jl. R. Met. Soc.*, **81**: 63–71.

WATTS, A. J. (1955). Sea-breeze at Thorney Island. *Met. Mag.*, **84**: 42–8.

WEST, E. S. (1952). A study of the annual soil temperature wave. *Australian J. Sci. Res.*, Ser. A, **5**: 303–14.

WEXLER, R. (1946). Theory and observations of land and sea breezes. *Bull. Amer. Met. Soc.*, **27**: 272–87.

WILKEN, G. C. (1967). Snow accumulation in a manzanita brushfield in the Sierra Nevada. *Water Resour. Res.*, **3**: 409–22.

YOSHINO, M. (1964). Some local characteristics of the winds as revealed by wind-shaped trees in the Rhone Valley in Switzerland. *Erdkunde*, **18**: 28–39.

YOUNG, F. D. (1921). Nocturnal temperature inversions in Oregon and California. *Mon. Weath. Rev.*, **49**: 138–48.

ZINKE, P. J. (1967). Forest interception studies in the United States. In Sopper, W. E. and Lull, H. W. (eds) *Forest Hydrology*, Pergamon, Oxford, 137–61.

# 3. Man-made climates

The previous chapter concluded with some reference to the inadvertent climatic effects produced by certain rural land-use practices. Such consequences tend to be fairly isolated and short-lived when compared with the more significant modifications of the planetary boundary layer produced by the spread of urbanization and the associated development of manufacturing industry. Over extensive areas, these man-made effects can assume equal importance with regional factors or the influence of topography in controlling local climate, and for this reason, as well the distinctive nature of the physical processes at work, urban topoclimates are treated separately. For convenience, this chapter is divided into two main sections. The first section deals with the more localized consequences of the whole complex of man-made changes—including air pollution—within the urban area itself, while the second part examines the wider meteorological implications of atmospheric contamination in both space and time.

## Urban climates

### 3.1 Introduction

It has been known for many years that cities create their own climates. The earliest detailed demonstration of this fact came in the nineteenth century, when Howard (1818) published the first edition of his work on London based on comparative temperature measurements begun in 1806 at Somerset House and various sites in what was at that time open country beyond the urban fringe. These pioneer studies were followed-up by investigations in several European cities during the opening decades of this century, but it is only within the last twenty years that the accelerating growth of urbanization, industrialization, and motor transport, together with the associated air pollution, has necessitated a better appreciation of urban climates. The implications of such changes have been examined by Lowry (1967), who concluded that, although on the credit side cities enjoy fewer days with snow, a longer gardening season, and lower heating bills than the surrounding countryside, they also have contaminated air, poorer visibility, less sunshine, and possibly more rainfall.

The overall complex of climatic modifications resulting from mid-latitude cities has been summarized by Landsberg (1960) and is shown in Table 3.1.

| *Element* | *Comparison with rural environs* |
|---|---|
| Contaminants: | |
| dust particles | 10 times more |
| sulphur dioxide | 5 times more |
| carbon dioxide | 10 times more |
| carbon monoxide | 25 times more |
| Radiation: | |
| total on horizontal surface | 15 to 20% less |
| ultraviolet, winter | 30% less |
| ultraviolet, summer | 5% less |
| Cloudiness: | |
| clouds | 5 to 10% more |
| fog, winter | 100% more |
| fog, summer | 30% more |
| Precipitation: | |
| amounts | 5 to 10% more |
| days with 5 mm | 10% more |
| Temperature: | |
| annual mean | 0·5 to 0·8 °C more |
| winter minima | 1 to 1·5 °C more |
| Relative humidity: | |
| annual mean | 6% less |
| winter | 2% less |
| summer | 8% less |
| Wind speed: | |
| annual mean | 20 to 30% less |
| extreme gusts | 10 to 20% less |
| calms | 5 to 20% more |

**Table 3.1** Generalized urban features compared with rural environs. (After Landsberg, 1960.)

Despite the wide-ranging nature of the elements listed, the urban atmospheric environment depends on the local modification of three main climatic factors: the heat balance, the composition of the air, and surface roughness conditions. These factors influence more than one climatic element, as when changes in the heat balance have repercussions on urban visibility and rainfall in addition to thermal conditions, and it will be convenient to treat these major controls separately from the discussion of the climatic elements themselves.

So far, few studies of the urban heat balance have been attempted, but it is clear that the heat transfer processes are altered, since temperatures are generally higher within the city limits than outside. This feature is partially achieved by the back-radiation of heat from the overlying smoke haze and the surrounding buildings, plus the reduced windspeeds within the city which limit the turbulent transfer of heat from the urban area. On a diurnal time-scale, the high thermal capacity of most building materials, such as concrete and bricks, allows the nocturnal release of heat energy received and stored during the daylight hours. Although all these factors are significant from time to time, the chief modification of the urban heat budget arises from the artificial output of heat produced by the combustion of fossil fuels.

Most of this output is in the form of low-level radiant heat, and the first calculations relating to this were made by Eaton (1877), who estimated that the consumption of 5 million tons of coal in the London Metropolitan District would be sufficient to raise the mean temperature of an air layer 30 m thick by 1·2 degC. Together with the physiological heat generated by the 3·5 million inhabitants of London, Eaton calculated that the total temperature rise would be around 1·4 degC. Similar work was undertaken on several large German towns by Kratzer (1956). He was able to show that coal consumption liberated a heat equivalent of some 40 cal/ $cm^2$ day, and, since radiation measurements in Hamburg indicated that the sun and sky provide an average of 34 cal/ $cm^2$ day in December and 50 cal/$cm^2$ day in January, the inevitable conclusion was that, in mid-winter, the artificial heat release was about the same as the natural contribution from insolation.

More recently, Garnett and Bach (1965) have made calculations of artificial heat released within the urban area of Sheffield in 1952 from all fuel consumption and the heat generated by human and animal metabolism. The results indicated an artificial radiation heat amounting to some 14·6 kg/cal $cm^2$ yr, which represented nearly one-third of the net radiation balance or about one-fifth of the total solar radiation received during that year. Expressed another way, it counter-balanced about half of the heat lost by net longwave radiation exchange. In addition to radiant heat, cities also liberate sensible heat especially in the form of warm-water effluent from sewage works, factories, and power-stations. Lamb (1963) has described such thermal pollution of the river Thames at the end of a long, cold spell in January 1963, when the water temperature rose from 0 °C at the mouth of the river obstructed by ice up to as much as 10·3 °C in central London.

It is commonplace for atmospheric pollution to be at its most severe in urban areas as a result of the concentration of the combustion process for residential, commercial, and industrial requirements. The most important solid pollutants are smoke particles, which are caused mainly by the incomplete combustion of solid fuels in domestic appliances, while sulphur dioxide remains the greatest source of general gaseous pollution and arises from the burning of the sulphur content which most fossil fuels have. On average, the nature and intensity of urban air pollution can be related to the character and density of the built-up area (Landsberg, 1967). Thus Fig. 3.1, taken from Chandler (1965), shows the average smoke distribution in London from April 1957 to March 1958. Apart from the general tendency of smoke pollution to decrease towards the outer suburbs, it can be seen that low values were found in central London, partly because of the small resident population and the scarcity of smoke-producing industries in this inner zone. On the other hand, heavily polluted air occurred in the lower Lea valley and between Bermondsey and Lewisham south of the river. These areas were mainly composed of densely populated eighteenth- and nineteenth-century terraced housing, with its associated industry, and formed concentrated sources of smoke pollution.

However, it should not be thought that such average patterns are in any way static, because urban air pollution is a very dynamic feature dependent on both man-made and natural factors. Some of these factors will be examined more closely later in the chapter, but McCormick (1971) has noted that, among the buildings of a city, the pollution concentration range can cover as much as two or more orders of magnitude. For example, in the downwind vicinity of a moderate local source of sulphur dioxide, the observed ground-level concentration of the concentrations of atmospheric smoke and lead pollution were respectively up to five and six times greater in the congested main street than in other parts of the borough subject to less exposure to vehicle exhausts.

Anti-pollution legislation can create significant trends. Commins and Waller (1967) reported a 32 per cent

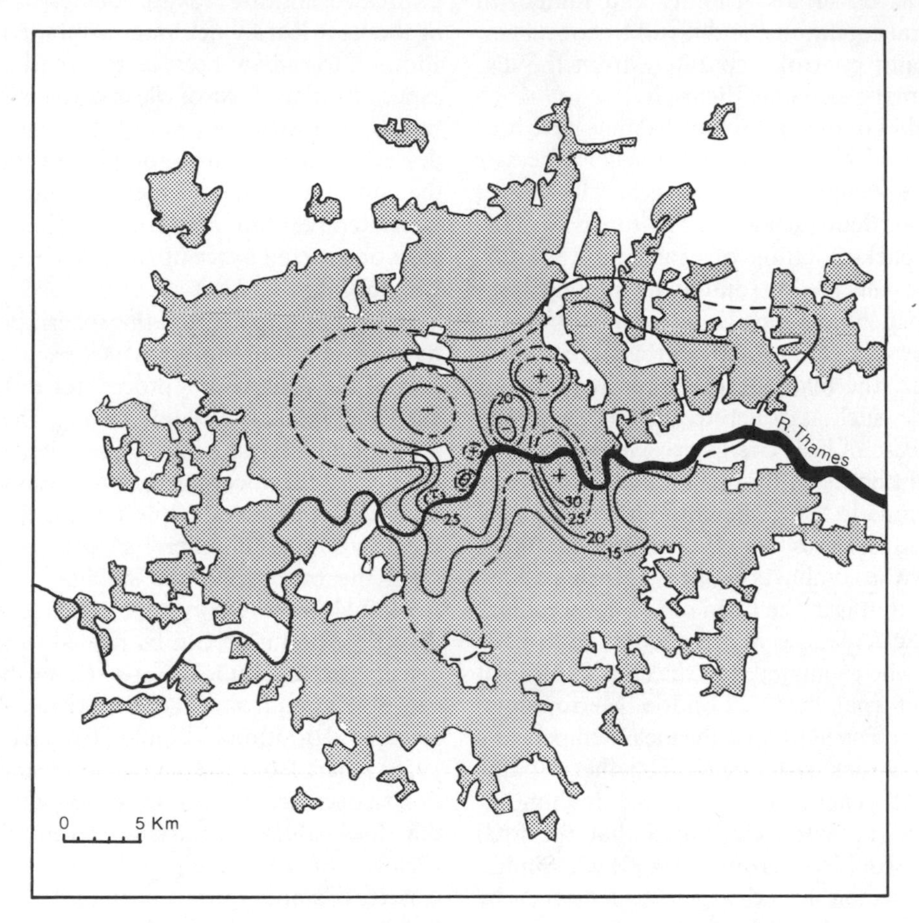


**Fig. 3.1 The average distribution of smoke concentration in London from April 1957 to March 1958 in mg/100m³. After Chandler (1965).**

gas may be around 3000 μg/m$^3$ compared with an urban average of the order of 300 μg/m$^3$ and a concentration in the clean air entering the city about 30 μg/m$^3$.

The movement of vehicular traffic within a town may well influence local pollution levels. This was demonstrated by Bullock and Lewis (1968) for the town of Warwick in the English Midlands, where measured reduction in smoke concentration within central London from 1959 to 1964 compared with the preceding five-year period as a result of declining coal consumption, although sulphur dioxide levels remained about the same owing to large outputs from power stations and new multi-storey office blocks. In some urban areas, a significant proportion of the atmospheric pollution may be imported from external sources. For example, in a study of smoke and sulphur dioxide patterns in the Reading area, Parry (1967a) found that, although the zone of maximum pollution was rarely displaced far from the closely built-up town centre, the highest smoke concentrations were invariably associated with light easterly winds which implied the transfer of some pollution from the Greater London area about 38 km to the east.

The major natural variable affecting urban air pollution is, of course, the weather itself. Thus, Commins and Waller (1967) have shown that the average smoke concentration in the City of London in winter is about three times that of summer, with a sulphur dioxide level about twice that of summer. Short-term variations are such that a ten-fold increase in smoke concentration has been observed within one hour and, in a study of the steel town of Windsor in Canada, Munn *et al.* (1969) noted that the daily prevailing wind was not necessarily the most reliable air-quality criterion, since, for example, most of the air pollution could come with only 4 h of westerly winds although the remainder of the day might have easterlies. Similar importance has been attached to local wind conditions in the industrialized town of Hamilton at the western end of Lake Ontario where Weisman *et al.* (1969) found the smoke pollution pattern to be complicated during the summer by a well-developed lake breeze.

Many authors have used multiple regression techniques to analyse the relationships between air pollution and the weather. In a study of sulphur dioxide concentrations in central Stockholm, Bringfelt (1971) obtained a good correlation between daily values and air temperature, wind speed, and mixing height variables over two winters, and this allowed daily sulphur dioxide levels to be predicted with a standard error of about 25 per cent.

The varied geometry of a city skyline introduces a high degree of surface roughness into the landscape which exceeds that of most rural areas. The chief impact of this roughness is on the air flowing over the urban surface, so that mechanical turbulence is set up. The total effect of this frictional drag is to reduce windspeeds within the built-up area, although localized turbulence and eddying leads to marked increases and gustiness under certain conditions. Furthermore, the mechanical turbulence defines, in part, the depth of the mixing layer into which pollutants will be dispersed and therefore is also an index of the severity of pollution likely to result under stated conditions of pollution output.

### 3.2 Radiation and energy balance

As might be expected, contaminated urban air is effective in reducing incoming solar radiation, and differences in albedo and other surface characteristics also modify the energy balance over cities. The interception of insolation is usually most apparent during the winter, when not only is the haze layer thicker, but in addition the sun's rays strike the overlying blanket at the lowest angles. According to Chandler (1965), the loss of direct solar radiation by suspended matter, principally smoke, on cloudless days at Kew averages 8·5 per cent when the sun's elevation is 30° and 12·8 per cent with an elevation of only 14·3°. On the gloomiest winter days, 90 per cent of all radiation may be intercepted and Meetham (1945) estimated that the city of Leicester lost 30 per cent of the incoming radiation in winter, compared with only 6 per cent in summer. Monteith (1966) showed that, during the 1957–63 period, the average smoke density in central London decreased by 10 $\mu g/m^3$ and that this led to a total solar radiation increase of around 1 per cent. Nevertheless, the mean smoke concentration of 200–300 $\mu g/m^3$ in the city centre still made the total receipt of solar energy some 20 to 30 per cent lower than that recorded in nearby rural areas.

No complete energy budget has yet been attempted in urban climatology but some preliminary work has been undertaken in the US as a first step towards the formulation of causal physical models of the thermal environment of cities. A heat budget study of Cincinnati city and its environs in summer anticyclonic weather by Bach and Patterson (1969) has revealed the 'greenhouse effect' of the local smoke pall, with shortwave radiation fluxes 6 per cent smaller and longwave radiation fluxes from 1 to 14 per cent larger within the city than outside. On a cloudy September day in Los Angeles, Terjung *et al.* (1970) found that, although there were marked areal patterns in the energy balance according to land-use contrasts, about 80 per cent of the net radiation received was disposed of by sensible heat flux into the air above the dry urban surfaces with the remainder entering pavements and other similar surfaces to be stored as heat flux into the ground.

Much of the modification of urban climates occurs in the lowest layers of the atmosphere and Bach *et al.* (1970) have claimed that the lowest 1000 m contribute on 'clean' days up to 35 per cent, and on 'polluted' days up to 70 per cent, of the total attenuation of the solar beam. Similarly, in the Greater Cincinnati area, Bach (1971) reported that air layers only 30 m deep may be responsible for as much as 15 per cent of the total attenuation of the solar beam and was also able to demonstrate some of the spatial variations of the Cincinnati haze layer.

The largest areal differences in daytime urban energy budgets tend to occur during clear summer weather and Terjung (1970) has presented maps which depict something of the spatial pattern in downtown Los Angeles during the early afternoon of 24 May 1968. Figure 3.2A shows the variation of incident solar radiation (direct beam plus diffuse solar radiation) on a horizontal surface in langleys per minute. It can be seen that high values occur to the north and south-east, whereas lower values in the centre coincide with the busy Santa Monica Free-

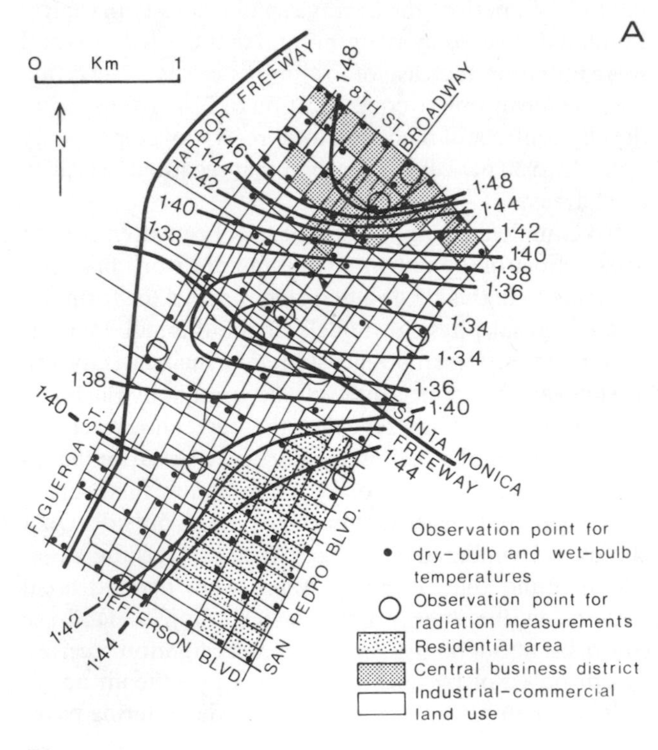


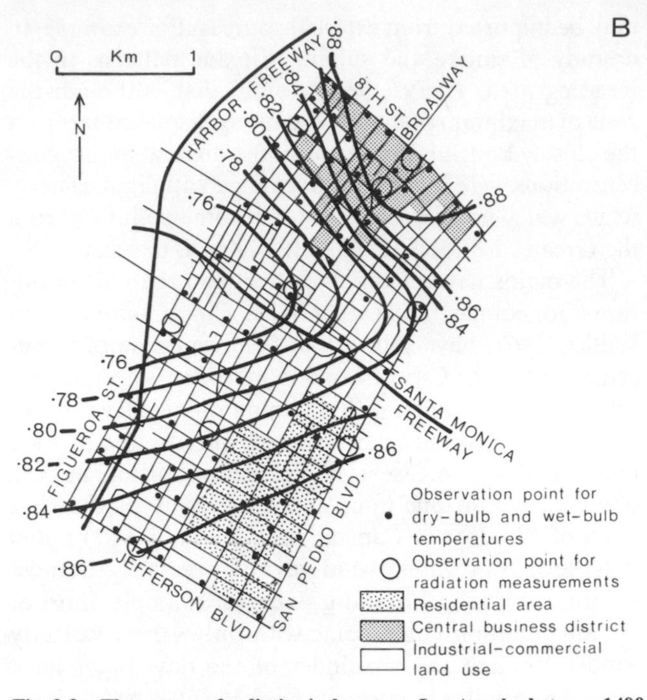


**Fig. 3.2 The pattern of radiation in downtown Los Angeles between 1400 and 1515 h on 24 May 1968.**
**(A) Direct plus diffuse solar radiation ($Q+q$) in langleys/min.**
**(B) The net radiation balance.**
**After Terjung (1970). Copyrighted by the American Geographical Society of New York.**

way and with the industrial areas where the pollution haze may well be thickest. The pattern of net radiation in Fig. 3.2B indicates a broadly comparable picture. According to Terjung, the residential zones in the south-east have a high net radiation because of lower albedo due to more vegetation, lower surface temperatures, and generally lower terrain temperatures, which together result in lower rates of emission of terrestrial radiation.

## 3.3 Sunshine and visibility

Like incoming radiation, the duration of bright sunshine shows significant reductions over urban areas, and in large cities a progressive decline can be traced towards the city centre. During the period 1921–50, the mean daily sunshine duration outside London was 4·33 h compared with 4·07 h in the outer suburbs, 3·95 h in the inner, low-lying suburbs and 3·60 h in the central area (Chandler, 1965). This amounted to an average loss ranging from

16 min/day in the outer suburbs to 44 min in the centre; the monthly values, expressed as a percentage of bright sunshine outside London, are shown in Fig. 3.3. This diagram indicates that, in mid-winter, the central district

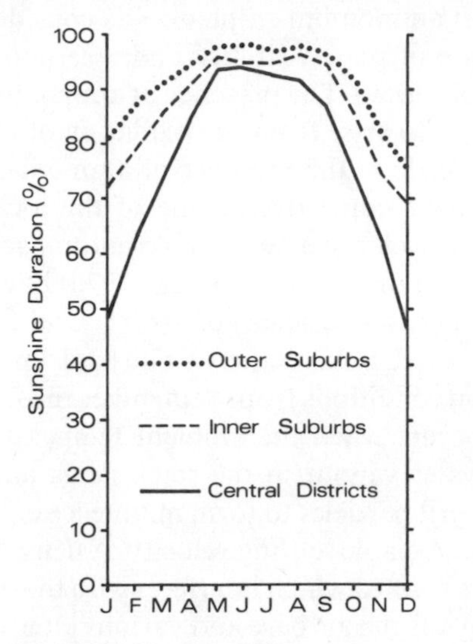


**Fig. 3.3 Mean monthly duration of bright sunshine at selected London stations as a percentage of the sunshine recorded outside London from 1921 to 1950. After Chandler (1965).**

had sunshine values which were less than half those recorded outside London and even in the summer months the duration was often 10 per cent below that of the surrounding countryside.

Such effects have been largely attributed to urban smoke pollution which, during recent years, has been progressively reduced over many cities as a result of statutory controls on smoke emission. The subsequent improvement in sunshine duration in the London area has been noted by Jenkins (1969), who examined average monthly sunshine between 1958 and 1967 at three sites as a percentage of the monthly normals for the 1931–60 period. Figure 3.4 shows that the greatest increases have taken place between September and March at the London Weather Centre, which represents a city-centre location compared with the suburban and rural sites of Kew and Wisley respectively. For the winter months from November to January, the average increase of bright sunshine has been about 50 per cent above the long-term mean and such changes are not untypical of other British cities.

In addition to reduced sunshine duration, most urban areas are also associated with poor visibility as a result of the local pollution and, in a survey of conditions near Liverpool, Reynolds (1957) estimated that the reduction in visibility due to Merseyside industry was about 25 per cent in the mornings when the air was fairly stable compared with only 10 per cent during the afternoons. Traditionally, London has held an unenviable inter-

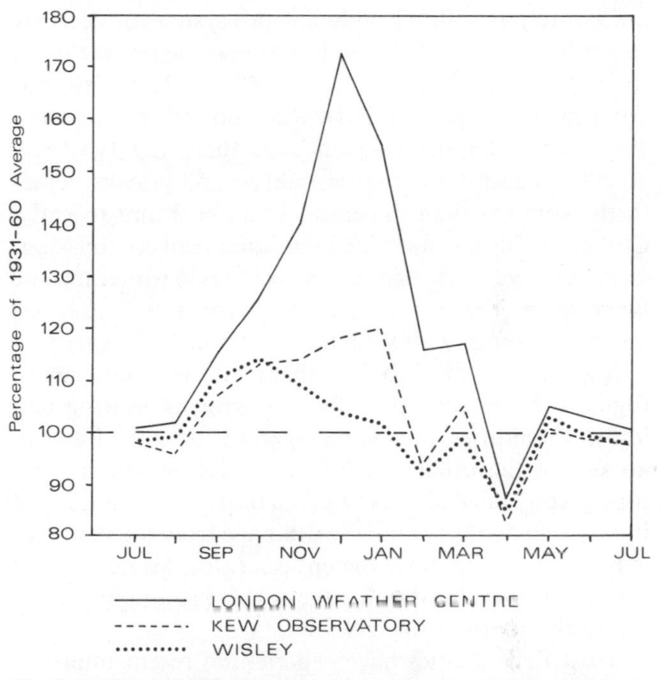


**Fig. 3.4 Mean monthly duration of bright sunshine at three stations in the London area during 1958–67 expressed as a percentage of the average for the 1931–60 period. After Jenkins (1969).**

national reputation for the prevalence of thick, smoky fogs as described by Lawrence (1953). Such 'smogs' have been a particular feature of the winter months but it would be wrong to attribute the decreased visibility to the *direct* effects of smoke pollution. Thus, in the severe fog outbreak of December 1952, although visibility was often less than 9 m and smoke concentration up to 15 times the normal December level, Douglas and Stewart (1953) have claimed that only 2 per cent of the reduced visibility was due to the suspended smoke particles themselves. This means that 98 per cent of the fog resulted from suspended water droplets, which were admittedly made smaller and more numerous by the *indirect* action of smoke particles operating as condensation nuclei.

Similarly, it should not be thought that urban areas always suffer the densest fogs: Shellard (1959) suggested that visibilities less than 200 m are more frequent in the rural parts of East Anglia than in central London. According to Chandler (1962a), although overall fog frequencies, defined as visibilities of less than 1000 m, are greatest in central London, with a systematic decrease towards the limit of the built-up area, the reverse tends to be true for visibilities below 400 m. This view was confirmed by Brazell (1964), who showed that between 1947 and 1962 the frequency of thick fog (visibility < 200 m) and dense fog (visibility < 50 m) was higher in the suburbs than in central London. More recently, Collier (1970) has described the same features for Manchester where, although city-centre fogs in winter may be twice as frequent as in the rural environs, the latter has more dense fogs with visibility < 500 m.

It would appear, therefore, that really dense city-centre fogs are often prevented by the artificial heating and lower humidities prevailing over the heavily built-up areas. On the other hand, local pollution sources and topography must always be taken into consideration and it may well be that katabatic drainage from the Pennines helps to limit fog development in central Manchester at the same time that small valleys are encouraging thick fog to the south of the city.

Most British cities have experienced recent improvements in visibility as a result of pollution control. In London, both Wiggett (1964) and Freeman (1968) have noted a significant decline in the frequency of visibilities in the range of 200–1000 m, although the incidence of thick fog (< 200 m) has been little changed. This suggests that, although smoke pall conditions have been alleviated, there has been little effect on dense water fogs. In Manchester, during the winter half-year, the incidence of hourly visibilities of < 2000 m under calm conditions has fallen from 71 per cent in 1955–59 to 61 per cent in 1960–64. The restriction of such analysis to calm conditions eliminates the problem of comparability of windspeed between one selected period and another (Atkins, 1968).

In some areas, the emission of airborne industrial effluent has been specifically implicated in the reduction of visibility. On Teesside in north-eastern England, ammonium and sulphate ions were found to comprise 65 per cent of the total soluble ion content of the atmosphere and ammonium sulphate was considered to be a major cause of poor visibility under certain conditions by Eggleton (1969). The presence of ammonium sulphate was believed to arise from the oxidation of atmospheric sulphur dioxide in the presence of ammonia gas, which may well have come from some of the local chemical industries as well as some possible natural sources.

In Alaska, Porteous and Wallis (1970) have described how moist gases discharged from coal-fired power stations at the US Air Bases at Fairbanks produce low-level ice fog conditions from September through to May. The fog occurs when the ambient temperature is such that the water vapour in the stack gases nucleates and freezes on grit particles to form minute ice crystals which have no appreciable settling velocity. A dense fog blanket is therefore formed which interferes with the operation of the runways at the air base and various alternatives have been suggested for drying the stack gases before emission.

### 3.4 Temperature

The characteristic increase of temperature towards the centre of large built-up areas is probably the most publicized aspect of urban climates. There is a long record of research into the so-called 'urban heat-island' effect, starting with comparative studies of fixed station sites in the nineteenth century, but a major breakthrough can be traced to the work of Schmidt (1929), who pioneered the use of motorized traverses in Vienna as a technique for understanding the spatial variations of minimum temperatures within the city. This method has been adopted by various workers, including Chandler (1960 and 1961a) in London, and new instrumentation is continually being developed to minimize thermal interference from small-scale convection from hot roads or car exhausts (Conrads and Van der Hage, 1971).

The results of such traverses can provide a clear indication of the cumulative horizontal influence of towns on temperature. For example, Fig. 3.5 depicts the mean temperature pattern at about 23.30 h on a roughly north–south route across London during the anticyclonic night of 11–12 October 1961 (Chandler, 1962b).

Despite the irregular thermal variations, which are related to the density of the urban fabric, the progressive rise and fall of temperature across the urban area is quite marked.

tures, are weaker and less persistent than those produced at night, and in London the maximum temperature differential to be expected in the early afternoon of a summer's day is about 4 degC. In contrast, night-time

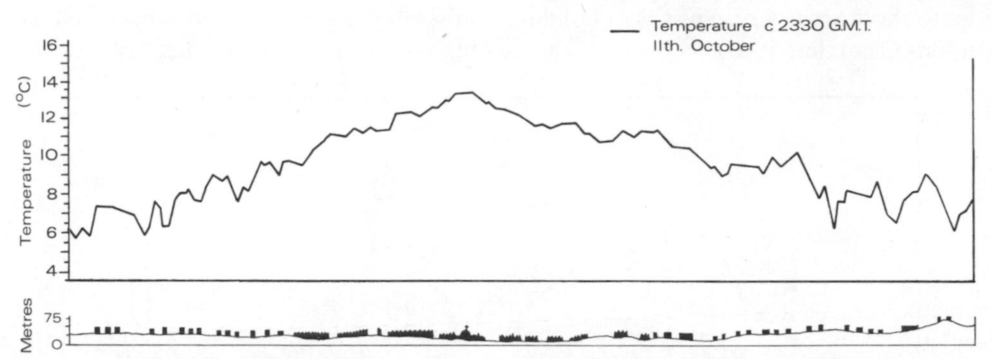


**Fig. 3.5 Nocturnal temperature across London as revealed by a traverse on the night of 11–12 October 1961. After Chandler (1965).**

Large cities such as London produce a thermal environment which is sufficiently distinct to be recognizable on the basis of mean annual data and Table 3.2 indicates that, although all values provide evidence for high city-centre temperatures, the largest increase is found with mean minima, which can be expected to be about 2 degC higher in central London than in the surrounding rural areas.

However, mean values tend to convey a false impression of stability, since the form and intensity of heat-islands are highly variable. In London, for example, although the heat-island generally follows the margin of the built-up area, the centre is usually displaced towards the north-east of the city as a result of the prevailing south-westerly winds and the high thermal capacity of the closely spaced terraced housing in that area (Chandler, 1962c). Heat-islands which develop during daylight hours, as identified by analysis of maximum tempera-

| | *Mean height (m)* | *Maximum (°C)* | *Minimum (°C)* | *Mean (°C)* |
|---|---|---|---|---|
| Surrounding country | 87·5 | 13·7 | 5·5 | 9·6 |
| Margins, high level | 144·2 | 13·4 | 6·2 | 9·8 |
| Suburbs, high level | 137·2 | 13·4 | 5·9 | 9·7 |
| Suburbs, low level | 61·9 | 14·2 | 6·4 | 10·3 |
| Central districts | 26·5 | 14·6 | 7·4 | 11·0 |

**Table 3.2** Mean annual temperatures in London, 1931–60. (After Chandler, 1965.)

heat-islands are stronger and simpler in form especially when they occur under conditions of a winter anticyclone. This is because not only is there a greater artificial release of heat during the winter, but also turbulence is normally weaker than in the day and local variations in cloud cover are less regionally differentiating.

Figure 3.6 depicts the sort of intense nocturnal heat-island which can develop over London with a strong anticyclone. On this occasion, the temperature increase was 6·7 degC with a typically steep thermal gradient along the margin of the built-up area and a slight displacement of the heat-island centre towards the north-east. It is interesting to note the particularly sharp temperature contrast in the east of the city, where cool air lying over lower Thames-side abuts more or less directly against the warm air created by the dense urban development in the east end of London. The result is a thermal discontinuity which is stronger than that produced by most frontal systems crossing Britain, and Chandler (1965) has suggested that such marked contrasts may be sufficient to generate localized thermal winds moving irregularly inwards towards the warmer central districts.

Although the large size of the London conurbation ensures ample illustration of urban heat-island phenomena, many lesser settlements, including small villages, show comparable features (Harrison, 1967). Indeed, Chandler (1964) has specifically suggested that the intensity of heat-islands is not necessarily related to city growth. In a later comparative study between London and Leicester, the same author concluded that, despite the large discrepancy in the size of the two cities, the nocturnal heat-islands created by Leicester revealed similar intensities to those areas with equivalent building densities in London (Chandler, 1967).

distributed by chinook winds and Preston-Whyte (1970) has shown how the centre of the summer daytime heat-island in Durban, South Africa, can be displaced from the central business district by the local sea breeze. Cities in sheltered valley locations may well have an unusually high frequency of intense heat-islands and Garnett (1967)

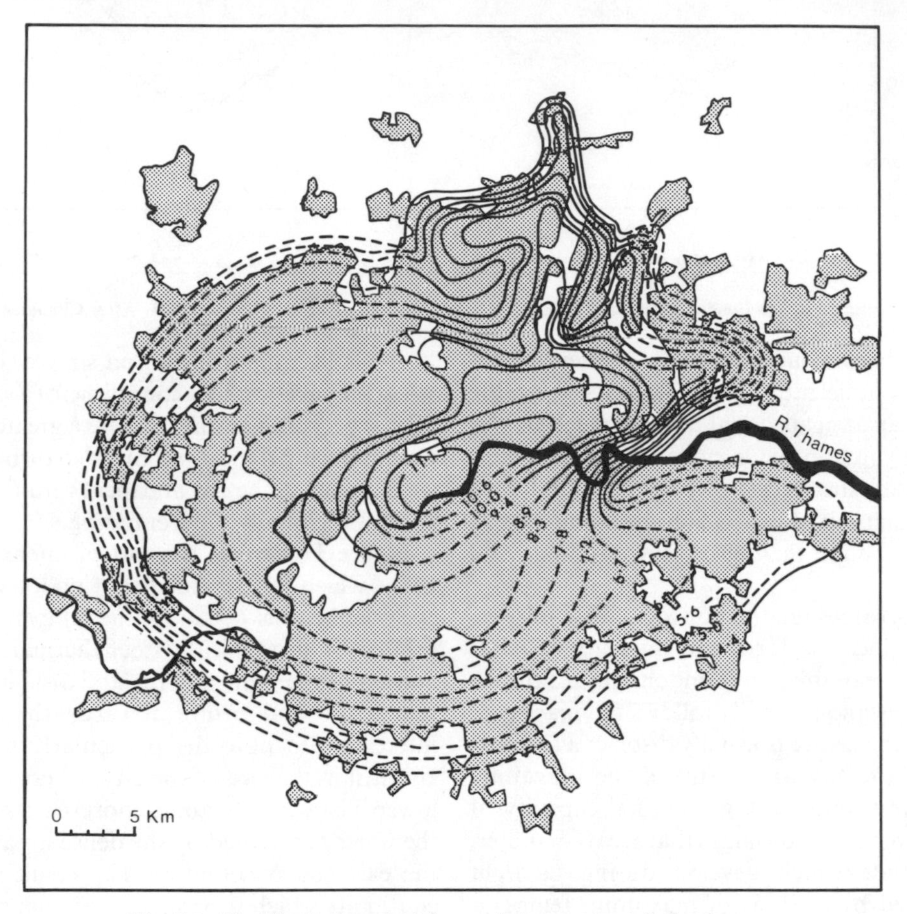


**Fig. 3.6 The distribution of minimum air temperatures in London on 14 May 1959. Broken lines indicate some uncertainty of position. After Chandler (1965).**

Certainly there is now plenty of evidence that heat-island conditions occur in small cities, such as that of Corvallis, Oregon, described by Hutcheon *et al.* (1967), and it appears that local factors other than settlement size may be more important influences. For example, Yudcovitch (1966/7) has reported how the urban temperature excess at Calgary may occasionally be re-

has documented a nocturnal temperature difference of up to 7·8 degC between the central core of Sheffield and the city margins when the latter were snow-covered during winter anticyclonic conditions. On the other hand, more recent work by Oke (1973) has demonstrated a recognizable relationship between city size, as measured by the population total, and the intensity of the urban heat-island. Survey data obtained under uniform conditions by automobile traverses for 10 comparable settlements ranging from 1000 to 2 million inhabitants in the St Lawrence lowlands of Canada showed that, on cloud-free nights, the magnitude of the urban temperature excess was related to the inverse of the regional windspeed and the logarithm of the population total. Overall, the heat-island intensity appeared to be roughly proportional to the fourth root of the population. In an attempt to validate the relationship more widely, Oke plotted the maximum observed differences between

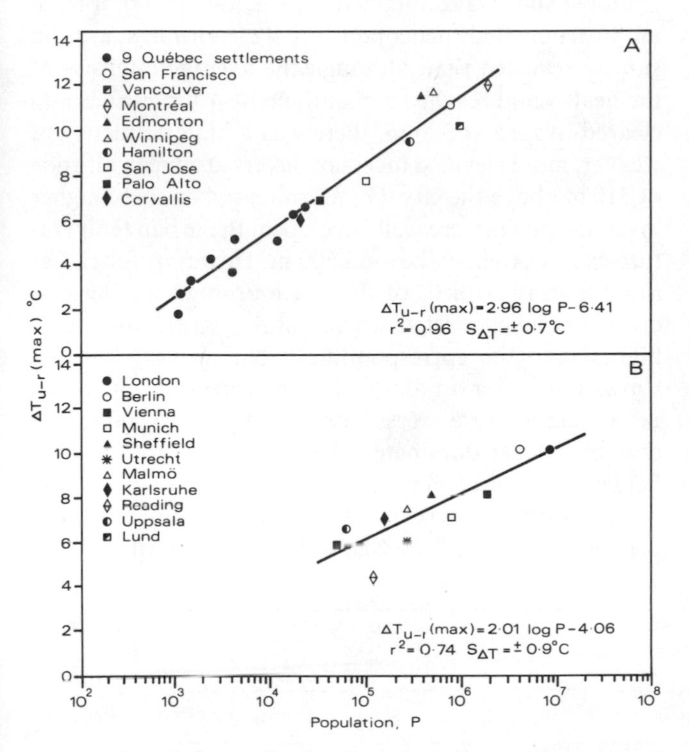


**Fig. 3.7 The relationship between observed maximum urban heat-island temperature and the logarithm of the population size for (A) North American settlements and (B) European settlements. After Oke (1973).**

background rural and urban temperatures $T_u - T_r$ (max) as reported in the literature for North American and European settlements, against the logarithm of the population.

As shown in Fig. 3.7, good correlation exists for both sets of data and the chief difference lies in the slope of the regression line so that, for a given population size, the European heat-island is apparently smaller than its North American counterpart. This result was somewhat unexpected in view of the generally higher population densities in European cities but may be partially due to functional or structural differences between settlements of similar size. In any event, the basic logarithmic relationship indicates that, for a given population increase, the heat-island effect is greater for a town than a large city. This helps to reinforce the physical theory that any heat island must have a finite limit which is governed partly by the amount of artificial energy released and partly by the presence of a convergent thermal breeze circulation which arises in large cities and helps to dissipate strong temperature gradients between urban and rural areas. Indeed, Parry (1967b) has gone so far as to state that the urban climate is most realistically viewed as a collection of varied microclimates.

On the other hand, the heat-island effect has been observed in tropical as well as temperate latitudes, as indicated by the work of Nieuwolt (1966) in Singapore, and there is a real need for the development of a physical model with wide applications in order to assess the magnitude of the various mechanisms involved. Myrup (1969) applied an energy-budget model to the problem and concluded that the most important factors determining the size of the heat-island were the reduced evaporation, the increased surface roughness, the thermal properties of building and paving materials, and the windspeed. Altogether, the results suggested that, although reduced evaporation was the dominant daytime feature, the thermal properties of the substrate became most important during the night.

Within recent years, an increasing amount of attention has been devoted to the vertical rather than the purely horizontal pattern of heat-islands. Early studies of the temperature profile overlying cities were confined to settlements located within valleys, where an observation network extending up the surrounding hillsides allowed some analysis of the effect of urban temperatures during inversion conditions. In southern England, Balchin and Pye (1947) claimed that the retained heat of Bath was sufficient to restrict the development of valley inversions for some 25 per cent of the occasions when such conditions were recorded outside the built-up area, and Parry (1956a) has drawn attention to similar destruction of low-level inversions by the town of Reading in the Thames valley.

More information has become available from balloon ascents and helicopter surveys. In a study of three cities in the San Francisco Bay area of California, Duckworth and Sandberg (1954) measured the vertical temperature gradients in the lowest 305 m under varying weather conditions during evening surveys and found that the built-up areas often caused atmospheric instability in otherwise stable air up to about three times the height of the buildings. In addition, a so-called 'crossover' point was sometimes located, where the air above the urban centre became cooler than that over the peripheral open areas, and it was suggested that this feature might be explained by variations in wind velocity between 30 and 300 m.

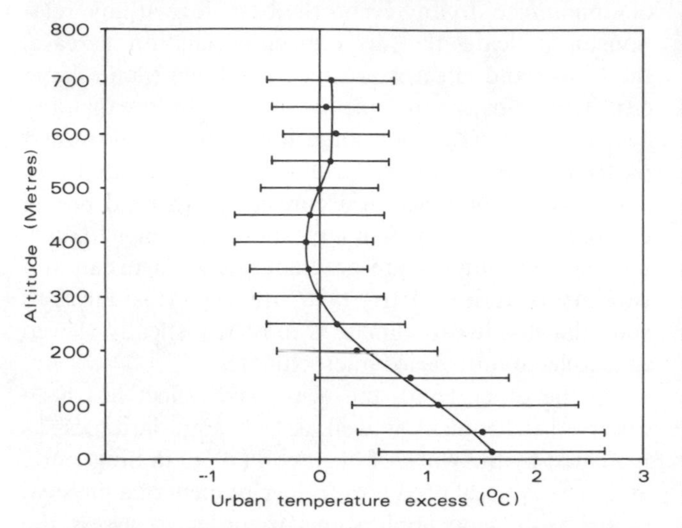


**Fig. 3.8 Height variation of the magnitude of the urban heat island above New York City during the hours near sunrise. The range of plus and minus one standard deviation is indicated. After Bornstein (1968).**

A similar survey was undertaken by Bornstein (1968) in New York, where a strong heat-island might be expected because combustion during winter in Manhattan releases 250 per cent more heat than reaches the surface from the sun. Data obtained for the lowest 700 m from an instrumented helicopter on 42 mornings around sunrise revealed that, although the average intensity of the heat-island reached a maximum near the surface and cleared to zero at 300 m, there was a high frequency of weaker, more elevated inversion layers at a mean altitude of 310 m above the city. On mornings when these higher inversion layers were well developed, the urban temperature excess extended beyond 500 m. However, for rather more than two-thirds of the test mornings, an elevated crossover layer existed where rural temperatures were higher than the corresponding urban values; Fig. 3.8 shows that, after an almost linear decrease in the mean urban temperature excess from the surface to 300 m, the crossover layer dominated the urban temperature field between 300 and 500 m.

Data from motorized traverses, pilot balloons, and a helicopter were used by Clarke (1969a) to illustrate how

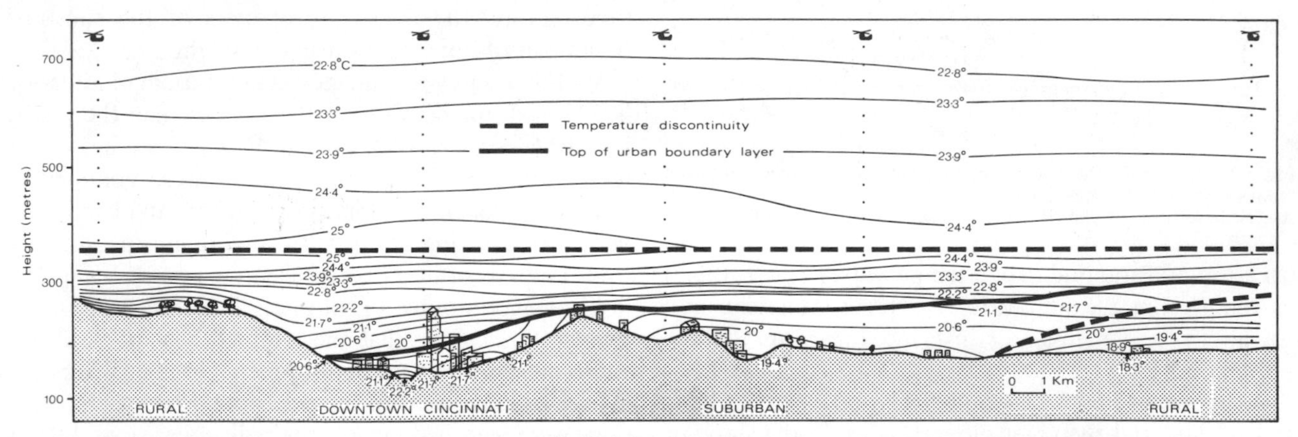


**Fig. 3.9 Cross-section of the temperature structure above Cincinnati on 13 June 1967. The dots aligned vertically indicate the locations where temperature observations were made by helicopter. The schematic representation of the buildings is roughly to scale. After Clarke (1969a).**

the urban boundary layer over Cincinnati becomes modified to a progressively greater vertical extent with increasing distance downwind. Figure 3.9 shows that a strong surface-based inversion extending to 366 m in the rural area upwind of the city gave way to a superadiabatic layer involving the lowest 45 m of the atmosphere over the central business district. In the downwind suburban areas, the boundary layer consisted of a weak inversion that gradually increased in height with downwind distance, while a stronger inversion was found above the total extent of the urban boundary layer. Beyond the downwind limit of the city, the relatively unstable layer formed by the outflowing plume of warm urban air was sandwiched between the more stable layers near the rural surface and in the air aloft.

### 3.5 Wind

Urban heat-islands are capable of modifying local air circulations and evidence presented for Leicester by Chandler (1961b), as well as that provided by Clarke (1969a) for Cincinnati, indicates the existence of thermally induced near-surface winds flowing across the boundaries of the built-up areas. During the winter in Toronto, Canada, Findlay and Hirt (1969) found an example of a weak centripetal wind system generated by the urban heat-island in the daytime as well as at night; similarly, Schmidt (1963) measured the airflow around an oil refinery covering only 4 km$^2$ in the Netherlands and demonstrated the existence of a minor cyclonic circulation with ascending air at the centre reaching vertical velocities of around 15 cm/s.

However, the major modification of airflow is created by the surface morphology of towns, as the irregular skyline presents a rough surface which, in turn, acts as a mechanical brake. This reduces mean wind speeds within the city complex and Chandler (1965) has claimed that average speeds within central London are about 5 per cent lower than in the rural environs. On the other hand, this extra frictional resistance is not applied uniformly and local eddying and turbulence can increase speeds at some sites, especially where the street plan favours the channelling of airflow down a particular thoroughfare. Further complications may arise from variations in the underlying topography. Croydon in south London has a lower average wind speed than might be expected from either its altitude or its position relative to the edge of the built-up area but, since the station is situated on the north-facing slope of the North Downs, it seems likely that the shelter afforded by the escarpment is as effective as that provided by the conurbation.

Urban wind speeds are also significant in that, above certain threshold values, airflow can lead to either the dislocation or the complete elimination of heat-islands through the combined effects of upward turbulence and lateral advection in the lower atmosphere. Chandler's work suggests that the thermal consequences of London are destroyed by wind speeds in excess of about 11 m/s and that lighter air movements displace the centre of the heat-island downwind with a resulting asymmetrical outline of the warm air. In smaller settlements, the thermal effects are more easily eliminated and Parry (1956a) has proposed a threshold wind speed of only 4·5–6·7 m/s for Reading.

### 3.6 Precipitation

In contrast to the well-documented heat-island effects, the influence of cities on local precipitation is much less clearly established. However, although urban humidities are usually lower than in the surrounding countryside owing to the lack of open water surfaces and transpiring vegetation, the prevailing view is that large built-up areas are instrumental in increasing at least certain types of precipitation from the atmosphere. This is probably due to a specific combination of the surplus condensation nuclei arising from pollutants, the increased upward component of air motion above cities due to both low-level turbulence and thermal updraughts, plus the possible local addition of water vapour from combustion processes.

A significant contribution to the study of urban precipitation was made by Ashworth (1929). He examined a 30-year rainfall record for part of industrial Lancashire and concluded that, in the manufacturing town of Rochdale, Sunday had less rainfall than any other day of the week whereas at Stonyhurst, with a lesser concentration of factories, the incidence of rainfall was similar on all days. Furthermore, more rain fell at Rochdale during working hours and the daytime generally than at night, and this diurnal pattern was not only reversed at Rochdale on Sundays but was also the direct opposite of the pattern found at all stations lying upwind of the main industrial district.

Other workers, such as Carter (1931), commented on apparent weekly rhythms in the weather, but it was not until Nicholson (1965 and 1969) used the rainfall record from Teddington near London to argue that Thursdays were notably wetter than other days during summer that a more rigorous appraisal was made. The reality of a weekly rainfall cycle has been largely accepted by Scorer (1964 and 1969) and equally rejected by Glasspoole (1969) on the basis of statistical analysis of available British data, but Lawrence (1971a and b) has emphasized the possible importance of other weekly patterns in the urban climatic environment. The cumulative evidence for London indicates that the weekly cycle of domestic and industrial activity causes weekly patterns of pollution with minimum smoke and sulphur dioxide levels on Sunday, which is also the sunniest day. Lawrence also showed that over the 20-year period 1949–69 the mean daily maximum temperature in central London during summer had less excess over the surrounding rural values on Thursdays than on Sundays. This relationship was significant at the 0·1 per cent level and led to the view that the apparent weekly variation in urban heat-island effect might produce a weekly pattern of convective rainfall caused by thermal instability, and that the rain pattern could also lead to the observed lowering of daily maximum temperatures during the week.

So far, such views on weekly events have not been substantiated for other towns and the most convincing evidence for urban rainfall in Britain has come from individual summer thunder storms for which urban areas appear to act as a trigger mechanism. Thus, Parry (1956b) has described a localized rainstorm in the Reading area, where the heaviest precipitation coincided with the centre of the town. More recently, Atkinson (1968 and 1969) has located a maximum of thunder rainfall over the central part of the London conurbation. The precipitation excess appears to be largely a summer feature associated with warm frontal storms and was attributed to higher daytime temperatures and increased air turbulence over the city centre.

Long-period changes in precipitation regime have proved more difficult to establish, not least because local rainfall variations are often dominated by topography and require dense observer networks for detailed analysis (Balchin and Pye, 1948). In addition, there is often a lack of long-period urban gauges, and the natural variability of rainfall itself provides further problems, as illustrated by Spar and Ronberg (1968) for New York City. They found that the records for Central Park showed a significant decreasing trend from 1927 to 1965, but this feature was not substantiated by data from nearby sites, some of which had experienced small rainfall increases over the same period. However, Barrett (1964), working in the Manchester region, has interpreted an unusual increase in both summer and winter rainfall at four stations on the north-eastern edge of the city from 1890 and 1960 as a downwind effect caused by the expanding urban area.

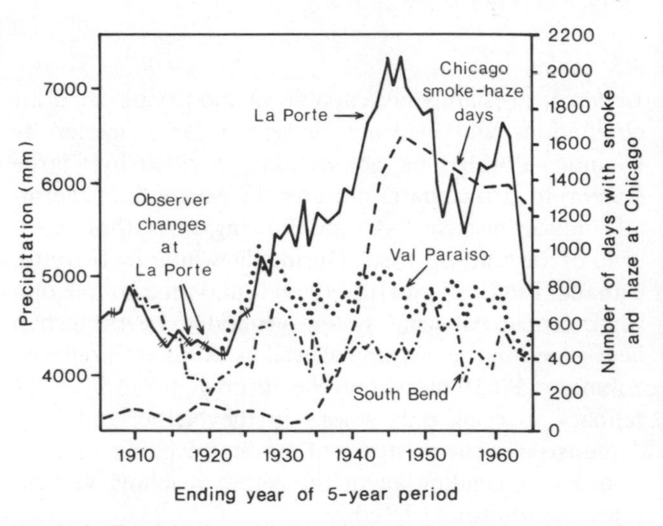


**Fig. 3.10 Precipitation values at selected Indiana stations and smoke-haze days at Chicago, both plotted as five-year moving totals. After Changnon (1968).**

A much clearer example has been provided from the US by Changnon (1968), who has concentrated on the situation at La Porte, Indiana, which lies 48 km downwind of the large heavy industry complex around Chicago. Since 1925, the variations in annual and warm-season precipitation at La Porte have fluctuated with steel output in the Chicago area and between 1951 and 1965 La Porte had 31 per cent more precipitation, 38 per cent more thunderstorms, and 246 per cent more hail-days than the surrounding stations. Something of this anomaly is illustrated in Fig. 3.10, which shows the related upward trend between La Porte precipitation and smoke-haze days at Chicago compared with the seemingly unchanged precipitation regimes at Valparaiso and South Bend which lie outside the path of downwind pollution.

More recently, Huff and Changnon (1972) have investigated the rainfall regime around the typical mid-Western city of St Louis, where the urban effect was strongest during late spring and summer leading to increases in average summer rainfall ranging from 6 to 15 per cent for distances up to 40 km downwind of the city. The mechanism appeared to be most active on days of moderate to heavy natural intensities, with the most pronounced effects during wet summers, and it was concluded that the thermal influences of the city were the most important causes. On the other hand, Ogden (1969) studied the data from 90 rainfall stations within a 100 km radius of the Port Kembla steelworks in Australia and, despite a high output of condensation nuclei from the plant together with large emissions of heat and water vapour, was unable to find evidence which supported a steelworks influence greater than 5 per cent on total rainfall. In the US, Schaefer (1969) has used aircraft observations to establish the presence of large numbers of condensation nuclei downwind of several cities and has also claimed that these have led to ice crystal clouds and snow flurries in some instances.

Snow is a particular type of precipitation which might be expected to occur less frequently in towns than in comparable areas outside, since heat-island influences are likely to reduce both the frequency of snowfall and the subsequent period during which the snow lies. Manley (1958) has looked at some aspects of snowfall in metropolitan England but some of the best data have been assembled by Lindqvist (1968) during the 1964–65 winter for the small city of Lund in southern Sweden. Although some attempt was made to measure differences in initial snow accumulation, a greater degree of success was achieved by the use of aerial photographs to study variations in snow cover due largely to melting on house roofs. Figure 3.11 shows the pattern on 1 February 1965 when snow had been lying for four days. Despite unavoidable complications such as the varying slope and exposure of the roofs, and differences in roofing materials and the amount of heat released in the buildings beneath, it can be seen that in the peripheral parts of the town the snow frequency on roofs was more than 80 per cent compared with less than 30 per cent in the central areas.

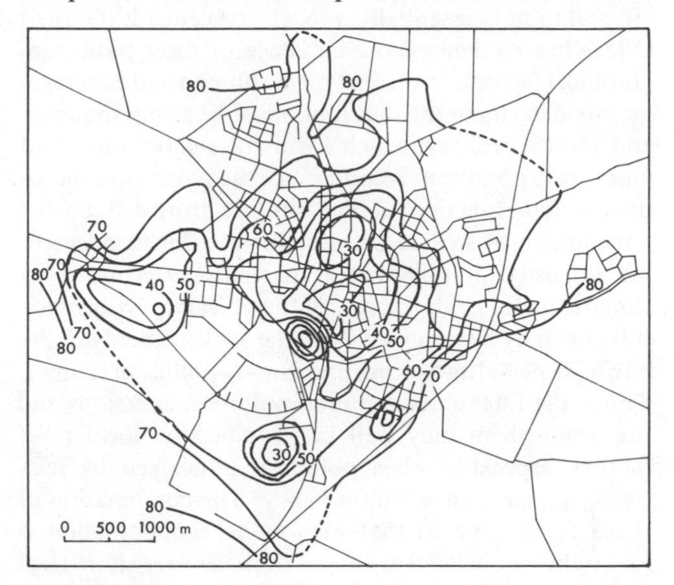


**Fig. 3.11 Isarithms of equal snow frequency on house roofs in Lund, Sweden, estimated from aerial photographs on 1 February 1965. After Lindqvist (1968).**

## Air pollution

### 3.7 The nature of air pollution

Although air pollution is mainly associated with built-up areas, its significance extends far beyond that of urban climates. Indeed, as we shall see, the nature, distribution, and intensity of atmospheric contamination has, in some circumstances, reached the point of global implications. Atmospheric pollutants can exist in the form of particles, liquid droplets, or gases, and the complexity of air pollution ensures that most definitions are both arbitrary and unsatisfactory. Broadly speaking, however, most pollutants comprise substances present in the atmosphere in sufficient concentrations to interfere with man's well-being or health, including the optimum and continued use of his material possessions.

Although pollutants may come from either natural or man-made sources, the term pollution is often restricted to considerations of air quality as modified by human actions, particularly as when effluents are emitted from nucleated urban or industrial sources at rates in excess of the natural diluting and self-purifying processes currently prevailing in the atmosphere. Interpreted in this sense, air pollution is essentially a local problem and Garnett (1957) has recognized the influence of three basic geographical factors. In the first place, human and economic factors determine the distribution of housing, industry, and motor vehicles, which form the major sources of man-made pollution. Second, since the pollution concentration is a function of the dilution provided by the atmosphere as well as the amount of effluent released, the intensity of pollution is partially determined by climatic factors, which control the efficiency with which effluents may be removed from the air together with the rate of dispersal away from the primary pollution sources. Third, the interaction between pollution emissions and the atmosphere may well be modified by local relief factors, especially when pollution is trapped by relatively stagnant air within a valley. The combination of these factors means that man-made contamination is invariably associated with the most severe, short-term pollution episodes and these have inevitably attracted the most attention, not only because of the serious consequences, but also because they offer the best possibilities for improvement through technological advances in effluent quality control, backed up where necessary by statutory regulations.

On the other hand, local man-made pollution is not the complete picture. In some parts of the world, the continuous emission of the more persistent pollutants has given rise to a regional as well as a local problem. This has happened in western Europe, where the release of sulphur dioxide in Britain has been blamed for the appearance of acidic rainfall over Sweden. At Barbados in the West Indies, Seba and Prospero (1971) have monitored pesticides believed to have originated in either Europe or North America. A few pollutants, such as the carbon dioxide resulting from the combustion of fossil fuels, may have world-wide significance and be implicated as possible causal factors leading to global fluctuations of climate. In addition, as shown in Table 3.3, many atmospheric pollutants originate from natural sources. Some natural air pollutants are particulate in form, including dust produced from the mechanical disruption of land surfaces by wind action, salt particles from the oceans, smoke particles from forest fires started by lightning strikes, and pollen blown from vegetated areas.

| *Type* | *Source* |
|---|---|
| $CO_2$ | Volcanoes<br>Burning fossil fuels<br>Animals |
| CO | Internal combustion engine<br>Volcanoes |
| Sulphur compounds | Bacteria<br>Burning fossil fuels<br>Volcanoes<br>Sea spray |
| Hydrocarbons | Internal combustion engine<br>Bacteria<br>Plants |
| Nitrogen compounds | Bacteria<br>Combustion |
| Particles | Volcanoes<br>Wind action<br>Combustion<br>Industrial processing<br>Meteors<br>Sea spray<br>Forest fires |

**Table 3.3** Types and sources of atmospheric pollutants. (After Varney and McCormac, 1971.)

Other natural pollutants are gaseous, such as hydrocarbons emitted by decomposing vegetation, whereas volcanic eruptions—like the oceans—are a source of both particulate and gaseous matter. With very few exceptions, these natural releases are either isolated and temporary (forest fires, dust storms) or produce undesirable substances slowly from widely dispersed sources (hydrocarbons from decomposing vegetation) and are therefore absorbed by the atmosphere without the more obviously adverse effects associated with excessively high man-made effluent concentrations. Furthermore, it must be appreciated that most of the natural gaseous releases are essential for the maintenance of life on the earth and only become pollutants when the concentrations become too high to be beneficial.

Nevertheless, it will be evident from Table 3.3 that many pollutants are both natural and man-made and, although it is true that man-made emissions are usually responsible for adversely affecting air quality by raising local pollution levels above some acceptable standard, the natural sources may well be more significant on a world scale for the determination of so-called background concentrations of the same pollutant. For example, although nitrogen can be released into the atmosphere by the combustion of fossil fuels and from vehicle exhausts, Robinson and Robbins (1970a) have shown that the vast majority is emitted by bacteria. Similarly, Stevenson (1968) has revealed that the annual concentration of chloride measured in rainwater over the British Isles (see Fig. 3.12) indicates a dominantly ocean source for this pollutant, which can also occur as a result of industrial processes. A seasonal variation was noted and the higher winter chloride concentrations were attributed to the increased storminess of the sea at this season, which adds to both the spray content of the air and the salt content at cloud level. In the case of particulate pollution, Abelson (1971) has claimed that the volcanic eruption of Krakatoa in 1883 injected more material into the atmosphere than has resulted from the smoke emitted by man-made fires throughout history.

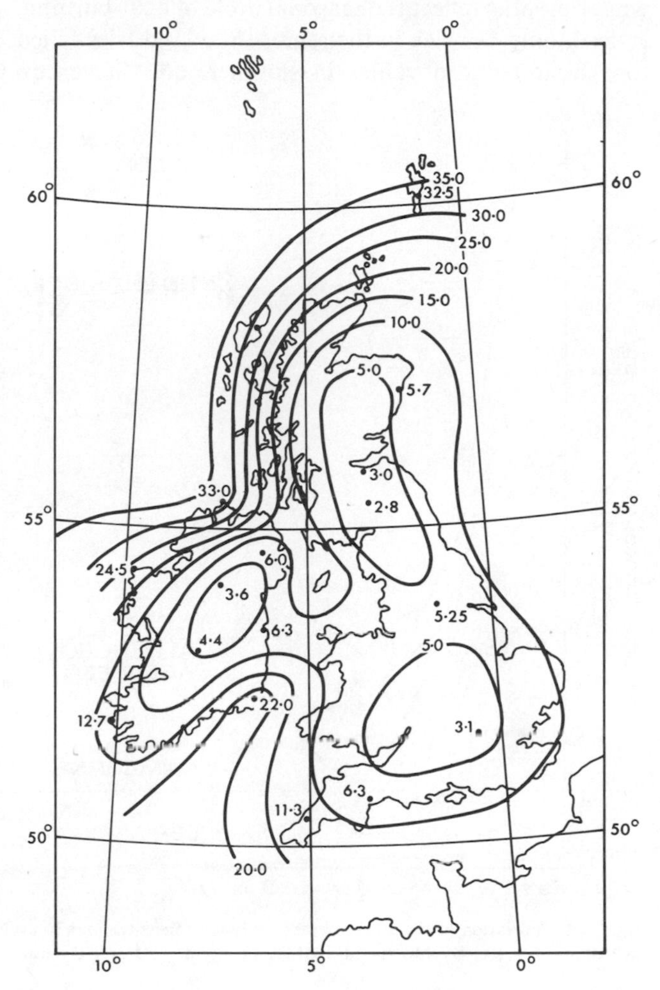


**Fig. 3.12 Annual chloride medians over the British Isles in mg/l. Isopleths drawn at 5 mg/l intervals. After Stevenson (1968).**

For a comprehensive classification of air pollutants, the reader should consult Chambers (1962), who recognized both primary emissions and also secondary pollutants due to chemical reactions in the atmosphere. Perhaps the best known form of secondary pollution occurs as a consequence of photochemical reactions in the Los Angeles area of California. For convenience, a brief account of some of the more important pollutants is given below.

*Particulates.* According to Varney and McCormac (1971), only about 10 per cent of the mass of all pollutants is emitted as particles and liquids compared with 90 per cent for gaseous compounds; however, particulate matter has important consequences for the turbidity and scattering properties of the atmosphere. It has also been estimated that over 90 per cent of all atmospheric particles come from natural sources and that the overwhelming volume remain in the troposphere, with some 80 per cent concentrated within the lowest kilometre depth of the atmosphere.

Particulate matter embraces a wide range of solid and liquid particles extending from more than 100 μ to less than 0·1 μ in diameter. Most of the smaller particles, approximately 10 μ or less in diameter, are of smoke and are produced by the incomplete combustion of fossil fuels. According to Meetham (1964), about 85 per cent of the suspended matter found near British towns is smoke and, since most of this arises from coal consumption, it contains a high proportion of carbon together with tarry hydrocarbons. The average diameter of a smoke particle is only 0·075 μ, which means that smoke does not fall readily out of the atmosphere under its own weight and may be blown for considerable distances in suspension. These smaller particles are more properly referred to as *aerosols* and can exist in the troposphere for perhaps a week before breaking up and, if introduced into the less turbulent conditions of the stratosphere, this type of particle may persist for years (Junge, 1963).

In contrast, coarser particles of ash or grit soon fall out. Ash is the incombustible material resulting from the burning of solid fuel and it has been estimated that in Britain over 80 per cent of this atmospheric ash comes from industrial sources with the remainder issuing from domestic chimneys. If particulate matter is ejected into the stratosphere in large quantities it is likely to have an influence on global climates but, on a world scale, stratospheric aerosols are almost entirely derived from natural sources and mainly comprise ammonium sulphate, volcanic dust, and meteoric dust.

*Sulphur and sulphur compounds.* Sulphur is one of the most widespread forms of atmospheric pollution and exists in the air mainly as hydrogen disulphide ($H_2S$), sulphur dioxide ($SO_2$), or as sulphate compounds such as sulphuric acid ($H_2SO_4$). Robinson and Robbins (1970b) have outlined the major sources of sulphur emissions and it can be seen from Table 3·4 that about two-thirds of the

| *Source* | *kg/yr* |
|---|---|
| $SO_2$—coal combustion | $4{\cdot}7 \times 10^{10}$ |
| $SO_2$—petroleum combustion | $1{\cdot}3 \times 10^{10}$ |
| $SO_2$—smelting | $7{\cdot}0 \times 10^{9}$ |
| $H_2S$—bacteria | $8{\cdot}8 \times 10^{10}$ |
| $SO_4$—sea spray | $4{\cdot}0 \times 10^{10}$ |
| Total | $19{\cdot}5 \times 10^{10}$ |

**Table 3.4** World-wide sulphur emissions. (After Robinson and Robbins, 1970b.)

total come from natural sources. The most important single source is the $H_2S$ emitted by organic processes, which comprises over 45 per cent of the total. However, the next most significant emission is sulphur dioxide, which is largely man-made. This comes from burning the sulphur contained in fossil fuels and accounts for one-third of all the sulphur emitted. According to Blokker (1970), some 80 to 90 per cent of the total atmospheric sulphur content over certain land areas may be $SO_2$.

Approximately two-thirds of all sulphur emissions occur in the northern hemisphere and this has been attributed to the fact that probably over 90 per cent of all the man-made releases are located north of the equator. Although natural sources dominate on a world scale, Fig. 3.13 shows the mean monthly concentrations measured over the 1959–64 period at various sites in Britain, where the levels appear highest near to the densely populated areas and the marked increases in the winter months reflect the seasonal cycle of coal-burning. Indeed, only Lerwick in the remote Shetland Isles failed to exhibit reduced values in summer, and Stevenson

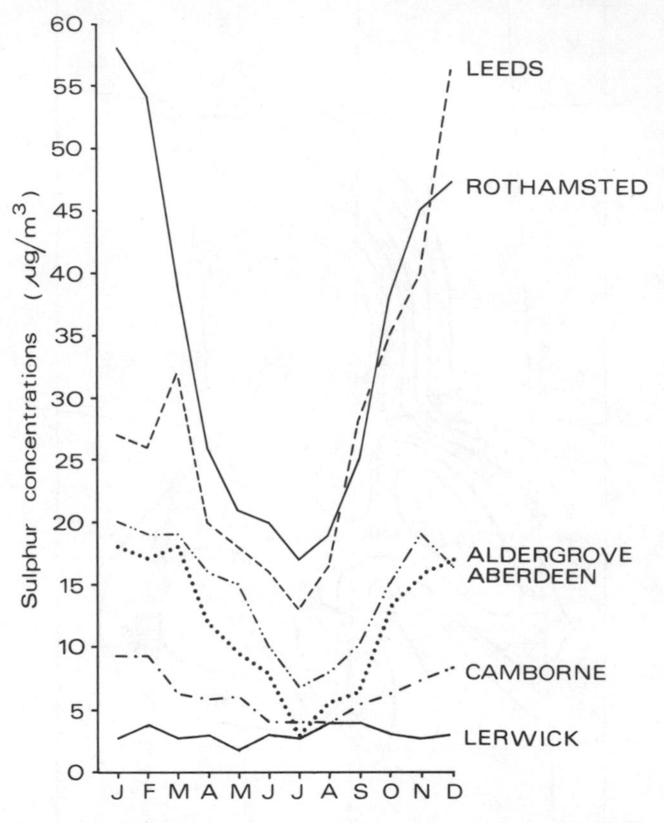


**Fig. 3.13 Variation of the mean monthly sulphur concentrations in air measured at selected stations in Britain in μg/m³. After Stevenson (1968).**

(1968) attributed this partly to lower fuel consumption and a more seasonally uniform climate plus the possible effects of relatively low sulphur contents in the peat which is burned in the area as a fuel. Significant differences in sulphur concentrations occur on much smaller scales and Andersson (1969) has demonstrated that the amount of sulphur in rainwater is higher within Uppsala than in the surrounding Swedish countryside.

Sulphur dioxide is a particularly serious pollutant. It is soluble in water and chemically attacks many materials in this form, but the most corrosive conditions occur when this atmospheric solution has been oxidized to form weak sulphuric acid. It has been claimed by Junge and Werby (1958) that all acids and sulphates are flushed from the atmosphere by precipitation within about 40 days, but this still allows considerable time for downwind transport and perhaps one-fifth of all the $SO_2$ produced in Britain is blown out to sea. On the other hand, measurements of $SO_2$ conducted by Brossett and Marsh (1970) during 1969–70 on a ship travelling between London and Gothenburg suggested that virtually all the atmospheric $SO_2$ had been dispersed some 50 km downwind of London.

*Carbon dioxide ($CO_2$).* Carbon dioxide occurs naturally in the atmosphere and the chief source is the respiratory loss from growing plants. Unlike the complementary process of $CO_2$ assimilation by plants through photosynthesis, respiration is not dependent on the receipt of light energy and so the process continues through both the day and the night. Since the highest release rates are found in summer, when plants have sufficient heat for growth, and since the maximum near-ground concentrations occur at night owing to restricted vertical mixing in the atmosphere, it follows that high values will be achieved at rural sites on summer nights. This has been confirmed by Clarke (1969b) working in Ohio, and the observations shown in Fig. 3.14 led him to conclude that $CO_2$ emissions from natural sources may be sufficient to exert an appreciable influence on the measured concentrations in urban areas during summer nights.

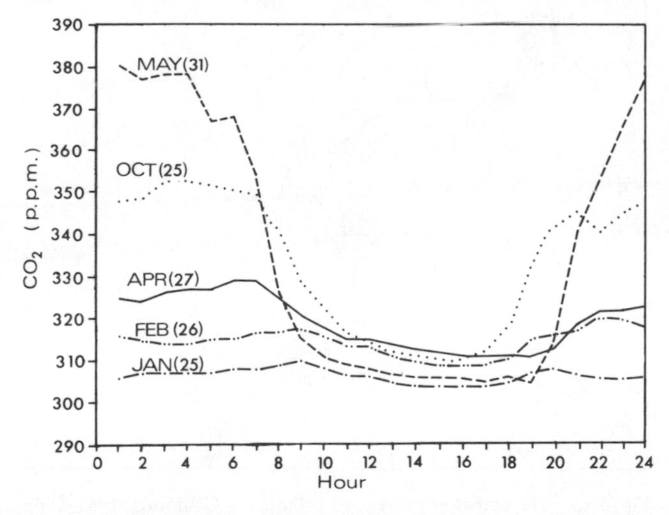


**Fig. 3.14 The diurnal variation of $CO_2$ concentration by months measured at a rural site 47 km south-east of Cincinnati, Ohio. The figures in parentheses indicate the number of days of valid data. After Clarke (1969b).**

Despite this, it has been estimated that the burning of fossil fuels produces about 20 per cent of the $CO_2$ emissions, or approximately $1{\cdot}4 \times 10^{13}$ kg/yr (Varney and McCormac, 1971). According to Bach (1972), about 330 billion tons of $CO_2$ have been added to the atmosphere by combustion since the Industrial Revolution and in some areas concentrations have already reached 1000 ppm compared with an average level of only 318 ppm for so-called clean air. Furthermore, a proportion of the carbon dioxide released remains in the atmosphere for up to several hundred years, as reported by Keeling (1960), and the possible consequences of this accumulation on long-term climatic trends are considered later in this chapter.

*Carbon monoxide (CO).* Carbon monoxide occurs through the incomplete burning of carbonaceous fuels and 80 per cent of the global production is from motor vehicles (Bach, 1972). CO is a highly toxic gas and 50 ppm is the safe upper limit recommended for British factory atmospheres. Fortunately, measured concentrations in Britain are usually much lower. Reed and Trott (1971) monitored CO values at street-level in various towns and found that the proportion of time each month when the concentration exceeded 30 ppm was always less than 2 per cent, and that values in excess of 50 ppm only occurred for a few minutes during any month.

*Radioactive substances.* Radioactive contamination causes concern because of the long period necessary before the disintegration of an isotope produces a stable element and because of the human genetic disturbances produced by radiation levels increased beyond the background exposure, as stressed by Mason (1970). Natural or background radioactivity exists largely through the diffusion of the radioactive gases *radon* and *thoron* from the soil into the air and the consequent decay of these gases. At Ghent in Belgium, van Cauwenberghe and

Bosch (1969) have shown that these natural sources were far in excess of the artificial radioactivity which occurs as a result of fall-out and radioactive debris from nuclear reactions either in deliberate explosions or from reactor accidents. Natural radioactivity varies with the local weather conditions and is reduced after heavy rainfall both by the 'wash-out' effect of the precipitation and the much lower rate of diffusion through the water-filled pores of the soil (Israel *et al.*, 1966; Malakhov *et al.*, 1966).

On the other hand, both global and local concentrations increase after artificial releases. Dyer and Hicks (1967) have monitored the radioactive fall-out in rain collected over a 10-year period at Aspendale, Australia, and have been able to relate increased concentrations to nuclear-weapons testing. More locally, Crabtree (1959) examined the dispersion of radioactive material emitted by the Windscale reactor in northern England in 1957 when dairy pastures were sterilized over a downwind strip of land some 80 km long and 16 km wide.

*Photochemical products.* Most hydrocarbons in the atmosphere originate from the natural decomposition of vegetation and are harmless, but the gasoline fractions from vehicle exhausts and other sources are highly reactive and create a special problem. When these pollutants are exposed to solar radiation, a number of photochemical reactions take place and nitric oxide, which is present from most combustion operations, is converted to nitrogen dioxide ($NO_2$). This photochemical oxidation may lead to the formation of particulates and may be accompanied by a characteristic odour, haze formation, eye irritation, and plant damage (Haagen-Smit, 1962). Another important feature is the excessive production of ozone which, although only present at about 1–3 parts per hundred million in unpolluted air, can reach concentrations twenty to thirty times higher. Further reaction gives rise to peroxide compounds, such as peroxyacetylnitrate (PAN), which act as eye irritants and as oxidizing pollutants at concentrations of 0·1 ppm or more (Haagen-Smit, 1952).

Owing to a particular combination of high emission rates and favourable climatic conditions resulting essentially from large inputs of sunlight, this type of air pollution is worst in California, although comparable areas, such as parts of Greece, may experience the same problems on a smaller scale (Papaioannou, 1967). The detailed meteorological background to air pollution in California has been presented by Bell (1967), and Leighton (1966) has stated that the observed symptoms are related to the intensity of the pollution with visibility reduction coming first, followed by plant damage and eye irritation as the pollution builds up.

As shown in Fig. 3.15, most of California is affected by air pollution, most of which is photochemical in character. The plant-damage areas, which are specifically caused by photochemical effects, have grown from a few square kilometres near Los Angeles in 1942 to cover some 30 000 km$^2$ of the State. Leighton further claimed that,

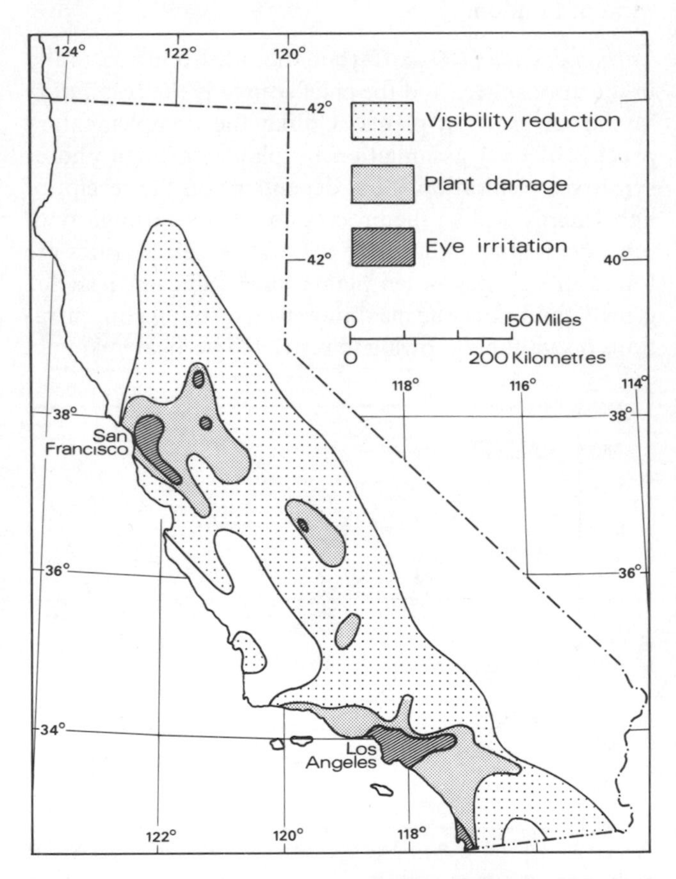


**Fig. 3.15 The extent of general air pollution in California, 1961–1963. The plant-damage areas are specific but the eye irritation and visibility reduction may be partly due to pollution other than the photochemical type. After Leighton (1966). Copyrighted by the American Geographical Society of New York.**

at a cruising speed of almost 100 km per hour, the average American automobile emits approximately 3 litres of nitrogen oxide every minute. In order to dilute this emission below 0·05 ppm, more than $6 \times 10^7$ litres of air per minute is required, which is sufficient to supply the normal breathing needs of five to ten million people over the same period of time.

### 3.8 Sources and measurement of air pollution

The major sources of air pollution are implicit in what has already been said about the nature of pollutants, but there are important differences in relative and absolute importance from place to place. Thus, figures quoted by Bach (1972) show that, in 1965, motor vehicles were responsible for 60 per cent of air pollution emissions by weight in the US compared with 17 per cent from industry and 14 per cent from power-plant sources. Carbon monoxide was easily the largest single polluting substance, with an emission rate of about $65 \times 10^9$ kg/yr, and hydrocarbons from motor vehicles and sulphur oxides from power plants were other important pollution sources. However, the most meaningful statements are probably those concerning specific pollutants in individual areas: Table 3.5, for example, shows that about 70 per cent of the particulates in New York originate from apartment heating and incineration (Eisenbud, 1970).

| *Source* | *per cent* | *kg/yr* |
|---|---|---|
| Space heating (apartments) | 32·3 | $2{\cdot}03 \times 10^7$ |
| Municipal incineration | 19·3 | $1{\cdot}22 \times 10^7$ |
| Apartment incineration | 18·4 | $1{\cdot}16 \times 10^7$ |
| Mobile source | 14·3 | $0{\cdot}90 \times 10^7$ |
| Power generation | 9·2 | $0{\cdot}58 \times 10^7$ |
| Industrial | 6·5 | $0{\cdot}41 \times 10^7$ |
| | Total | $6{\cdot}3 \times 10^7$ |

**Table 3.5** Sources of particulates in New York City during November 1969. (After Eisenbud, 1970.)

In Britain, air pollution from traffic is small compared with other sources (Sherwood and Bowers, 1970), and remains essentially a local problem restricted to certain city streets for fairly short periods of time. On the other hand, smoke from domestic chimneys and sulphur dioxide from industrial premises have caused concern for many years and a National Survey of these pollutants embracing about 1200 measuring sites was started in 1961–62. Over 80 per cent of smoke pollution arises from the use of coal in open domestic fires but Craxford and Weatherley (1971) have shown that, over the period 1951 to 1968, smoke emissions in the UK have fallen from 2·36 to 0·93 million tonnes. This represents a decline of 60 per cent and has been achieved despite a 10 per cent increase in population and a 17 per cent increase in annual gross energy consumption. Figure 3.16 provides a breakdown of this trend in terms of domestic and industrial sources together with forecast values up to 1975.

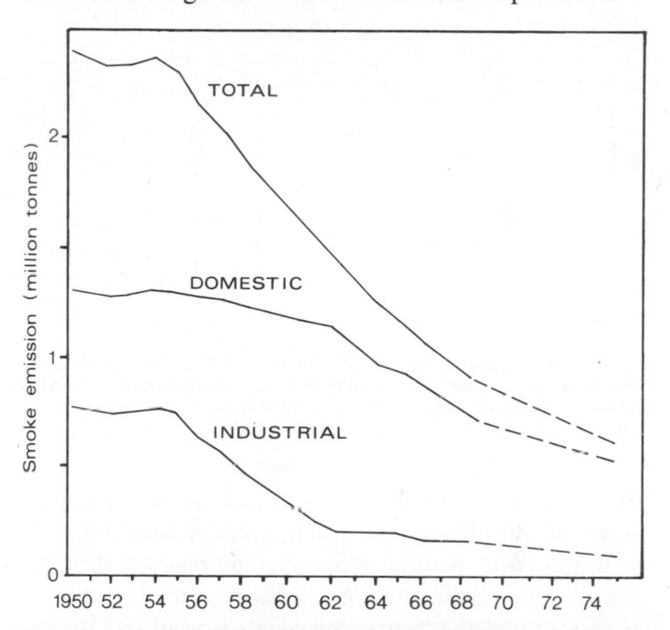


**Fig. 3.16 Emissions of smoke in the UK 1950–1968 from both domestic and industrial sources with projections up to 1975. After Craxford and Weatherley (1971). Crown Copyright, reproduced by permission of the Director, Warren Spring Laboratory.**

The statutory provisions of the Clean Air Act of 1956 are largely responsible for the progressive reduction in smoke pollution. For industrial sources, an immediate improvement can be detected, while the introduction of smoke-control zones in the urban areas has accelerated the decline of the traditional open fire and its replacement with cleaner, labour-saving fuels such as oil, gas, and electricity. The highest regional concentrations of smoke are found in the north with values in excess of 100 $\mu g/m^3$ compared with concentrations of only one-third of this level in the south-west. Particularly striking improvements have been achieved with smoke control in London owing to careful implementation of the Clean Air Act and substantial re-housing programmes.

Figure 3.17 indicates that $SO_2$ emissions reached a peak of 6·34 million tonnes in 1963 and then declined by 6 per cent to 6·01 million tonnes in 1968. A continuing decline of 13 per cent is expected by 1975 as a result of the increasing use of natural gas and nuclear energy.

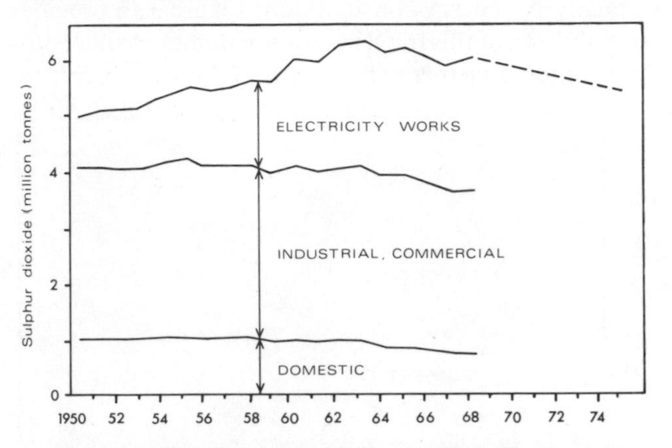


**Fig. 3.17 Emissions of sulphur dioxide in the UK 1950–1968 from the three major sources with a projection of total emissions up to 1975. After Craxford and Weatherley (1971). Crown Copyright, reproduced by permission of the Director, Warren Spring Laboratory.**

Domestic emissions have decreased more than those from industrial sources which are dominated by the pollution from electricity works. The regional distribution of sulphur dioxide conforms, in general, to that of smoke, except that very high levels prevail in London. This is attributed by Craxford and Weatherley (1971) to the high population density and the large number of commercial premises in the central area and they suggest that the burning of oils with a lower sulphur content may become a necessity in Inner London.

The successful measurement of air pollution depends on the nature of the local contaminants, but the three major types of pollutant emitted from chimneys are deposited matter, smoke, and sulphur dioxide. In Britain, all deposited matter—solid and liquid, including rain—is collected inside a 305 mm diameter glass collecting bowl known as a *deposit gauge*. The material passes into a glass bottle below and at regular intervals, usually monthly, the contents are chemically analysed to isolate the main pollutants.

For smoke and sulphur dioxide, daily measurements are obtained using the combined standard instrument shown in Fig. 3.18. The electrically driven suction pump continuously draws in a sample of air from outside the building by means of an inverted glass funnel. The total volume of air drawn through the apparatus is measured by the gas meter. First, a measurement of suspended smoke content is achieved by passing the air through a filter paper, which is renewed each day; the smoke is retained on the paper and forms a stain (Ministry of Technology, 1966). The darkness of the stain depends largely on the amount of the smoke collected and can be measured with a reflectometer. A calibration is available to relate the darkness of the filter stain to the weight of retained smoke for a standard smoke composition, and so the concentration of smoke per unit volume of air drawn through the filter can be calculated in relation to this standard.

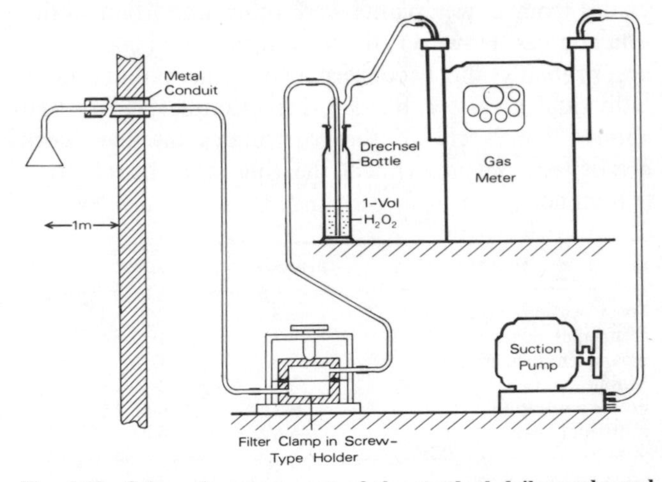


**Fig. 3.18 Schematic arrangement of the standard daily smoke and sulphur dioxide sampling apparatus. After Ministry of Technology (1966). Crown Copyright, reproduced by permission of the Director, Warren Spring Laboratory.**

The amount of sulphur dioxide in the air is measured by bubbling the same sample of filtered air through a dilute solution of hydrogen peroxide in a Drechsel bottle. The sulphur dioxide is then converted to sulphuric acid and the strength of acid is then taken as an index of the atmospheric concentration of $SO_2$.

## 3.9 Air pollution meteorology

It will be apparent from what has already been said that there are some highly effective natural cleansing processes at work in the atmosphere which prevent the continuing build-up of many pollutant concentrations. For example, Meetham (1950) has stated that over Britain the average atmospheric life of a smoke particle is only 1 or 2 days, and that of a molecule of $SO_2$ is less than 12 h. Pollutants are removed from the troposphere by a variety of processes, of which the action of rain and snowfall in flushing out both particulate and gaseous pollutants into the earth and soil is one of the most efficient. Other things being equal, this means that air pollution is likely to be more persistent in arid rather than humid regions and Greenfield (1957) has noted that a uniform rainfall of 1 mm/h for 15 min can scavenge 28 per cent of the 10 μ particulates from the air through which the rain passes. For particles of 2 μ or less, the watering action of rain has little effect.

Other pollutants may be filtered out by the foliage of trees and grasses, as shown by Martin and Barber (1971), and a further removal is effected by the breathing action of animals and humans. Some natural materials show a high capacity for pollution absorption and Braun and Wilson (1970) have reported that limestone building stones exposed for 500 years could still absorb atmospheric sulphur as fast as a new sample of the same stone. It was suggested that the surface reactivity was maintained by the removal of calcium sulphate in solution by rainwashing and that this outdoor deposition may be important in explaining reduced indoor concentrations of $CO_2$.

Oxygen in the air combines with many pollutants and changes them into forms which are more easily removed. The effect of sunlight may be significant here in producing particulate matter from gases and in this sense the development of a Californian smog is actually part of the atmospheric cleansing process (Air Conservation Commission, 1965). On the other hand, pollution which enters the stratosphere is much longer-lasting and Cramer (1959) has claimed that, of the original fission load of material injected into the stratosphere by a thermonuclear explosion, only 1 to 5 per cent falls out during the first 30 days after the blast. Even in the troposphere, it may take several days for the pollution from large urban concentrations to be completely cleared and, in the meantime, that body of air may well have reached another pollution source. The effect of this downwind transfer on air pollution concentrations across the US is shown in Fig. 3.19 (Neiburger, 1969). Clean air from the Pacific is heavily polluted by high emissions and from diffusion around Los Angeles, but pollution concentrations fall eastwards with the passage of the air over the Rockies and desert areas. Isolated cities such as Denver, Kansas City, and St Louis provide pollution peaks as the air traverses mainly agricultural intervening areas, but a cumulative increase occurs as the air crosses the metropolitan areas further east giving the highest pollution concentrations near the East Coast before further cleansing takes place over the Atlantic.

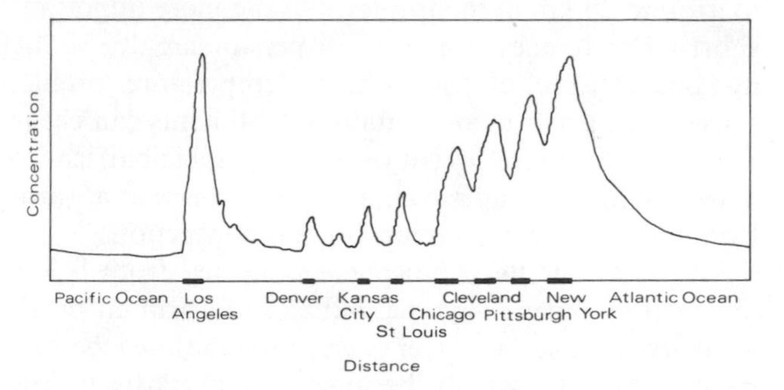


**Fig. 3.19 A schematic representation of the pollution concentration in air crossing the US. After Neiburger (1969).**

Figure 3.19 illustrates the general principle that air-pollution concentrations depend largely on the dilution available from relatively clean air and, potentially, this means the entire tropospheric layer which is about 10 km thick and contains some $5 \times 10^{18}$ $m^3$ of air (Pack, 1964). However, the actual dilution available at any one time is determined by the prevailing meteorological conditions as these control the rates of dispersal and diffusion of airborne pollutants from their sources.

Atmospheric diffusion theory is a large and complex field of study (Pasquill, 1962), but all dispersion formulae are common to the extent that the concentration downwind of a continuous emission is directly proportional to the rate of emission and inversely proportional to the product of the wind speed, the crosswind spread, and the vertical spread (Pasquill, 1972). For our purposes, the main factors are those controlling either the vertical or the horizontal transport of material. The type of pollution source may also be important. Thus, pollution from *point sources*, which may be either *instantaneous* (nuclear explosion) or *continuous* (factory chimney), tends to be dissipated by local dispersion processes and for a single chimney stack the travel distance may be less than 10 km downwind. Alternatively, the pollution from large *area sources* such as a city may travel up to 100 km from the emission area and the city is dependent on the ventilation provided by the movement of a large volume of air.

*Vertical dispersion* within the troposphere is highly variable in extent. The average thickness of the tropospheric layer itself varies from about 8 km over the poles to around 20 km in the tropics, but the more important short-term changes in upward dispersion are due to the dynamic nature of the vertical temperature profile. Large-scale uplift of air containing pollutants can occur as a result of forced ascent over topographic barriers or when warmer air slides over a colder airmass at a warm front, but the most common cause is convection.

The fact that the atmosphere is heated from below means that, during daytime, bubbles of warm air break away from the surface layer overlying some local thermal source and rise rapidly because of the relatively low density. As the parcel of air rises through the atmosphere, it undergoes adiabatic expansion and cooling. As long as the temperature of the air parcel remains above dew point and no water vapour condenses, the lapse rate of temperature with height is known as the *dry adiabatic lapse rate* (DALR) and remains constant at 1 degC per 1000 m irrespective of the original surface temperature of the air. Such an air parcel will continue to climb if it remains warmer and more buoyant than the surrounding atmosphere, and this will occur only if the *environmental lapse rate* (ELR) is greater than the lapse rate of the rising body of air. In turn, a steep, positive ELR naturally results from high daytime radiation inputs when super-adiabatic conditions prevail near the earth's surface and a vigorous upward dispersion of pollutants takes place in the unstable air. The depth of such a well mixed layer depends on the thermal characteristics of the underlying surface as well as the intensity of the solar radiation, and according to Pack (1964) may vary from more than 3 km over deserts to only 100 to 200 m over forested lake country.

Just as daytime heating tends to favour upward motion, so the nocturnal surface cooling of the earth's surface and the immediately overlying air layers produces a *temperature inversion* or increase of temperature with altitude. As shown in Fig. 3.20, such a situation

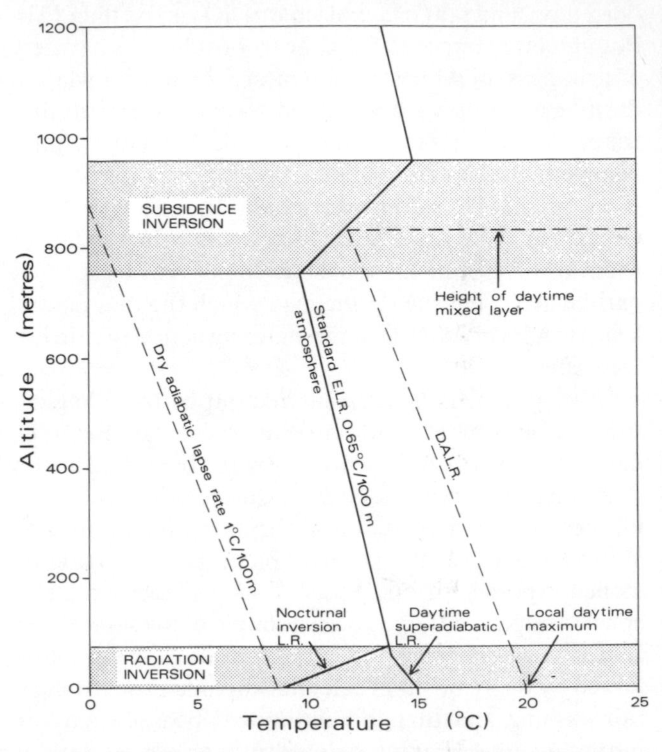


**Fig. 3.20 The effect of the vertical temperature profile and the presence of inversion conditions on the depth of atmospheric mixing. Outside the inversion layers, the environmental lapse rate is that of the International Standard Atmosphere, with a sea-level temperature of 15°C and a uniform lapse rate of 0.65 deg. C per 100m.**

means that the theoretical adiabatic path taken by a rising air parcel would always make that parcel cooler than the surrounding environment and in this highly stable context there can be little effective convectional activity. Thus, even if an effluent emerges from a chimney or other source much warmer than the near-surface air, its temperature is rapidly modified by the surrounding air so that, after rising only a few metres, it has no further buoyancy and remains trapped within the inversion layer.

Nocturnal or *radiation inversions* are formed when

clear skies permit the maximum longwave radiation exchange, and long winter nights with light winds represent the most favourable conditions. All such inversions are shallow and most are less than 100 m deep, although on extreme occasions the top may be at 200 to 300 m (Pedgley, 1962). Not surprisingly, this can lead to very high ground-level concentrations of pollutants, especially when the chilling and stagnation of air is increased within a valley.

In addition to the surface-based inversions of radiation nights, so-called *subsidence* or *high-level inversions* may occur in the middle troposphere (Fig. 3.20). According to Scorer (1968a), such inversions may be caused by radiation from the top of a cloud layer, by subsidence between convection, or, more likely, by air subsiding within developing or semi-permanent anticyclones between about 500 and 5000 m above the earth's surface. During the descent, the air is compressed and warmed adiabatically, thus producing a fairly thick inversion layer, and such features are most commonly found in association with the semi-permanent high-pressure cells located off the west coast of the continents, e.g., the Pacific high off the Californian coast.

Both surface and high-level inversions are often found together in high-pressure systems because anticyclones frequently provide the clear-sky low-windspeed conditions for radiation inversions to develop as well as the slow descent of air necessary for subsidence inversions. Not surprisingly, therefore, many authors, such as Absalom (1954) and Meetham (1955) have pointed to the significance of anticyclones for atmospheric pollution. Meade (1954) has shown that the highest air-pollution intensities over Britain occur beneath the central areas of anticyclones, and Lawrence (1967) working in southern England has found a relationship between mean daily concentrations of $SO_2$ and the height of the low-level inversions over a period of five winter half-years. In summer the convection processes are normally strong enough to break down the nocturnal surface inversion and high concentrations of summer pollution are found in the vicinity of shallow depressions or cols rather than directly beneath high-pressure areas (Lawrence, 1969a). Nevertheless, the presence of a low-level inversion at night and an inversion base below 2000 m during the day were claimed to be important factors.

In the US, forecasts of air-pollution potential have been prepared since the early 1960s on the basis of the frequency occurrence of large anticyclones, since Holzworth (1962) has noted that poor air quality appears to be most usually associated with quasi-stationary high-pressure cells accompanied by a warm ridge aloft. Figure 3.21A shows the frequency of low-level (surface to 150 m) inversions in the fall season and may be related to Fig. 3.21B which illustrates the number of days of forecast high-pollution potential. This indicates quite clearly that

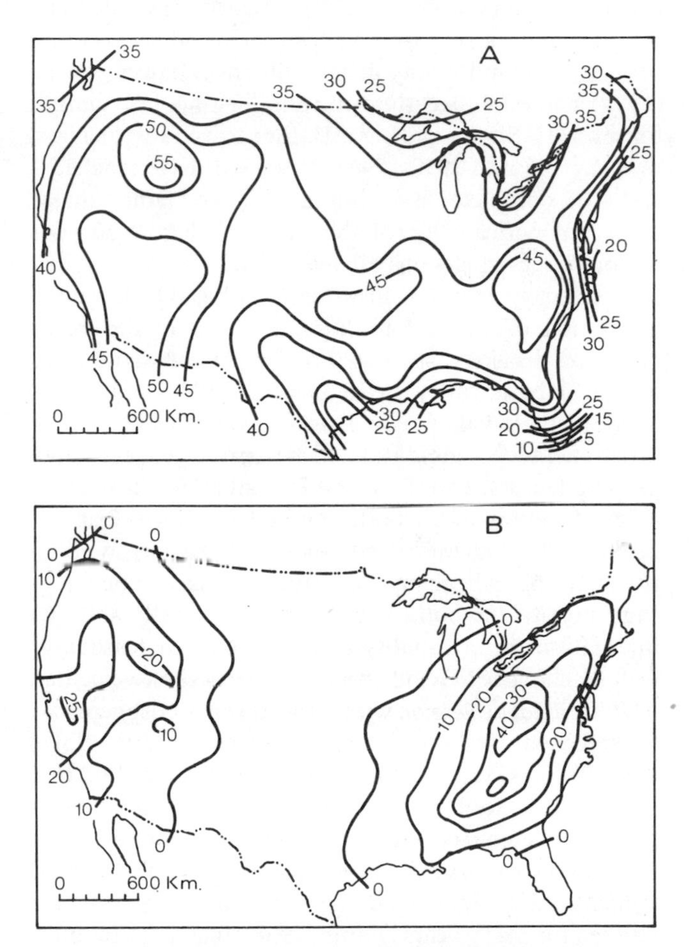


**Fig. 3.21 The general pattern of air pollution potential in the US.**
**(A) Frequency of low-level (surface to 150 m) inversions in the fall season. Isopleths represent average percentage of hours of inversion per day. After Pack (1964).**
**(B) Distribution of forecast days of high air pollution potential. After Leighton (1966). Copyrighted by the American Geographical Society of New York.**

California and the Appalachian region experience the highest pollution potential irrespective of the emission of pollutants. On the other hand, such regional analysis disguises the effect of local conditions which may either increase or decrease the significance of large-scale anticyclones. For example, the urban heat island is likely to reduce the impact of nocturnal temperature inversions over built-up areas, and Baker and Enz (1969), using thermistor data collected from a television tower in the central metropolitan area of St Paul-Minneapolis, have taken an observed decrease in inversion frequency and intensity over a seven-year period as an indication of a progressively modifying urban influence. Indeed, Ewing (1972) has estimated the future night-time heat output for several US cities and has used the projections to suggest that, by the mid-1990s, New York will have reached a state of complete inversion avoidance. Other major cities in America will probably need another 20–30 years to counteract inversions all year round.

Conversely, steep-sided valleys may lead to increased stagnation of anticyclonic air and some of the major pollution disasters, such as those in the industrial areas of the Meuse valley near Liège in 1930 and in the Monongahela valley near Donora, US in 1948, may be attributed to a combination of meteorological and topographic factors. Panofsky and Prasad (1967) have made a specific study of air pollution at Johnstown, Pennsylvania, which is an industrial town sited in a narrow valley through the Allegheny plateau, and have concluded that most of the pollution is produced locally and that fluctuations in air quality can be explained reasonably well by changes in wind speed and vertical air velocities.

*Horizontal dispersion* within the lower atmosphere may be achieved over considerable distances by the global wind systems. Wanta (1962) has described the wind rose as the basic tool in air pollution climatology and, in general, the concentration of pollution emitted from a continuous source is inversely proportional to the wind velocity. The downwind drift of pollution has been studied for many years (Lamb, 1938) and, more recently, experimental evidence has been collected on the transport of pesticide using fluorescent trace particles (Murray and Vaughan, 1970) and the movement of ragweed pollen (Raynor, Ogden, and Hayes, 1970).

Prevailing wind conditions must be seen within the context of the synoptic situation and weather type. For example, observed decreases in winter $SO_2$ concentrations at Rotterdam over a six-year period were found by Schmidt and Velds (1969) to be not entirely due to improvements in industrial processes and partially attributable to an increase in a circulation type which promoted better dispersion. The degree of instability or turbulence within a wind is important and Rouse and McCutcheon (1970) have reported that pollution levels in Hamilton, Ontario, are twice as great with stable easterly winds as for winds from all other sectors. Similarly, Scorer (1968b) investigated the siting of a proposed chemical fertilizer plant on Merseyside, England, and concluded that, although the prevailing wind was directed towards a nearby town, it was sufficiently fresh and unstable to disperse the pollution. The effect of low-turbulence flow aloft on pollution dispersal has been studied by Slade (1969).

Over the years, most attention has probably been given to the behaviour of individual stacks (Scorer, 1955a; Priestley, 1956) and, as shown by Panofsky (1969), a large number of formulae have been developed to relate plume dispersion to meteorological and other factors. The highest ground-level concentrations of pollution from single stacks, even under neutral or unstable conditions, usually occur within a downwind distance equal

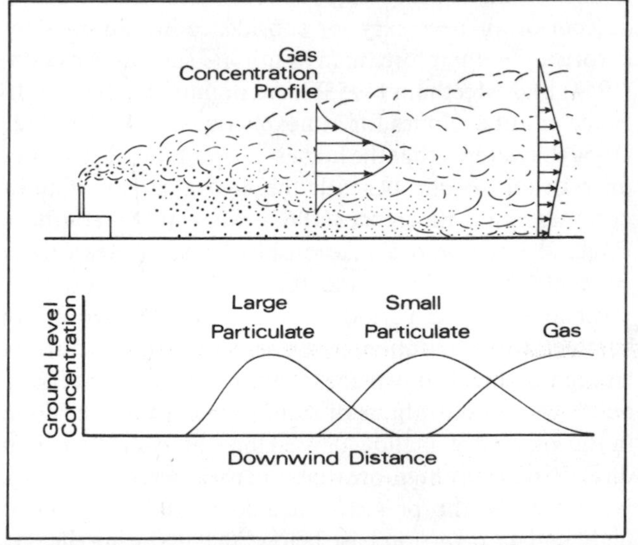


**Fig. 3.22 A diagrammatic comparison of the concentration of gaseous and particulate plumes at a given windspeed. After Strom (1962).**

to 20 times the stack height, and gas diffusion theories show that such concentrations decrease with the inverse of the plume height squared (Strom, 1962). Figure 3.22 illustrates the typical progressive downwind dilution and ground-level concentrations of effluent composed of both gases and particulate matter emitted from a single stack. It can be seen that the gas concentration is highest at the axis of the plume and decreases towards the edges, and that the greatest ground level concentrations are found at some distance from the stack. On the other hand, the particulates have greater free-settling velocities which lead to maximum ground-level concentrations nearer the stack.

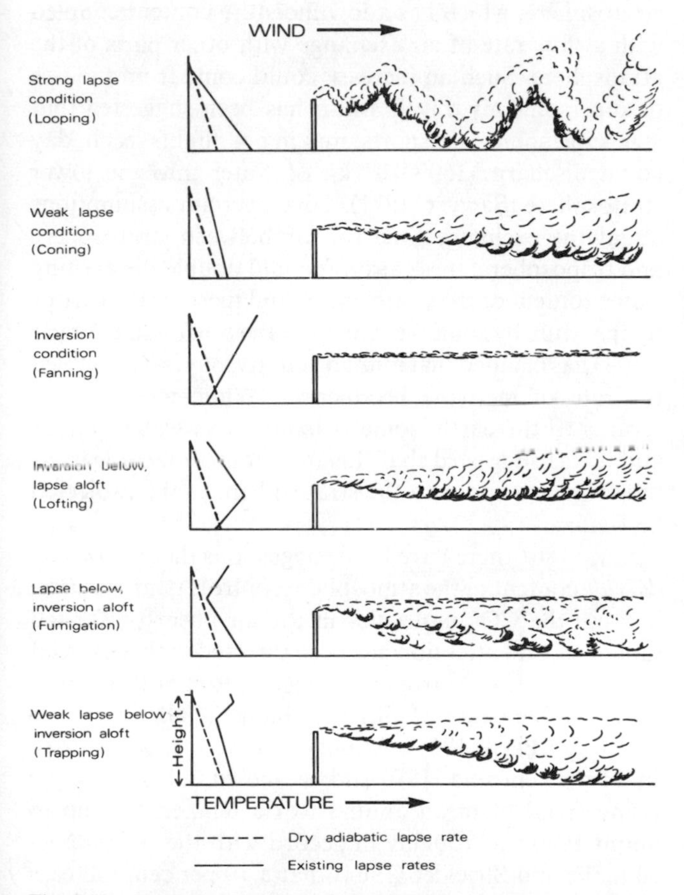


**Fig. 3.23 Characteristic types of plume behaviour under various conditions of atmospheric stability and instability. Broken lines at left represent dry adiabatic lapse rates and solid lines indicate existing lapse rates. After Bierly and Hewson (1962).**

The actual trajectory of a plume is dependent on a variety of factors, including ejection speed and temperature of the effluent, but is mainly controlled by the vertical temperature and wind profiles. Six characteristic types of plume behaviour, after Bierly and Hewson (1962), are presented in Fig. 3.23.

*Looping* occurs only during daylight hours, when superadiabatic conditions and light winds produce strong thermal eddies which can cause high ground-level concentrations of effluent. Another restricting meteorological condition is *fumigation* which occurs when, for a short period after sunrise, solar radiation creates an unstable air layer near the ground, which mixes with the plume developed aloft under a nocturnal inversion and brings pulses of effluent down to the surface. Plume *trapping*, by comparison, may occur at any time when the stack effluent is restricted by a high-level inversion of one type or another. More favourable dispersion from high stacks is achieved by *coning*, *fanning*, and *lofting*. *Coning* can occur at any time when weak lapse rates produce an evenly shaped and diluted plume such as that illustrated in Fig. 3.23. It produces relatively low ground-level concentrations as does *fanning*, which is principally a nocturnal feature caused by a radiation inversion, and the thin plume tends to remain at the stack height as it drifts slowly downwind. Finally, *lofting* is a particularly favourable condition for plume dispersal and may occur at any time other than the middle of the day. The stack height penetrates the shallow surface inversion while a change to unstable conditions above the stable layer ensures that the plume continues to disperse upwards.

## 3.10 The effect of air pollution on the atmosphere

Mention of the atmospheric consequences of air pollution around towns has already been made in this chapter, and it is the purpose of the present section to consider more closely some of the longer-term global effects of such contamination. This is not, of course, to imply that urbanization itself fails to create some continuing local anomalies. For example, Lawrence (1968) and Moffitt (1972) have attributed temperature increases of about 1 degC at Manchester airport from 1942 to 1966 and at Kew, London, from the 1880s to urban growth.

Changes in pollution patterns may introduce different trends, however, and Lawrence (1969b) has detected a

1 degC *decrease* in the mean daily winter minimum in central London compared with surrounding stations and an *increase* in mean maximum temperature of approximately 0·5 degC from 1930 to 1950, which could reflect the increasing use of cleaner and more efficient heating methods. On the other hand, such effects are predominantly local and we must turn to more widespread effects. Unfortunately, not enough is known about the mechanisms of global climate which produce natural changes, and there is, therefore, still considerable uncertainty about the role played by man-made pollutants. Most theories of climatic change, whether naturally or artificially induced, are concerned with temperature trends and, according to Bryson (1968), the mean temperature of the earth depends on the intensity of solar radiation together with the emissivity and reflectivity of the globe.

*Changes in radiation energy*. Variations in solar output are not thought to be influenced by atmospheric pollution, although at least one investigator has sought a relationship between sunspot activity and smog conditions (Lawrence, 1966). However, the most likely change in total atmospheric energy is from *thermal pollution*. According to Cole (1969), the total energy now produced by man is equivalent to only $5 \times 10^{19}$ erg/s, or less than 1 per cent of the $2 \times 10^{24}$ erg/s total radiation from the earth. But, assuming that energy production continues to grow at 7 per cent per year, it is estimated that an overall warming of 1 degC will be achieved in 91 years and Waggoner (1966) believes that this could be sufficient to cause a re-location of plant community boundaries.

A global temperature increase of 3 degC, which would melt the polar icecaps, would take about 780 years from the present and Cole concludes that in less than 1000 years the mean temperature of the earth would be doubled from 15 °C to 30 °C and the earth would be effectively uninhabitable. Rather more pessimistically, Chapman (1970) has calculated that thermal pollution could raise the globe's surface temperature by 3·5 degC within 60 years and has suggested that non-equilibrium sources of power (fossil fuels and nuclear fission) should be limited to 5 per cent of the total energy output, the rest coming from hydroelectric, geothermal, and solar sources. Not surprisingly, the main problem occurs in the US, which consumes about 36 per cent of the world's energy. On the other hand, recent experiments with a numerical model conducted by Washington (1972) have suggested that the thermal pollution effect is small when compared with the natural fluctuations of climate.

*Changes in global emissivity*. The emission of radiation from the atmosphere back into space is determined by a number of factors. Of these, water vapour and carbon dioxide are perhaps the most important since their presence restricts the passage of longwave radiation and creates the so-called 'greenhouse effect'. A climatically significant increase of *water vapour* is most likely in the stratosphere, which has a low moisture content coupled with a slow rate of air exchange with other parts of the atmosphere. Such an increase could come from the rise of supersonic air travel, and it has been suggested that 400 supersonic transports making 4 flights each day could discharge $150 \times 10^6$ kg of water into the lower stratosphere (Sawyer, 1971). Given certain assumptions about the exchange time for air between stratosphere and troposphere, these aircraft could double the existing water content of the stratosphere and increase the earth's temperature by some 0·6 degC. On the other hand, Singer (1971) has claimed that human activity may have doubled the rate of methane production. While most of this returns to the earth, some is oxidized to water vapour and it was suggested that this may contribute at least as much water vapour to the stratosphere as the projected SST effect.

Since 1861 there have been suggestions that the *carbon dioxide* content of the atmosphere controls temperatures, because the $CO_2$ molecules in the air absorb infra-red radiation, and attention has been directed at the artificial production of $CO_2$ from the consumption of fossil fuels as a possible cause of climatic change (Callendar, 1938 and 1958; Plass, 1956). Atmospheric $CO_2$ has increased by 11 per cent since 1870 and, as shown in Fig. 3.24, the rising trend of mean annual world temperature up to about 1940 was roughly in accord with the estimate by Manabe and Strickler (1964) that a 10 per cent increase in $CO_2$ concentration would increase the world mean temperature by 0·3 degC. More recently there is evidence that this trend has been overshadowed by other factors, but this may be only a temporary event. Certainly, not all the $CO_2$ increase may be directly the result of man's activities, and we still know insufficient about the natural $CO_2$ cycle, although there is very little loss of atmospheric carbon dioxide through precipitation and estimated average residence periods range from 5 up to 15 years (Sawyer, 1971).

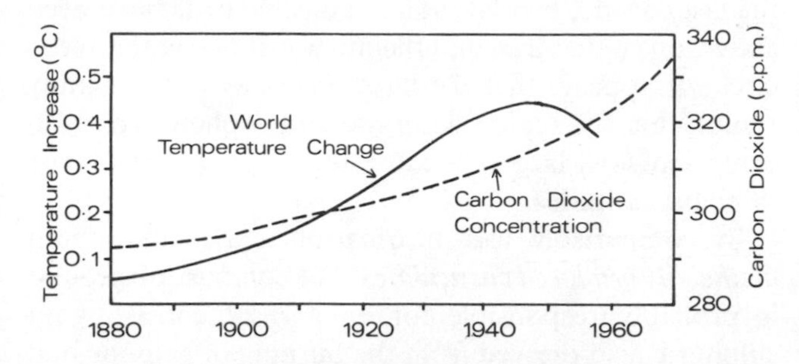


**Fig. 3.24 Trends in mean annual world temperature and carbon dioxide concentration in the atmosphere. After Lovelock (1971).**

Nevertheless, even in the central Pacific, remote from pollution sources, there is a recorded increase of 0·72 ppm/yr of $CO_2$ (Pales and Keeling, 1965). A more recent detailed analysis of the annual tropospheric $CO_2$ budget for the northern hemisphere by Bolin and Bischof (1970) revealed a comparable, well defined increase of $0{\cdot}7 \pm 0{\cdot}1$ ppm each year. Assuming that the emission increase is 4 per cent annually, rising to 5 per cent after 1980, it was predicted that the $CO_2$ concentration will rise to between 418 and 450 ppm by the year 2010 compared with just over 290 ppm in 1880 (Figure 3.24). Sometime next century, therefore, it seems possible that the $CO_2$ content of the atmosphere may well have doubled and Manabe (1970) has calculated that this would lead to a rise in global temperature of 2 degC assuming an unchanged absolute humidity and an increase of 3 degC assuming that the relative humidity remains unchanged.

*Changes in global reflectivity*. The mean temperature of the atmosphere is sensitive to changes in the albedo of the earth; according to Bryson (1968), a 1 per cent increase in reflectivity from the present 37 per cent would lower mean temperatures by 1·7 degC. Positive or negative changes could occur through a variety of human actions. *Land use* modifications, for example, may be locally important and Sawyer (1971) has suggested that the largest effect may occur during spring in snowy areas as a result of deforestation. On the other hand, such alterations are unlikely to have world-wide repercussions. *Cloud cover* influences the earth's reflectivity by increasing the albedo. Schaefer (1969) has described how artificially produced ice crystals, originating in vehicle exhausts or ploughed land, may overseed supercooled clouds, thereby reducing the amount of local precipitation but creating extensive sheets of cirrus cloud. The condensation trails from aircraft have been similarly implicated by Scorer (1955b), and Lamb (1970a) has quoted evidence that cirrus cloud cover has increased by 1/20 to 1/10 of the sky in certain areas since the 1950s.

Within recent years, however, most attention has been paid to the role of *particulates,* since there is evidence that atmospheric turbidity is increasing. Thus, McCormick and Ludwig (1967) obtained some early turbidity data, depicted in Table 3.6 for eastern US and Switzerland, which represent increases of approximately 57 per cent over 60 years and 88 per cent over 30 years respectively.

| | *Period* | *Mean Turbidity* |
|---|---|---|
| Washington, DC | 1903–1907 | 0·098 |
| | 1962–1966 | 0·154 |
| Davos, Switzerland | 1914–1926 | 0·024 |
| | 1957–1959 | 0·043 |

**Table 3.6** Mean values of turbidity. (After McCormick and Ludwig, 1967.)

It was considered that the Davos measurements, at an altitude of 1600 m, might represent global background conditions, but more recent data on solar radiation on Hawaii over a 13-year period have led Ellis and Pueschel (1971) to conclude that short-term fluctuations in turbidity over the Pacific appear to be related to natural tropospheric aerosols—especially volcanic dust. On the other hand, Flowers *et al.* (1969) made turbidity measurements over the US during the early 1960s and found generally low values over the western plains and the Rockies but high values in the east, which suggested the importance of man-made contributions.

Munn and Bolin (1971) have stressed that, although an increasing dust load in the atmosphere will probably result in a temperature decrease, little is still known about the radiative properties of particulate matter and there is even less information about the role played by human activity in generating particulates. Figure 3.25, from Lovelock (1971), illustrates global trends in the combustion of carbon and sulphur from fossil fuels and the relationship to carbon dioxide content and atmospheric turbidity. It can be seen that, although the $CO_2$ increase is roughly equivalent to the rate of carbon combustion, turbidity is increasing much faster at some 30 per cent per decade, and it has been predicted that within a few decades the mean temperature of the northern hemisphere could approach that of the last ice age.

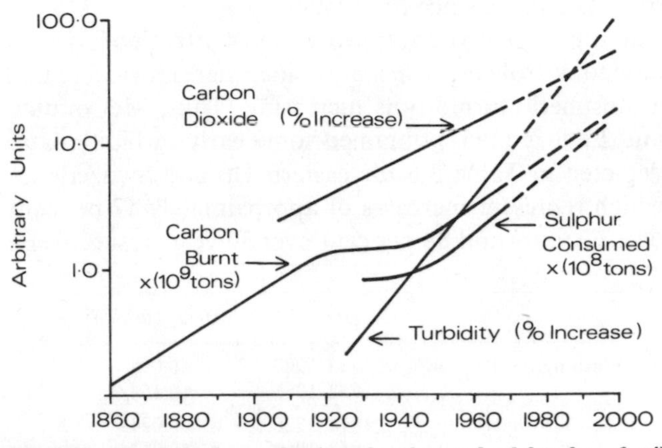


**Fig. 3.25 Trends in the combustion of carbon and sulphur from fossil fuels in relation to the carbon dioxide content of the atmosphere and atmospheric turbidity. After Lovelock (1971).**

*Naturally produced aerosols* include the blue haze commonly found over forests as a result of the photochemical conversion of the hydrocarbons released from the vegetation and the soil particles found above deserts, but climatologists have given most emphasis to volcanic dust (Wexler, 1952). The injection of volcanic material into the atmosphere can produce large increases in turbidity, as when the eruption of Krakatoa, Indonesia, in 1883 caused an ejection of some $10^{13}$ kg of dust into the stratosphere and a 10 per cent reduction in solar radiation at Montpellier Observatory, France, over the following three years (Lamb, 1970b).

Short-term effects may result from tropospheric injections (Lamb and Parker, 1971) but the less frequent release of volcanic dust and gases into the stratosphere is more significant since there is no removal by precipitation. Bigg (1971) has used photographic evidence from the southern hemisphere to show that, after the eruption of Mt Agung, Indonesia, in March 1963, solid particles composed mainly of ammonium sulphate were replaced by liquid droplets of sulphuric acid and $H_2SO_4$ particles. An increase in global volcanism affecting the stratosphere since 1963 has been reported by Cronin (1971). Within the last decade, two latitudinal volcanic belts have been active, one equatorial and the other just below the arctic circle. It appears that the latter may have been mainly responsible for replenishing the stratosphere with volcanic aerosols, largely because the tropopause is lower in higher latitudes.

By comparison, less information is available about *artificially produced particulates*. The combustion process is probably responsible for an aerosol consisting of sulphuric acid derived from the burning of sulphur but, in Britain, smoke concentration appears more significant. Over Britain the effect of smoke control, based on a value of 1·4 per cent radiation loss per 10 $\mu g/m^3$ of smoke, should have increased average insolation by approximately 1 per cent per year (Monteith, 1962). On a global scale, however, it might well be that the extension of mechanized agriculture, particularly into semi-arid areas, is the major factor explaining the increasing atmospheric dust load. Davitaya (1969) has studied the decreasing albedo of glaciers associated with the accumulation of surface dust, and taken dust samples from the firn layers of a Caucasian glacier at an altitude of 4600 m. As shown in Fig. 3.26 there has been a marked increase

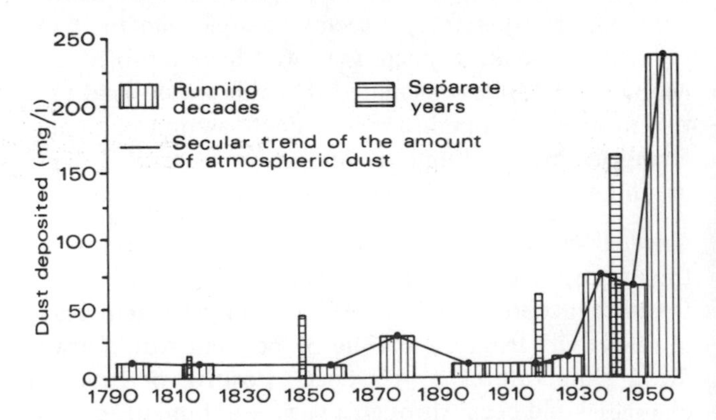


**Fig. 3.26 Amounts of dust deposited in firn layers during decades at the Maili Plateau, Kazbek, USSR. After Davitaya (1969). Reproduced by permission of the Association of American Geographers.**

of dust concentration from 1930 since when the rise has paralleled that of economic activity in the Soviet Union and other Eurasian countries.

There seems little doubt that increasing turbidity is one of the most important changes currently taking place in the atmosphere. According to Bryson (1968), a rise of 3–4 per cent averaged over the world could lower global temperatures by 0·4 degC and several authorities suggest that increasing turbidity has recently overtaken the influence of $CO_2$ additions and is responsible for the apparent trend towards world-wide cooling. Indeed, Rasool and Schneider (1971) have stated that an increase in the global aerosol background concentration by a factor of four may be sufficient to reduce surface temperatures by 3·5 degC and, if sustained over several years, this might possibly trigger off an ice age.

### 3.11 Other effects of atmospheric pollution

The social and economic implications of air pollution are wide-ranging and in this section only a brief account of pollution costs together with the damage to human health, plant life, and other materials will be possible.

*Costs of air pollution.* The economic analysis of air pollution is difficult because, in addition to the direct losses caused to material property, there are substantial costs incurred by indirect factors such as lost production through ill-health and the reduced efficiency of transport facilities. Ogden (1966) has emphasized that an accurate economic assessment should also include a cost: benefit appraisal of control measures but, however deficient existing estimates are, there is little doubt about the substantial economic consequences of pollution.

In Britain, Beaver (1954) estimated that the total cost of air pollution, excluding the loss of health, rose from about £45 million in 1924 to £250 million in 1954. Of this, some £100 million was attributed to indirect losses and, for example, it was claimed that one-third of the cost of track replacement on the railway system was due to atmospheric corrosion. More recently, the Royal College of Physicians (1970) has noted that 30 million working days are lost every year in Britain through the partially pollution-dependent disease of bronchitis, and that the total cost of this amounts to £65 million per year in lost production and medical care.

According to Bach (1972), estimated total economic losses arising from air pollution in the US vary between $2 billion and $12 billion per year. The health costs alone are near to $1 billion, while crop damage involves some $500 million each year of which about $125 million occurs in California (Low, 1968). In urban areas, the cost of cleaning and restoring public buildings may be large. Thus, several years ago it cost over $2 million to re-face the New York City Hall and Benline (1965) has estimated that the economic burden within the city due to the deterioration of public and private property is no less than $520 million per year. There is accumulating evidence of the effect of air pollution on residential property values. Nourse (1967) quoted studies in the St Louis Metropolitan Area, where house prices declined by $245 for every increase of 0·5 mg of sulphur trioxide per 100 cm$^2$ per day and some properties decreased in value by $1000 with deteriorating air quality following the establishment of a metal fabricating plant. Similarly, Anderson and Crocker (1971) have used data from St Louis, Kansas City, and Washington, DC to develop a qualitative model showing reductions in property values of between $30 and $70 per unit of pollution in each city.

*Effects on human health.* The clearest illustrations of the adverse influence of air pollution on human health occur during the acute outbreaks of pollution associated with cold, foggy weather in highly industrialized urban centres in the northern hemisphere. A number of such early episodes in Britain have been described by Marsh (1947) but probably the most serious and best documented event anywhere took place in London between 5 and 9 December 1952, as shown in Fig. 3.27, when a dense anticyclonic smog was responsible for the deaths of approximately 4000 people (Ministry of Health, 1954; Wilkins, 1954). The largest relative increases in mortality were for lung diseases such as bronchitis, pneumonia, and respiratory tuberculosis, together with heart disease. Most of the fatalities were amongst the elderly and the high mortality was largely responsible for the passing of the national Clean Air Act in 1956.

Some measure of the part played by smoke pollution in 1952 can be gained, since comparable meteorological conditions produced a fog which lasted for the same length of time in December 1962. Although sulphur dioxide concentrations were very similar to those a decade earlier, there was much less smoke and the number of excess deaths above normal rose to only 700 on this occasion (Scott, 1963). A similar winter fog took place at Donora, Pennsylvania, in 1948. It was also accompanied by a rise in the incidence of chronic bronchitis and subsequent investigation showed that 43 per cent of the population experienced some illness during the event (McDermott, 1961; Goldsmith, 1962).

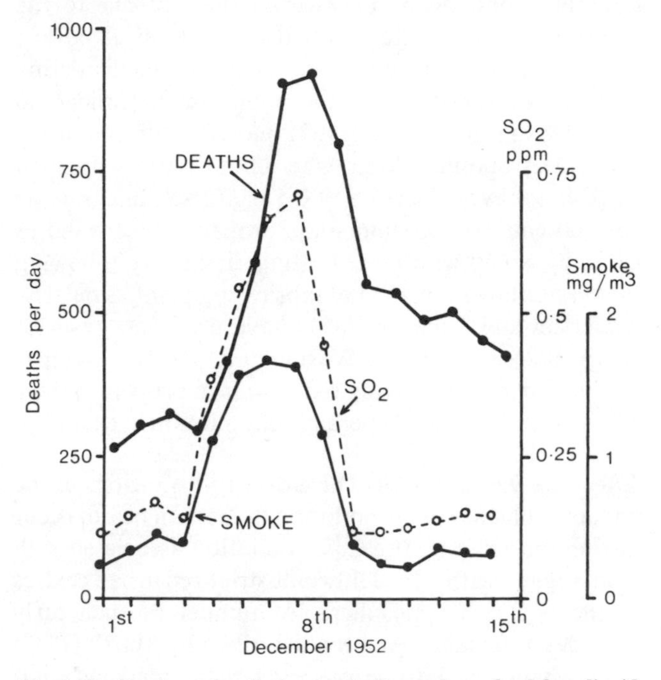


**Fig. 3.27 Deaths and pollution levels for smoke and sulphur dioxide during the London fog of December 1952. After Royal College of Physicians (1970).**

The US also suffers from the photochemical smogs of the West Coast, where the action of strong ultraviolet light on a mixture of nitrogen oxide, hydrocarbons, and ozone in stable air produces substances that irritate the eyes and respiratory membranes. Although no direct fatalities have been confirmed, the smog interferes with lung function and produces eye irritation that affects nearly three-quarters of the metropolitan population of Southern California, according to Goldsmith (1962).

It is rare for specific air pollutants to reach toxic levels. For example, the dangerous gas carbon monoxide, emitted from petrol engines, is thought to reach concentrations of only 10–20 ppm in British cities, with occasional peaks approaching 100 ppm (Waller *et al.*, 1965), compared with the prescribed industrial limit for continuous exposure of 50 ppm. It is, therefore, much more common for air pollution to be implicated in the exacerbation of existing diseases, notably those of the respiratory tract. In this context, smoke and sulphur dioxide are thought to be significant pollutants. Lawther (1958) has found that the deterioration of bronchitics coincided with an increase in the atmospheric concentration of both smoke and $SO_2$, and Gore and Shaddick (1958) have claimed that mortality from lung cancer in London is highest in the areas mainly associated with these two pollutants.

Bronchitis and other forms of respiratory disease are much more common in Britain than the US, especially among older people living in the large towns (Reid *et al.*, 1964; Holland *et al.*, 1965). The incidence of low-temperature fog has long been blamed for this phenomenon and Russell (1924 and 1926) was able to report an inverse relationship between weekly death rates and air temperatures recorded during the preceding week. However, the relative importance of such separate factors as air pollution, fog, and low temperature is still uncertain.

Although Reid and Fairbairn (1958) found that the monthly absenteeism of London postmen with bronchitis rose sharply when visibility fell below 1000 m, a study of weekly death rates from bronchitis and pneumonia in London and East Anglia showed that mortality was more closely related to low temperatures than fog frequency (Boyd, 1960). Nevertheless, in periods of cold, foggy weather the death rates were higher in the polluted London air. More recently, Gregory (1970) has examined absences from a Sheffield steelworks due to chronic bronchitis and discovered that, although smoke pollution appears to influence the incidence of exacerbations, low temperatures affect the period of absence by delaying the return to work.

Like bronchitis, lung cancer is a largely British disease and, when emigration takes place, the death rates of British-born migrants in countries such as New Zealand and the US are found to be intermediate between the lower rate for the native-born population and the higher rate prevailing in Britain (Eastcott, 1956; Haenszel, 1961). Carcinogenic elements are found in urban and industrial smoke and Stocks and Campbell (1955) have suggested that Liverpool air has 8–12 times as many carcinogens as occur in the surrounding rural area. Human exposure to radioactive pollution may be responsible for the development of a wide range of body cancers together with other disorders. In addition to the immediate, short-term effects there are long-term genetic consequences. For a more comprehensive review of human health aspects, the reader should consult Masters (1971).

*Effects on vegetation.* Comprehensive reviews of the effects of air pollution on plants have been presented by Brandt (1962), Thomas (1965), and Linzon (1971). Plant growth may be restricted by soot clogging leaf stomata and thereby reducing the absorption of $CO_2$ from the atmosphere while, in Los Angeles smog conditions, excessive nitrogen dioxide and leaf flecking induced by high ozone concentrations may reduce growth by up to 30 per cent (Koritz and Went, 1956). In turn, it has been claimed that photochemical pollution has lowered citrus fruit production around Los Angeles by around 20 per cent (Anon., 1964).

Injury to forest stands has been summarized by Hepting (1964) and includes the early smelter fume damage to conifers in the upper Columbia valley downriver from Trail, British Columbia. The more recent chloritic decline of ponderosa pine, first noted during the early 1950s in the San Bernardino Mountains of California, has been attributed to a combination of smog and drought despite a distance of over 50 miles from Los Angeles itself (Miller *et al.*, 1963). Monteith (1970) has considered the possible long-term global consequences of atmospheric pollution on plant growth and has estimated that the rate at which green plants produce dry matter could increase by 12 per cent before the end of the present century owing to increasing levels of atmospheric carbon dioxide and the attendant rise of photosynthetic activity. On the other hand, the increase could be much smaller and the trend even reversed if mean air temperatures are reduced by added turbidity.

*Effects on materials.* Air pollution greatly accelerates the natural atmospheric weathering of exposed materials and Munn (1959) recognized soiling, erosion, and corrosion as the three principal methods of deterioration. Although soiling and abrasion by particulates are important, it is the corrosive effects, often aided by the presence of moisture, which are most widespread. For example, $CO_2$ attacks limestone building materials through the conversion of calcium carbonate to the soluble bicarbonate form, the sulphur oxides damage ferrous metal, leather, paper, and textiles, and ozone produces extensive cracking of rubber products and reduces the colour-fastness of dyes (Yocom, 1962). The deterioration of works of art is an especially difficult problem. For example, sulphur dioxide has been found by Thomson (1969) to cause damage to Italian frescoes composed of pigmented lime plaster which is converted to gypsum, with a resulting increase in volume and subsequent loss of coherence.

## References


ABELSON, P. (1971). The environmental crisis. *Trans. Amer. Geophys. Un.*, **52**:124–9.

ABSALOM, H. W. L. (1954). Meteorological aspects of smog. *Q. Jl. R. Met. Soc.*, **80**:261–6.

AIR CONSERVATION COMMISSION (1965). *Air conservation* Pub. No. 80, Amer. Ass. Adv. Sci., Washington, D.C. Reprinted in T. R. Detwyler *Man's impact on environment*, McGraw-Hill, 1971.

ANDERSON, R. J. and CROCKER, T. D. (1971). Air pollution and residential property values. *Urban Studies,* **8**:171–80.

ANDERSSON, T. (1969). Small-scale variations of the contamination of rain caused by washout from the low layers of the atmosphere. *Tellus*, **21**:685–92.

ANONYMOUS (1964). Citrus study under way: smog answer sought. *Air in the News*, **1**:12.

ASHWORTH, J. R. (1929). The influence of smoke and hot gases from factory chimneys on rainfall. *Q. Jl. R. Met. Soc.*, **55**: 341–50.

ATKINS, J. E. (1968). Changes in the visibility characteristics at Manchester/Ringway airport. *Met. Mag.*, **97**:172–4.

ATKINSON, B. W. (1968). A preliminary examination of the possible effect of London's urban area on the distribution of thunder rainfall. *Trans. Inst. Brit. Geog.*, **44**:97–118.

ATKINSON, B. W. (1969). A further examination of the urban maximum of thunder rainfall in London, 1951–60. *Trans. Inst. Brit. Geog.*, **48**:97–119.

BACH, W. (1971). Atmospheric turbidity and air pollution in Greater Cincinnati. *Geogrl. Rev.*, **61**:573–94.

BACH, W. (1972). *Atmospheric pollution.* McGraw-Hill Book Co., New York, 144 pp.

BACH, W., HAGEDORN, T., and MATHEWS, E. (1970). Variation of atmospheric turbidity with height over an urban area. *Proc. Ass. Amer. Geogrs.*, **2**:4–8.

BACH, W. and PATTERSON, W. (1969). Heat budget studies in greater Cincinnati. *Proc. Ass. Amer. Geogrs.*, **1**:7–11.

BAKER, D. G. and ENZ, J. W. (1969). Frequency, duration, commencement time and intensity of temperature inversions at St. Paul-Minneapolis. *J. Appl. Met.*, **8**:747–53.

BALCHIN, W. G. V. and PYE, N. (1947). A micro-climatological investigation of Bath and the surrounding district. *Q. Jl. R. Met. Soc.*, **73**:297–323.

BALCHIN, W. G. V. and PYE, N. (1948). Local rainfall variations in Bath and the surrounding district. *Q. Jl. R. Met. Soc.*, **74**:361–78.

BARRETT, E. C. (1964). Local variations in rainfall trends in the Manchester region. *Trans. Inst. Brit. Geog.*, **35**:55–71.

BEAVER, H. (1954). *Report of Committee on Air Pollution.* HMSO, London.

BELL, G. G. (1967). Meteorological effects on California air pollution. In Tromp, S. W. and Weihe, W. H. (eds) *Biometeorology Vol. 2, Pt. 2.* Pergamon Press, Oxford, 628–40.

BENLINE, A. J. (1965). Air pollution control problems in the city of New York. *Trans. N.Y. Acad. Sci.*, **27**:916–22.

BIERLY, E. W. and HEWSON, E. W. (1962). Some restrictive meteorological conditions to be considered in the design of stacks. *J. Appl. Met.*, **1**:383–90.

BIGG, E. K. (1971). Stratospheric pollution and volcanic eruptions. *Weather*, **26**:13–18.

BLOKKER, P. C. (1970). The atmospheric chemistry and long-range drift of sulphur dioxide. *J. Inst. Petroleum*, **56**:71–9.

BOLIN, B. and BISCHOF, W. (1970). Variations of the carbon dioxide content of the atmosphere of the northern hemisphere. *Tellus*, **22**:431–42.

BORNSTEIN, R. D. (1968). Observations of the urban heat island effect in New York City. *J. Appl. Met.*, **7**:575–82.

BOYD, J. T. (1960). Climate, air pollution and mortality. *Brit. J. Soc. Prev. Med.*, **14**:123–35.

BRANDT, C. S. (1962). Effects of air pollution on plants. In Stern, A. C. (ed) *Air Pollution* 1, Academic Press, New York: 255–81.

BRAUN, R. C. and WILSON, M. J. G. (1970). The removal of atmospheric sulphur by building stones. *Atmos. Envir.*, **4**:371–8.

BRAZELL, J. H. (1964). Frequency of dense and thick fog in central London as compared with frequency in outer London. *Met. Mag.*, **93**:129–35.

BRINGFELT, B. (1971). Important factors for the sulphur dioxide concentration in central Stockholm. *Atmos. Envir.*, **5**:949–72.

BROSSET, C. and MARSH, K. J. (1970). Measurements of sulphur dioxide over the North Sea. *Atmos. Envir.*, **4**:225.

BRYSON, R. A. (1968). All other factors being constant. . . . A reconciliation of several theories of climatic change. *Weatherwise*, **21**:56–61.

BULLOCK, J. and LEWIS, W. M. (1968). The influence of traffic on atmospheric pollution. The High Street, Warwick. *Atmos. Envir.*, **2**:517–34.

CALLENDAR, G. S. (1938). The artificial production of carbon dioxide and its influence on temperature. *Q. Jl. R. Met. Soc.*, **64**:223–40.

CALLENDAR, G. S. (1958). On the amount of carbon dioxide in the atmosphere. *Tellus*, **10**:243–8.

CARTER, H. E. (1931). Week-end weather in 1931. *Met. Mag.*, **66**:163–4.

CHAMBERS, L. A. (1962). Classification and extent of air pollution problems. In Stern, A. C. (ed.) *Air Pollution* 1, Academic Press, New York: 3–22.

CHANDLER, T. J. (1960). Wind as a factor of urban temperatures—a survey in North-east London. *Weather*, **15**: 204–13.

CHANDLER, T. J. (1961a). The changing form of London's heat island. *Geography*, **46**:295–307.

CHANDLER, T. J. (1961b). Surface breeze effects of Leicester's heat island. *East Mid. Geog.*, **2**:32–8.

CHANDLER, T. J. (1962a). London's urban climate. *Geogrl. Jl.*, **128**:279–98.

CHANDLER, T. J. (1962b). Temperature and humidity traverses across London. *Weather*, **17**:235–42.

CHANDLER, T. J. (1962c). Diurnal, seasonal and annual changes in the intensity of London's heat island. *Met. Mag.*, **91**:146–53.

CHANDLER, T. J. (1964). City growth and urban climates. *Weather*, **19**:170–4.

CHANDLER, T. J. (1965). *The climate of London.* Hutchinson, London, 292 pp.

CHANDLER, T. J. (1967). Night-time temperatures in relation to Leicester's urban form. *Met. Mag.*, **96**:244–50.

CHANGNON, S. A. (1968). The La Porte weather anomaly—fact or fiction. *Bull. Amer. Met. Soc.*, **49**:4–11.

CHAPMAN, P. F. (1970). Energy production—a world limit? *New Scientist*, **47**:634–6.

CLARKE, J. F. (1969a). Nocturnal urban boundary layer over Cincinnati, Ohio. *Mon. Weath. Rev.*, **97**:582–9.

CLARKE, J. F. (1969b). A meteorological analysis of carbon dioxide concentrations measured at a rural location. *Atmos. Envir.*, **3**:375–83.

COLE, L. C. (1969). Thermal pollution. *Bio Science*, **19**:989–92.

COLLIER, C. G. (1970). Fog at Manchester. *Weather*, **25**:25–9.

COMMINS, B. T. and WALLER, R. E. (1967). Observations from a ten-year study of pollution at a site in the City of London. *Atmos. Envir.*, **1**:49–68.

CONRADS, L. A. and VAN DER HAGE, J. C. H. (1971). A new method of air-temperature measurement in urban climatological studies. *Atmos. Envir.*, **5**:629–35.

CRABTREE, J. (1959). The travel and diffusion of the radioactive material emitted during the Windscale accident. *Q. Jl. R. Met. Soc.*, **85**:362–70.

CRAMER, H. E. (1959). A brief survey of the meteorological aspects of atmospheric pollution. *Bull. Amer. Met. Soc.*, **40**:165–71.

Craxford, S. R. and Weatherley, M-L. P. M. (1971). Air pollution in towns in the United Kingdom. *Phil. Trans. R. Soc., Lond. Series A*, **269**:503–13.

Cronin, J. F. (1971). Recent volcanism and the stratosphere. *Science*, **172**:847–9.

Davitaya, F. F. (1969). Atmospheric dust content as à factor affecting glaciation and climatic change. *Ann. Ass. Amer. Geogr.*, **59**:552–60.

Douglas, C. K. M. and Stewart, K. H. (1953). London fog of December 5–8, 1952. *Met. Mag.*, **82**:67–71.

Duckworth, F. S. and Sandberg, J. S. (1954). The effect of cities upon horizontal and vertical temperature gradients. *Bull. Amer. Met. Soc.*, **35**:198–207.

Dyer, A. J. and Hicks, B. B. (1967). Radioactive fallout from the French 1966 Pacific tests. *Austral. J. Sci.*, **30**:168–70.

Eastcott, D. F. (1956). The epidemiology of lung cancer in New Zealand. *Lancet*, **1**:37–9.

Eaton, H. S. (1877). Presidential Address. *Q. Jl. R. Met. Soc.*, **3**:309–17.

Eggleton, A. E. J. (1969). The chemical composition of atmospheric aerosols on Tees-side and its relation to visibility. *Atmos. Envir.*, **3**:355–72.

Eisenbud, M. (1970). Environmental protection in the City of New York. *Science*, **170**:706–12.

Ellis, H. T. and Pueschel, R. F. (1971). Solar radiation: absence of air pollution trends at Mauna Loa. *Science*, **172**:845–6.

Ewing, R. H. (1972). Potential relief from extreme urban air pollution. *J. Appl. Met.*, **11**:1342–5.

Findlay, B. F. and Hirt, M. S. (1969). An urban-induced meso-circulation. *Atmos. Envir.*, **3**:537–42.

Flowers, E. C., McCormick, R. A. and Kurfis, K. R. (1969). Atmospheric turbidity over the United States, 1961–66. *J. Appl. Met.*, **8**:955–62.

Freeman, M. H. (1968). Visibility statistics for London/Heathrow Airport. *Met. Mag.*, **97**:214–18.

Garnett, A. (1957). Climate, relief and atmospheric pollution in the Sheffield region. *Adv. Sci.*, **13**:331–41.

Garnett, A. (1967). Some climatological problems in urban geography with special reference to air pollution. *Trans. Inst. Brit. Geog.*, **42**:21–43.

Garnett, A. and Bach, W. (1965). An estimation of the ratio of artificial heat generation to natural radiation heat in Sheffield. *Mon. Weath. Rev.*, **93**:383–5.

Glasspoole, J. (1969). Wet Thursdays. *Weather*, **24**:241–2.

Goldsmith, J. R. (1962). Effects of air pollution on humans. In Stern, A. C. (ed.) *Air Pollution* 1, Academic Press, New York: 335–86.

Gore, H. T. and Shaddick, C. W. (1958). Atmospheric pollution and mortality in the County of London. *Brit. J. Prev. Soc. Med.*, **12**:104–13.

Greenfield, S. M. (1957). Rain scavenging of radioactive particulate matter from the atmosphere. *J. Meteor.*, **14**:115–125.

Gregory, J. (1970). The influence of climate and atmospheric pollution on exacerbations of chronic bronchitis. *Atmos. Envir.*, **4**:453–68.

Haagen-Smit, A. J. (1952). Chemistry and physiology of Los Angeles smog. *Ind. Eng. Chem.*, **44**:1342–6.

Haagen-Smit, A. J. (1962). Reactions in the atmosphere. In Stern, A. C. (ed.) *Air Pollution* 1, Academic Press, New York: 41–64.

Haenszel, W. (1961). Cancer mortality among the foreign-born in the United States. *J. Nat. Cancer Inst.*, **26**:37–132.

Harrison, A. A. (1967). Variations in night minimum temperatures peculiar to a valley in mid-Kent. *Met. Mag.*, **96**:257–65.

Hepting, G. H. (1964). Damage to forests from air pollution. *J. Forestry*, **62**:630–4.

Holland, W. W., Reid, D. D., Seltser, R. and Stone, R. W. (1965). Respiratory disease in England and the United States: studies of comparative prevalence. *Arch. Envir. Health*, **10**:338–45.

Holzworth, G. C. (1962). A study of air pollution potential for the western United States. *J. Appl. Met.*, **1**:366–82.

Howard, L. (1818). *The climate of London*. Vol. 1. London.

Huff, F. A. and Changnon, S. A. (1972). Climatological assessment of urban effects on precipitation at St. Louis. *J. Appl. Met.*, **11**:823–42.

Hutcheon, R. J. *et al.* (1967). Observations of the urban heat island in a small city. *Bull. Amer. Met. Soc.*, **48**:7–9.

Israel, H., Horbert, M. and Israel, G. W. (1966). Results of continuous measurements of radon and its decay products in the lower atmosphere. *Tellus*, **18**:638–41.

Jenkins, I. (1969). Increase in averages of sunshine in Greater London. *Weather*, **24**:52–4.

Junge, C. E. (1963). *Air Chemistry and Radioactivity*. Academic Press, New York: 382 pp.

Junge, C. E. and Werby, R. T. (1958). The concentration of chloride, sodium, potassium, calcium and sulphate in rain-water over the U.S. *J. Meteor.*, **15**:417–25.

Keeling, C. D. (1960). The concentration and isotopic abundances of $CO_2$ in the atmosphere. *Tellus*, **12**:200–3.

Koritz, H. G. and Went, F. W. (1956). Physiological action of smog on plants. I—Initial growth and transpiration studies. *Pl. Physiol.*, **28**:50–62.

Kratzer, A. (1956). *Das Stadtklima*. Verl. Vieweg, Braunschweig. (2nd Ed.)

Lamb, H. H. (1938). Industrial smoke drift and weather. *Q. Jl. R. Met. Soc.*, **64**:639–43.

Lamb, H. H. (1963). What can we find out about the trend of our climate? *Weather*, **18**:194–216.

Lamb, H. H. (1970a). Climatic variation and our environment today and in the coming years. *Weather*, **25**:447–55.

LAMB, H. H. (1970b). Volcanic dust in the atmosphere: with a chronology and assessment of its meteorological significance. *Phil. Trans. R. Soc.*, Lond. Series A, **266**:425–533.

LAMB, H. H. and PARKER, B. N. (1971). Volcanic eruption in Jan Mayen, September 1970. *Weather*, **26**:263–7.

LANDSBERG, H. E. (1960). *Physical climatology*. Gray Printing Co., Dubois, Pennsylvania. (2nd Ed.)

LANDSBERG, H. E. (1967). Air pollution and urban climate. In Tromp, S. W. and Weihe, W. H. (eds) *Biometeorology, Vol. 2, Pt. 2*, Pergamon Press, Oxford:648–56.

LAWRENCE, E. N. (1953). London's fogs of the past. *Weather*, **82**:367–9.

LAWRENCE, E. N. (1966). Sunspots—a clue to bad smog? *Weather*, **21**:367–70.

LAWRENCE, E. N. (1967). Atmospheric pollution during spells of low-level air temperature inversion. *Atmos. Envir.*, **1**:561–76.

LAWRENCE, E. N. (1968). Changes in air temperature at Manchester Airport. *Met. Mag.*, **97**:43–51.

LAWRENCE, E. N. (1969a). High values of atmospheric pollution in summer at Kew and the associated weather. *Atmos. Envir.*, **3**:123–33.

LAWRENCE, E. N. (1969b). Effects of urbanization on long-term changes of winter temperature in the London Region. *Met. Mag.*, **98**:1–8.

LAWRENCE, E. N. (1971a). Urban climate and day of the week. *Atmos. Envir.*, **5**:935–48.

LAWRENCE, E. N. (1971b). Day-of-the-week variation in weather. *Weather*, **26**:386–91.

LAWTHER, P. J. (1958). Climate, air pollution and chronic bronchitis. *Proc. Roy. Soc. Med.*, **51**:262–4.

LEIGHTON, P. A. (1966). Geographical aspects of air pollution. *Geogrl. Rev.*, **56**:151–74.

LINDQVIST, S. (1968). Studies on the local climate in Lund and its environs. *Geograf. Ann.*, **50A**:79–93.

LINZON, S. N. (1971). Effects of air pollutants on vegetation. In McCormac, B. M. (ed.) *Introduction to the scientific study of atmospheric pollution*. D. Reidel Publishing Co., Dordrecht: 131–51.

LOVELOCK, J. E. (1971). Air pollution and climatic change. *Atmos. Envir.*, **5**:403–11.

LOW, I. (1968). Smog over the fields. *New Scientist*, **28**:494.

LOWRY, W. P. (1967). The climate of cities. *Scientific American*, **217**:15–23.

MCCORMICK, R. A. (1971). Air pollution in the locality of buildings. *Phil. Trans. R. Soc., Lond. Series A*, **269**:515–26.

MCCORMICK, R. A. and LUDWIG, J. H. (1967). Climatic modification by atmospheric aerosols. *Science*, **156**:1358–9.

MCDERMOTT, W. (1961). Air pollution and public health. *Scientific American*, **205**:49–57.

MALAKHOV, S. G. *et al.* (1966). Diurnal variations of radon and thoron decay product concentrations in the surface layer of the atmosphere and their washout by precipitation. *Tellus*, **18**:643–54.

MANABE, S. (1970). The dependence of atmospheric temperature on the concentration of carbon dioxide. In *Global Effects of Environmental Pollution*, Reidel Pub. Co., Holland; 25–9 (quoted by Munn and Bolin, 1971).

MANABE, S. and STRICKLER, R. F. (1964). Thermal equilibrium in the atmosphere with convective adjustment. *J. Atmos. Sci.*, **21**:361–85.

MANLEY, G. (1958). On the frequency of snowfall in metropolitan England. *Q. Jl. R. Met. Soc.*, **84**:70–2.

MARSH, A. (1947). *Smoke. The Problem of Coal and the Atmosphere*. Faber and Faber Ltd, London, 306 pp.

MARTIN, A. and BARBER, F. R. (1971). Some measurements of loss of atmospheric sulphur dioxide near foliage. *Atmos. Envir.*, **5**:345–52.

MASON, P. F. (1970). Spatial variability of atmospheric radioactivity in the U.S. *Proc. Ass. Amer. Geogr.*, **2**:92–7.

MASTERS, R. L. (1971). Air pollution—human health effects. In McCormac, B. M. (ed.) *Introduction to the scientific study of atmospheric pollution*. D. Reidel Publishing Co., Dordrecht: 97–130.

MEADE, P. J. (1954). Smogs in Britain and the associated weather. *Int. J. Air. Pollut.*, **2**:87–91.

MEETHAM, A. R. (1945). Atmospheric pollution in Leicester. *DSIR* Tech. Paper No. 1, HMSO, London.

MEETHAM, A. R. (1950). Natural removal of pollution from the atmosphere. *Q. Jl. R. Met. Soc.*, **76**:359–71.

MEETHAM, A. R. (1955). Know your fog. *Weather*, **10**:103–5.

MEETHAM, A. R. (1964). *Atmospheric pollution*. Pergamon Press, London (3rd Ed.) 301 pp.

MILLER, P. R., PARMETER, J. R., TAYLOR, O. C. and CARDIFF, E. A. (1963). Ozone injury to the foliage of Pinus ponderosa. *Phytopathology*, **53**:1072–6.

MINISTRY OF HEALTH (1954). Mortality and morbidity during the London fog of December 1952. *Reports on Public and Medical Subjects*. No. 95, HMSO, London.

MINISTRY OF TECHNOLOGY (1966). *Instruction Manual*. National Survey of Smoke and Sulphur Dioxide, Warren Spring Laboratory, 142 pp.

MOFFITT, B. J. (1972). The effects of urbanisation on mean temperatures at Kew Observatory. *Weather*, **27**:121–9.

MONTEITH, J. L. (1962). Solar radiation: a climatological study. *Q. Jl. R. Met. Soc.*, **88**:508–21.

MONTEITH, J. L. (1966). Local differences in the attenuation of solar radiation over Britain. *Q. Jl. R. Met. Soc.*, **92**:254–6.

MONTEITH, J. L. (1970). Prospects for photosynthesis from AD 1970 to AD 2000. *Weather*, **25**:456–62.

MUNN, R. E. (1959). Engineering meteorology: The weathering of exposed surfaces by atmospheric pollution. *Bull. Amer. Met. Soc.*, **40**:172–8.

MUNN, R. E. and BOLIN, B. (1971). Global air pollution—metorological aspects. *Atmos. Envir.*, **5**:363–402.

MUNN, R. E., THOMAS, D. A. and COLE, A. F. W. (1969). A study of suspended particulate and iron concentrations in Windsor, Canada. *Atmos. Envir.*, **3**:1–10.

MURRAY, J. A. and VAUGHAN, L. M. (1970). Measuring pesticide drift at distances to four miles. *J. Appl. Met.*, **9**:79–85.

MYRUP, L. O. (1969). A numerical model of the urban heat island. *J. Appl. Met.*, **8**:908–18.

NEIBURGER, M. (1969). The role of meteorology in the study and control of air pollution. *Bull. Amer. Met. Soc.*, **50**:957–965.

NICHOLSON, G. (1965). Wet Thursdays again. *Weather*, **20**:322–323.

NICHOLSON, G. (1969). Wet Thursdays. *Weather*, **24**:117–19.

NIEUWOLT, S. (1966). The urban microclimate of Singapore. *J. Trop. Geog.*, **22**:30–7.

NOURSE, H. O. (1967). The effect of air pollution on house values. *Land Econ.*, **43**:181–9.

OGDEN, D. C. (1966). Economic analysis of air pollution. *Land Econ.*, **42**:137–47.

OGDEN, T. L. (1969). The effect on rainfall of a large steelworks. *J. Appl. Met.*, **8**:585–91.

OKE, T. R. (1973). City size and the urban heat island. *Atmos. Envir.*, **7**:769–79.

PACK, D. H. (1964). Meteorology of air pollution. *Science*, **146**:1119–28.

PALES, J. C. and KEELING, C. D. (1965). The concentration of atmospheric $CO_2$ in Hawaii. *J. Geophys. Res.*, **70**:6053–76.

PANOFSKY, H. A. (1969). Air pollution meteorology. *Amer. Sci.*, **57**:269–85.

PANOFSKY, H. A. and PRASAD, B. (1967). The effect of meteorological factors on air pollution in a narrow valley. *J. Appl. Met.*, **6**:493–9.

PAPAIOANNOU, J. (1967). Air pollution in Athens. *Ekistics*, **24**:72–80.

PARRY, M. (1956a). Local temperature variations in the Reading area. *Q. Jl. R. Met. Soc.*, **82**:45–57.

PARRY, M. (1956b). An 'urban rainstorm' in the Reading area. *Weather*, **11**:41–8.

PARRY, M. (1967a). Air pollution patterns in the Reading area. In Tromp, S. W. and Weihe, W. H. (eds) *Biometeorology. Vol. 2, Pt. 2.* Pergamon Press, Oxford: 657–67.

PARRY, M. (1967b). The urban 'heat-island'. In Tromp, S. W. and Weihe, W. H. (eds) *Biometeorology. Vol. 2, Pt. 2.* Pergamon Press, Oxford: 616–24.

PASQUILL, F. (1962). *Atmospheric Diffusion.* Van Nostrand, London and New York, 297 pp.

PASQUILL, F. (1972). Factors determining pollution from local sources in industrial and urban areas. *Met. Mag.*, **101**:1–8.

PEDGLEY, D. E. (1962). *A Course in Elementary Meteorology.* HMSO, London, 189 pp.

PLASS, G. N. (1956). The carbon dioxide theory of climatic change. *Tellus*, **8**:140–53.

PORTEOUS, A. and WALLIS, G. B. (1970). A contribution towards the reduction of ice fog caused by humid stack gases at Alaskan power stations. *Atmos. Envir.*, **4**:21–33.

PRESTON-WHYTE, R. A. (1970). A spatial model of an urban heat island. *J. Appl. Met.*, **9**:571–3.

PRIESTLEY, C. H. B. (1956). A working theory of the bent-over plume. *Q. Jl. R. Met. Soc.*, **82**:165–76.

RASOOL, S. I. and SCHNEIDER, S. H. (1971). Atmospheric carbon dioxide and aerosols: effects of large increases on global climate. *Science*, **173**:138–41.

RAYNOR, G. S., OGDEN, E. C. and HAYES, J. V. (1970). Dispersion and deposition of ragweed pollen from experimental sources. *J. Appl. Met.*, **9**:885–95.

REED, L. E. and TROTT, P. E. (1971). Continuous measurement of carbon monoxide in streets, 1967–1969. *Atmos. Envir.*, **5**:27–39.

REID, D. D., ANDERSON, D. O., FERRIS, B. G. and FLETCHER, C. M. (1964). An Anglo-American comparison of the prevalence of bronchitis. *Brit. Med. J.*, **2**:1487–91.

REID, D. D. and FAIRBAIRN, A. S. (1958). The natural history of chronic bronchitis. *Lancet*, **1**:1147–52.

REYNOLDS, G. (1957). Variations in visibility over urban and rural areas. *Weather*, **12**:314–20.

ROBINSON, E. and ROBBINS, R. C. (1970a). Gaseous nitrogen compound pollutants from urban and natural sources. *J. Air. Poll. Control Ass.*, **20**:303–7.

ROBINSON, E. and ROBBINS, R. C. (1970b). Gaseous sulphur pollutants from urban and natural sources. *J. Air. Poll. Control Ass.*, **20**:233–8.

ROUSE, W. R. and MCCUTCHEON, J. G. (1970). The effect of the regional wind on air pollution in Hamilton, Ontario. *Can. Geogr.*, **14**:271–85.

ROYAL COLLEGE OF PHYSICIANS (1970). *Air pollution and health.* Pitman Medical and Scientific Publishing Co. Ltd, London, 80 pp.

RUSSELL, W. T. (1924). The influence of fog on mortality from respiratory diseases. *Lancet*, **2**:335–9.

RUSSELL, W. T. (1926). The relative influence of fog and low temperature on the mortality from respiratory disease. *Lancet*, **2**:1128–30.

SAWYER, J. S. (1971). Possible effects of human activity on world climate. *Weather*, **26**:251–62.

SCHAEFER, V. J. (1969). The inadvertent modification of the atmosphere by air pollution. *Bull. Amer. Met. Soc.*, **50**:199–206.

SCHMIDT, F. H. (1963). Local circulation around an industrial area. *Int. J. Air Water Poll.*, **7**:925–9.

SCHMIDT, F. H. and VELDS, C. A. (1969). On the relation between changing meteorological circumstances and the decrease of the sulphur dioxide concentration around Rotterdam. *Atmos. Envir.*, **3**:455–60.

SCHMIDT, W. (1929). Die Verteilung der Minimumtemperaturen in der Frostnacht des 12 Mai 1927 in Gemeindegebeit von Wien. *Fortschritte der Landwirtschaft,* **2**:681–6.

SCORER, R. S. (1955a). Plumes from tall chimneys. *Weather,* **10**:106–9.

SCORER, R. S. (1955b). Condensation trails. *Weather,* **10**:281–4.

SCORER, R. S. (1964). A problem in rainfall statistics. *Weather,* **19**:131–2.

SCORER, R. S. (1968a). *Air pollution.* Pergamon Press Ltd, Oxford, 151 pp.

SCORER, R. S. (1968b). Air pollution problems at a proposed Merseyside chemical fertilizer plant: A case study. *Atmos. Envir.,* **2**:35–48.

SCORER, R. S. (1969). Wet Thursdays. *Weather,* **24**:336.

SCOTT, J. A. (1963). The London fog of December 1962. *Medical Officer,* **109**:250–2.

SEBA, D. B. and PROSPERO, J. M. (1971). Pesticides in the lower atmosphere of the northern Equatorial Atlantic Ocean. *Atmos. Envir.,* **5**:1043–50.

SHELLARD, H. C. (1959). The frequency of fog in the London area compared with that in rural areas of East Anglia and south-east England. *Met. Mag.,* **88**:321–3.

SHERWOOD, P. T. and BOWERS, P. H. (1970). Air pollution from road traffic—a review of the present position. *Min. of Transport, Road Res. Lab.* RRL Report LR352, 27 pp.

SINGER, S. F. (1971). Stratospheric water vapour increase due to human activities. *Nature,* **233**:543–5.

SLADE, D. H. (1969). Low turbulence flow in the planetary boundary layer and its relation to certain air pollution problems. *J. Appl. Met.,* **8**:514–22.

SPAR, J. and RONBERG, P. (1968). Note on an apparent trend in annual precipitation at New York City. *Mon. Weath. Rev.,* **96**:169–72.

STEVENSON, C. M. (1968). An analysis of the chemical composition of rain-water and air over the British Isles and Eire for the years 1959–1964. *Q. Jl. R. Met. Soc.,* **94**:56–70.

STOCKS, P. and CAMPBELL, J. M. (1955). Lung cancer death rates among non-smokers and pipe and cigarette smokers. Evaluation in relation to air pollution by benspyrene and other substances. *Brit. Med. J.,* **2**:923–9.

STROM, G. H. (1962). Atmospheric dispersion of stack effluents. In Stein, A. C. (ed.) *Air Pollution* 1, Academic Press, New York: 118–95.

TERJUNG, W. H. (1970). Urban energy balance climatology. A preliminary investigation of the city-man system in downtown Los Angeles. *Geogrl. Rev.,* **60**:31–53.

TERJUNG, W. H. *et al.* (1970). The energy balance climatology of a city-man system. *Ann. Ass. Amer. Geogrs.,* **60**:466–92.

THOMAS, M. D. (1965). The effects of air pollution on plants and animals. In Goodman, G. T., Edwards, R. W. and Lambert, J. M. (eds) *Ecology and the industrial society,* Oxford: 11–33.

THOMSON, G. (1969). Sulphur dioxide damage to antiquities. *Atmos. Envir.,* **3**:687.

VAN CAUWENBERGHE, A. R. and BOSCH, F. M. (1969). Natural and artificial radioactivity of the air in Ghent (Belgium). *Atmos. Envir.,* **3**:633–41.

VARNEY, R. and MCCORMAC, B. M. (1971). Atmospheric pollutants. In McCormac, B. M. (ed.) *Introduction to the scientific study of atmospheric pollution.* D. Reidel Publishing Co., Dordrecht: 8–52.

WAGGONER, P. E. (1966). Weather modification and the living environment. In Darling, F. F. and Milton, J. F. (eds) *Future Environments of North America.* Natural History Press, Garden City, N. York.

WALLER, R. E., COMMINS, B. T. and LAWTHER, P. J. (1965). Air pollution in a City street. *Br. J. Ind. Med.,* **22**:128–38.

WANTA, R. C. (1962). Diffusion and stirring in the lower troposphere. In Stern, A. C. (ed.) *Air pollution* 1, Academic Press, New York: 80–117.

WASHINGTON, W. M. (1972). Numerical climatic-change experiments: the effect of man's production of thermal energy. *J. Appl. Met.,* **11**:768–72.

WEISMAN, B., MATHESON, D. H. and HIRT, M. (1969). Air pollution survey for Hamilton, Ontario. *Atmos. Envir.,* **3**:11–23.

WEXLER, H. (1952). Volcanoes and world climate. *Scientific American,* **843**:3–5.

WIGGETT, P. J. (1964). The year-to-year variation of the frequency of fog at London (Heathrow) Airport. *Met. Mag.,* **93**:305–8.

WILKINS, E. T. (1954). Air pollution aspects of the London fog of December 1952. *Q. Jl. R. Met. Soc.,* **80**:267–71.

YOCOM, J. E. (1962). Effects of air pollution on materials. In Stern, A. C. (ed.) *Air pollution 1,* Academic Press, New York: 199–219.

YUDCOVITCH, N. (1966/67). Factors influencing the temperature variation at Calgary. *The Albertan Geographer,* **3**:11–19.

# 4. Climate and agriculture

## 4.1 Introduction

Agriculture is the most important of all the primary economic activities pursued throughout the world, and is also the most dependent on atmospheric conditions. The world food problem is rapidly emerging as one of the most critical contemporary issues and every effort will be necessary to raise production in order to meet the demands of the growing population. Landsberg (1968) has stressed that the optimization of output depends not only upon reducing agricultural hazards, such as hail, drought, frost, and disease, but also on realizing the full potential of land, which is currently under-used in terms of the total climatic energy available for agriculture.

Such under-utilization is not confined to the developing countries, and Fig. 4.1 illustrates the typical pattern of seasonal output from British farms in relation to the incidence of solar radiation during the year. It can be seen that the seasonal variation in crop production, for example, is poorly synchronized with the available light and heat energy, since cereals make little use of the insolation in the second half of the growing season, while root crops make only inadequate use in the early part of the year. Duckham (1963) attributed this irregular phasing to factors such as weather hazards in the form of spring frosts, which delay the planting of potatoes, and failing soil trafficability in August and September, which prevents the adoption of really late-maturing cereals, but there is little doubt that even advanced farmers do not make full use of their climatic environment. A similar view has been expressed by Smith (1967a), who commented that progress in agricultural climatology is dependent on closer co-operation between agriculturalists and meteorologists, with an emphasis on the application of existing knowledge rather than on ambitious research projects.

Several writers, such as Watson (1963), have made the point that the type of agricultural production, such as the crop range, is related to the climate, and that the annual yields and the profitability of farming are determined by the weather. On the other hand, it must not be assumed that any weather pattern can be recognized as either uniformly suitable or unsuitable. Jones (1964), taking the long view of the economic historian, was concerned with farm income and noted the essentially competitive relationship between arable and livestock farming. In this situation, high grain prices may well favour the former but be a burden to the livestock farmer when grain is purchased for feed, whereas an abundant harvest may depress prices enough to create distress amongst the agricultural community.

Furthermore, it is necessary to distinguish between the ecological optimum and the economic maximum since, according to Taylor (1967a), the former is a long-term concept implying biological stability as opposed to the shorter-term economic profit which may produce ecological deterioration through misuse such as monoculture and over-cropping. The need to tailor farming systems to the environment, in order to make full use of the climatic resources, has been outlined in the South African context by Whitmore (1957). In detail, therefore, most relationships between the atmosphere and agricultural production are highly complex, and it is particularly difficult to isolate the effect of human inputs such as fertilizers, drainage, and subsidies from the more direct weather influences. The varying assessments that individual farmers may make of the atmospheric resources in different areas are also difficult to rationalize, and some weather events may set up a world-wide chain of reactions. Thus, Smith (1967a) cited the case of a drought in Argentina which caused a drop in beef exports to Europe. In turn, this created a rise in the price of fat cows, which led to a drop in the British milk herd and a consequent decline in total milk production in the United Kingdom.

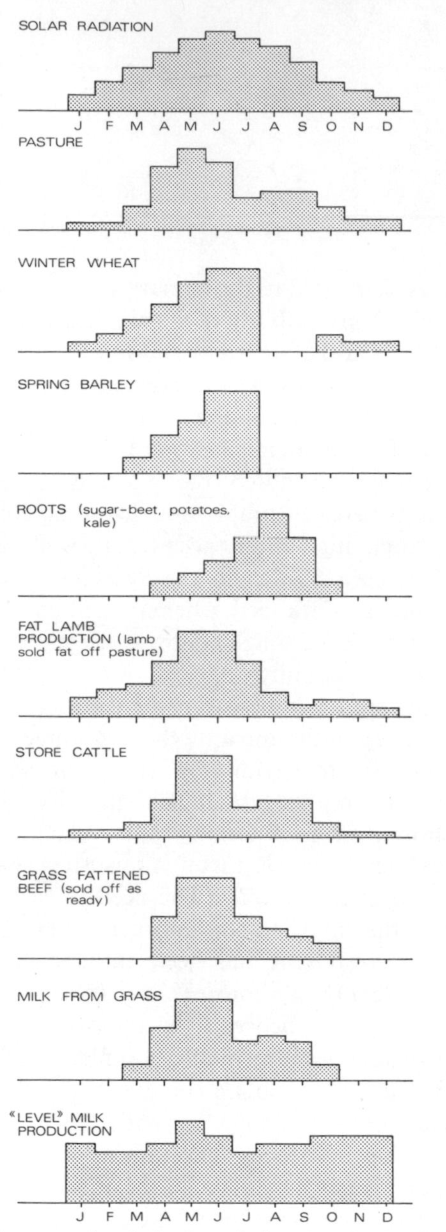


**Fig. 4.1 The seasonal variation in agricultural production in Britain in relation to the yearly cycle of solar radiation. After Duckham (1963).**

## 4.2 Some general relationships

Variations in agricultural production can often be related to deviations from the normal seasonal climate which is best described in terms of the *growing season* concept. In only a few parts of the world is farming output regular throughout the year and Duckham (1963) has drawn attention to the problem of seasonality in advanced societies, where man is an aseasonal animal requiring and expecting a constant diet throughout the year. One of the main problems facing agricultural production in developed economies is to minimize the variations in seasonal output as far as possible, but considerable seasonal dependence still exists. For example, as shown in Fig. 4.1, both livestock and milk production in the United Kingdom have a marked relationship with the growth cycle of the grass crop. For cattle, sheep, and lambs, weekly slaughterings reach a marked peak in autumn compared with the low output in spring when few livestock are available owing to the expense of winter fattening. Similarly, milk production from grass shows a seasonal maximum around May as a result of the 'spring flush' of grass with a generally declining production with the age of the grass crop. A more level output of milk through the year can be achieved, at a cost, only by the provision of feedstuffs and often the provision of animal shelters also.

It is clear that agricultural production is influenced to a greater or lesser extent by the growing season, but there are real difficulties in drawing up a wholly acceptable definition of this term. Taylor (1967b) has explored some of these difficulties and shown that there is much variation in usage with regard to the threshold temperature of 6°C, which is most usually taken as the minimum thermal requirement for any substantial growth in temperate crops.

In Britain, the growing season is most often taken as that seasonal period when the mean temperature exceeds 6 °C, although there is disagreement over whether daily, fortnightly, or monthly means should be employed, or even whether soil or air temperatures are involved. In the US, the growing season is held to exist between the last killing frost of spring and the first killing frost of autumn. Whether or not a frost 'kills' depends largely on the nature of the crop, but in this context a 'screen' frost is generally regarded as a 'killing' frost. The intensity of the growing period, as opposed to its length, is frequently represented by the accumulation of temperature units above the growth threshold. The basic unit is the *degree-day* which takes into account the amount by which the daily mean temperature exceeds the stated minimum. It is normally applied on the Fahrenheit scale so that a mean temperature of 53 °F for one day counts as 10 degree-days with a 43 °F (6 °C) threshold, as do temperatures of 48 °F for two days. However, Holmes and Robertson (1970) have pointed out that the growing degree-day concept contains a number of important assumptions, notably that there is only one significant base temperature operative throughout the life of the plant, that day and night temperatures are of equal importance for plant growth, and that the plant response to temperature is linear over the entire temperature range.

By the use of these criteria, it can be shown that large differences exist in the climatological potential of even small countries such as Britain. Figure 4.2 taken from Hogg (1965) indicates that the average duration of the growing season in England and Wales, based on a threshold temperature of 5·6 °C (42 °F), increases from a minimum of 210 days over the Pennines and the Welsh uplands to the entire year in the South-Western Peninsula and along narrow strips of the west and south coasts. In terms of the screen frost-free period, this lasts from as little as 10 weeks between 1 June and 15 August in the northern Pennines and the Scottish Highlands, to as long as 10 months between 15 February and 15 December in the coastal parts of Cornwall. Similarly, maps of accumulated temperature by Gregory (1954) show that degree-days per year in the British Isles above the growth temperature range from less than 500 in the northern uplands to over 3000 in restricted areas of the south coast.

It will be apparent from Fig. 4.2 that, excluding local climatic factors, the major influences are altitude and latitude, and in Britain it has been found that the length of the growing season decreases by about 4 or 5 days with every 30 m of altitude. If specific consideration is given to the occurrence of fairly high temperatures during spring and summer rather than values simply above the growth threshold, a strong latitudinal variation is revealed across the country. Thus, Hurst (1969) has noted that in spring Kinloss, near sea-level on the Scottish coast, has only 25 per cent of the season with hourly temperatures above 10 °C compared with almost 60 per cent at Croydon some 65 m above sea level in inland south-east England. The choice of higher base temperatures further magnifies the differences in favour of the inland, southern areas despite the influence of other factors such as coastal situation and altitude.

**Fig. 4.2 The average duration of the growing season in days over England and Wales. After Hogg (1965).**

All the indices of growing season so far discussed relate to the limitations imposed by low temperatures, but in many tropical areas growth is more likely to be inhibited by high temperatures, especially if these occur with hot, dry winds which cause plant dessication. In addition, the general assumption that plants will grow more quickly the higher the thermal environment rises above 6 °C is only valid, even in temperate regions, as long as soil moisture remains adequate.

Figure 4.3 shows the effective growing season for forage grasses in Mediterranean, Continental and Maritime climatic environments in terms of both thermal and moisture requirements. In these contrasting regimes, it can be seen that growth is limited by winter cold or summer drought, but Cooper (1965) stated that the growth limitations imposed in north and central Europe by low energy receipts are less important for potential production than the drought restrictions, as the latter coincide with the period of maximum radiation income. Indeed, the insolation which can actually be used at Algiers, in the absence of irrigation, is no greater than that received during the growing season in south-west England.

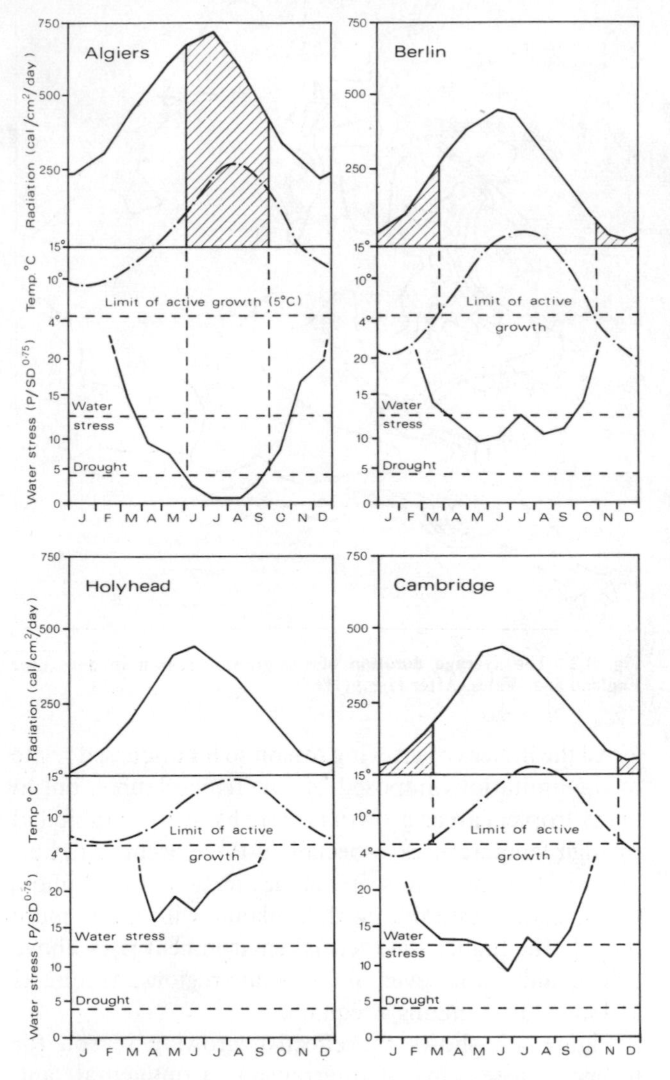


**Fig. 4.3 Climatic limitation of forage production in Mediterranean, Continental, and Maritime regimes. After Cooper (1965).**

In many parts of the world, therefore, rainfall rather than temperature restricts plant growth. Trumble (1937) defined the start of the growing season in South Australia when precipitation exceeded one-third of the open-water evaporation and went on to map the differing seasons for cereals and pasture on this basis. Pasture production is particularly dependent on a suitable moisture regime and Hurst and Smith (1967) have defined the grass growing season in Britain as the number of days between April and September inclusive when the soil moisture deficit does not exceed 50 mm. This value represents the approximate division between unhindered and drought-retarded growth and, in an average year, the number of days lost to grass production varies from less than 10 in northern and western areas to more than 40 days per year around the Thames estuary.

As in Britain, a moisture-based growing season often shows a reversed gradient from one dependent on temperature characteristics and may be closely related to the seasonal pattern of evapotranspiration, which has been proposed by Arkley and Ulrich (1962) as a plant growth index. Other workers have expressed dissatisfaction with the use of air temperature alone, even in areas where growth is determined largely by thermal conditions. The incidence of radiation is fundamental to plant growth, since it controls the rate of photosynthetic activity as well as transpiration opportunity, and Monteith (1965) has suggested that radiation is a more basic element than air temperature in the climatic environment of the plant. Certainly, radiation has wider ecological implications than temperature and is also more directly in phase with growth potential because air temperature lags behind the radiation income by some 4 to 6 weeks in mid-latitude areas.

This brief section on seasonality and growth would be incomplete without some reference to the fact that climatic fluctuations can create significant changes for agriculture. It is well known that agricultural patterns have differed through time with climatic oscillations and Lamb (1965) has depicted the wide distribution of vineyards which flourished in southern England during a warm epoch between AD 1000 and 1300, while the subsequent deterioration after 1300 appears to have led to the large-scale abandonment of wheat growing in east and south Scotland. Smith (1965) has analysed the possible effect of seasonal climatic changes on British agriculture and decided that alterations in spring and autumn weather would have the most impact on farming operations, especially if an adverse change limited the growing season. Nevertheless, the summer moisture balance is the largest individual element influencing land-use, since any trend to wetter conditions would favour dairying at the expense of cereal cultivation with drier weather producing the reverse effect.

In addition to influencing long-term agricultural land-use policies, the weather is also an important factor in the day-to-day operational decisions which are necessary on any farm. For example, Taggart (1970) has used a linear programming model in an attempt to isolate the effects of weather on the availability of labour on a farm near Edinburgh, and Donaldson (1968) and Smith (1970) have emphasized the significance of the weather factor in relation to the routine use of farm machinery. In this case, the weather must be considered right from the time that any investment in farm machinery is planned, in order to ensure that there will be enough suitable working days both to justify the purchase of expensive plant and to complete the work load on schedule.

Stansfield (1970) has stressed that many of the day-to-day weather-dependent managerial decisions may well be crucial for the profitability of the farm, since they are often directly related to the most effective timing of fundamental agricultural operations such as seeding or harvesting, although Duckham (1967) was equally convinced that, owing to the lack of quantified knowledge, such decisions are normally based almost entirely on practical experience and intuition.

Nevertheless, the growing sophistication of farming methods suggests the need for a closer liason between the farmer and the state weather service. Hogg (1964b) has shown that, in Britain, the official recognition of the farmers' needs by the Meteorological Office came with the establishment of the Crop Weather Service in 1924 but it was not until over 20 years later that a specialized Agricultural Branch was set up. At the present time, the Agricultural Branch provides a comprehensive service ranging from the compilation of weekly weather summaries to the initiation of complex research programmes as outlined by Hurst (1965 and 1970). In many cases, the most important information issued is that concerned with weather prediction and some of the agricultural implications of improved weather forecasting on various time scales have been considered by Hudson (1972) and Duckham (1972).

### 4.3 Weather and crop production

A large number of studies have been undertaken to show the direct effect of meteorological factors on crop yield but, as noted by Penman (1962), most workers have approached the various relationships involved from either a statistical/correlation or an environmental/physiological standpoint. The former method has been employed most frequently for yield forecasting but both approaches have been complicated by the rise of more scientific farming techniques. On a field scale it has so far only proved practical to modify the moisture relations of the crop through irrigation and Penman (1967) has stressed that even irrigation can only be fully exploited by the efficient farmer. The thermal environment of most crop plants still lies beyond effective modification and, for decades, the frontiers of cultivation have been extended more by the genetic adaptation of quick-ripening or drought-resistant varieties than any other technological impact.

The need for a reliable index for measuring weather influences on crop yields has been outlined by Stallings (1961), who also found that it was difficult to define even a theoretically ideal weather index. This was because potential users could not agree about the proposed use of the measure, some requiring a comprehensive index accounting for both direct and indirect weather influences and others wishing to know the effect of particular variables, such as precipitation, at specific times with the exclusion of various indirect influences, such as crop disease, associated with the weather. In general, the former approach would be more suitable for a retrospective analysis, whereas the latter method would find a role in yield forecasting. The employment of controlled laboratory environments has been a less successful approach, partly because of the expense involved, but also because of the difficulty of extrapolating the results to field conditions (Maunder, 1968).

Much of the work on weather-dependent crop production has been done on cereals, with much attention devoted to *corn* yields in the mid-western states of the US. This is probably due to the wide variations in yields which can occur and, in an early paper, Rose (1936), recognized climate as the major influence on output. In this study, temperature fluctuations were found to be more critical than variations in precipitation, although no one factor was entirely dominant. Later studies have also used both temperature and precipitation data. In the marginal area of South Dakota, Basile (1954) found that damaging dry periods reduced the yield both in the year of the drought and occasionally in the following year and concluded that, since yields were not reduced in even the wettest years, there was a deficiency of water for corn in every year.

More recently, Changnon and Neill (1968) have made a detailed study of a highly productive cash-grain farming area in central Illinois where the crop value in 1963 was \$16·5 million. The nine-year study period from 1955 to 1963 was characterized by steadily increasing yields due largely to technological improvements, such as the greater application of nitrogen fertilizers and higher planting rates, together with a trend to better corn-weather conditions (cooler July and August). There was some incentive, therefore, to isolate the effects of atmospheric and man-made inputs and the strongest relationships with yield were the negative correlations achieved with July and August mean temperatures and the cumulative number of degrees above 32·2 °C during July and August. The study showed that a reduction of July and August temperatures would be of much greater value than the provision of irrigation water, which would be beneficial only in late June and July, and, although several weather variables were more highly correlated with corn yields than any of the newer agricultural practices, no single parameter could explain more than 50 per cent of the yield variability.

Early surveys of the weather dependency of *wheat* yields were conducted in England notably by Hooker (1922), who correlated the yield of ten crops with rainfall and temperature in the eastern part of the country. He concluded that the summer climate of this area was too warm for all crops except potatoes and the winter climate too cold for the crops on the ground through that season, although he did concede that the late winter frosts helped to break up the land for the spring crops. Hooker stressed the equal importance for wheat of a dry seed-time and a dry subsequent winter and also pointed to the anomaly whereby, for the best seed development, a warm summer was necessary whereas ideally the early summer of the harvest year should be cool.

Tippett (1926) made a pioneer study of the influence of sunshine duration, although most subsequent workers have relied on rainfall and temperature. Thus, Frisby (1951) noted the significance of both adequate moisture during the growing season and early summer temperatures for wheat yields on the Great Plains. More specifically, Johnson (1964) has claimed that the amount of soil moisture available at seeding time in the Great Plains is so critical that it should determine whether the farmer seeds to wheat, summer fallows the land, or seeds later to another crop. As an aid to this decision-making process, Johnson found that the average contribution to yield of 25 mm increment of stored water was 2·44 bushels per acre for spring wheat and 2·74 bushels per acre for winter wheat. In Britain, Croxall and Smith (1965) have pointed to the importance of timing the autumn sowing of winter wheat in relation to falling soil temperatures at this season. For example, the *Capelle* variety should not be sown earlier than 10 days after the 100 mm soil temperature drops below 12·8 °C and not later than 45 days after this point in order to ensure optimum yields.

Wheat cultivation has been extended into highly marginal areas, such as parts of India, where the crop has to ripen under excessively high temperatures associated with drying winds. These adverse conditions reduce yields, probably because of increases in respiration at the expense of photosynthesis, and, experimenting in a controlled environment, Asana and Williams (1965) have shown that raising artificial 'day' temperatures from 25 to 31 °C brings about a decrease in mean yields of 16 per cent. In the US wheat yields are more commonly determined by the moisture balance and Hewes (1958) has reported that drought was the main cause of crop failure in western Nebraska between 1931 and 1937, when autumn precipitation less than 48 mm more or less guaranteed a failure of at least 30 per cent of the planted acreage. On the other hand, many other hazards such as hail and wind have important roles and the more reliable yields in this area since 1937 may be explained by a variety of factors including an increased use of summer fallow, better control of soil blowing as well as more adequate autumn precipitation.

Dry-land wheat farming takes place in parts of the northern Negev desert in Israel, where the mean annual rainfall varies from about 200 to 410 mm, and Lomas (1972) has estimated the rainfall-yield relationships in terms of probable profitability for five different stations in the area. If it is assumed that a yield of 1250 kg/ha is necessary to cover production costs and that farming operations require a 20 per cent gross return on investment, then a basic yield of about 1500 kg/ha will be necessary and Fig. 4.4 illustrates the probability of achieving such yields. Given that 70 per cent probability is accepted as a fair economic risk in this type of agriculture, it can be seen that a 20 per cent gross profit can be obtained only with a mean annual rainfall of at least 300 mm. Below 300 mm, irrigation becomes necessary and around 240 mm a 20 per cent profit margin can be reached only in 3 or 4 years out of 10.

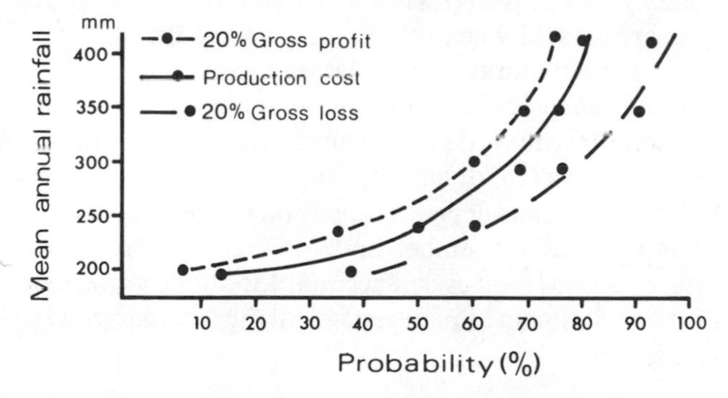


**Fig. 4.4 The probability of recovering production costs or making a 20 per cent gross profit or loss on wheat cultivation in the northern Negev, Israel. Data based on 1970/71 prices. After Lomas (1972).**

Moisture conditions are of particular significance in the extensive wheat-growing areas of the Canadian prairies and Williams (1972) has demonstrated that, in the driest and lowest-yielding areas, only one selected moisture parameter is needed to account for 40 per cent of the yield variance, whereas in the more humid parts at least three such variables are required. This has led to studies of the soil moisture regime associated with different cropping patterns, such as that undertaken in the upper Peace River district by Carder and Hennig (1966). More recently, Baier (1972) has used probability estimates of spring soil moisture in southern Saskatchewan to investigate the economics of spring wheat grown either continuously or on fallow in a two-year rotation programme. It was found that average wheat yields from the continuously cultivated areas were 71 per cent of those from the fallow-wheat areas and, based on an average farm wheat price of $1·65 per bushel, the mean annual net returns per cultivated acre were $13·23 for continuous wheat grown over the entire area and $9·53 for fallow-wheat which was seeded over only half the area under cultivation. The range of available moisture, comprising both spring soil moisture and growing season precipitation, varied from 150 to 355 mm and Table 4.1 indicates that each additional 25 mm above the 150 mm threshold yielded an additional net return of $5·05 for continuous wheat compared with only $2·53 for fallow-seeded wheat at the average farm price. Accordingly, it was concluded that, even in this semi-arid area, the available moisture was sufficient to justify continuous wheat cultivation, since it gave appreciably higher net returns.

| *Practice* | *Farm wheat prices in dollars/bushel* | | | | |
|---|---|---|---|---|---|
| | *1·35* | *1·50* | *1·65* | *1·80* | *1·95* |
| Fallow-seeded wheat ($) | 2·07 | 2·30 | 2·53 | 2·76 | 2·99 |
| Continuous wheat ($) | 4·13 | 4·59 | 5·05 | 5·51 | 2·97 |

**Table 4.1** Economic value of 25 mm of water from spring soil moisture and growing season precipitation in terms of average annual net returns from fallow-seeded wheat in two-year rotation and continuous wheat for various farm wheat prices. (After Baier, 1972.)

Canadian wheat production has also been expanded in the north over a long period by the introduction of fast-ripening varieties and Hopkins (1968) has indicated how the protein content of hard red spring wheat may be linked to weather factors. Below-average mid-summer rainfall and above-average July temperatures were significantly associated with above-average protein levels, whereas very heavy spring rainfall and hot dry weather during terminal ripening were correlated with low protein percentages.

Successful *barley* production is partly dependent on fairly dry weather near the beginning of the season, as shown by Wishart and Mackenzie (1930) at Rothamsted, and most workers in Britain, such as Hooker (1922) in eastern England and Grainger *et al.* (1954) in western Scotland, have commented on the need for low temperatures during May to August. However, Geddes (1922) claimed that, in the cooler climate of north-eastern Scotland, both barley and oats benefitted from temperatures rather above normal and that, although both crops required more than the usual summer rainfall amounts, the extra precipitation was needed earlier in the season for barley than for oats. In the US, temperatures soon become too high for optimum production and Weaver (1943) suggested that a reduction in mean temperature of 0·5 degC between mid-spring and mid-summer improved the barley yield by 1·6 bushels per acre. As already implied, the most suitable weather conditions for *oats* are similar to those for barley, although there has perhaps been slightly more emphasis on the need for sufficient moisture during the growing season and Zacks (1945) has demonstrated that optimum yields in southern Ontario are closely linked to water availability in the early summer.

After cereals, the *potato* is one of the most important mid-latitude field crops. The optimum growing conditions vary according to the stage of development reached but a generally equable temperature range is important for much of the time and Bodlaender (1963) has revealed how temperatures between 13·3 and 20·0 °C with high insolation can accelerate the initiation of tubers. Frost causes severe growth checks but low night minima are beneficial in raising tuber yields. The market price of potatoes normally declines with the delay in lifting the crop and early production is of some economic significance. In south-west Wales, early production is concentrated near the coast where crops may gain an advantage of up to one week over a site only 302 km inland (Tyrrell, 1970). Even along the coast, local site differences can delay lifting dates by four weeks and the average market price can vary by £45 per acre over very short distances.

As might be expected, the emphasis on crop-weather relationships in tropical regions has been placed on plantation agriculture and a major review has been presented by Russell (1967). The early growth of upland *cotton* has been studied under controlled environmental conditions by McQuigg and Calvert (1966), and a major investigation into *tea* yield in Assam was conducted by Sen *et al.* (1966). In this work, multiple regression techniques indicated that rainfall and temperature between January and March had most influence on the early crop yield, which led in turn to an increase in the main crop, and it was suggested that irrigation may not be economical in areas where the cold season rainfall exceeded 180 mm.

*Cane sugar* production is also dependent on adequate moisture (Hanna, 1971) and Oguntoyinbo (1966) has shown that moisture deficiency during the early stages of cane growth is critical in Barbados. On the other hand, it has been suggested by Oguntoyinbo (1965) that the high cost of irrigation schemes may make cane plantations uneconomic in Nigeria. In these circumstances, it may be better to lengthen the crop cycle so that the crop passes through a 17-month life cycle for plant cane and a 12-month cycle for the ratoons. In this way, the crop benefits from more than one season's rainfall and may possibly be cultivated without irrigation. Similar problems are created when tropical crops are cultivated near their thermal limits. Thus, Israel lies at the northern limit of *banana* cultivation, and one night with freezing temperatures may be sufficient to ruin a crop. Lomas and Shashova (1970) have demonstrated that the success of Israeli bananas depends largely on temperature conditions between November and March and that appropriate thermal indices can account for about 83 per cent of the variation in annual yield of all the banana-growing areas.

### 4.4 Grassland and pastoral production

Pastoral production embraces a wide range of commodities including milk (and its major derivatives, butter and cheese), meat (principally beef and lamb), and wool. In global terms, grassland farming is mainly located in the southern hemisphere and Maunder (1963) has claimed that about 94 per cent of the mutton and lamb exports, 86 per cent of the wool exports, 57 per cent of the veal and beef exports, and 38 per cent of the butter exports of the world come from the southern continents. Several countries, such as Australia and New Zealand, place an overwhelming reliance on these products as foreign exchange earners and it has been said that such countries are effectively exporting their climates in the form of these commodities. Maunder (1966) used statistical methods in an attempt to produce an agroclimatological model capable of quantifying the relationship between farming output and financial income for New Zealand. He was able to show the sympathetic fluctuation between moisture conditions in January and dairy farmers' income, and, where a significant correlation occurred between climate and agriculture, it was found that a departure of one standard deviation from the average climate produced price variations (at 1964 prices) ranging from £1 to £4 per acre for barley, peas, and maize, to over £40 per acre for potatoes.

Despite their ultimate diversity, all pastoral products owe their origins to favourable conditions for grass growth. In Britain, Smith (1960 and 1962) has worked on the assumption that potential evapotranspiration (PE) is the best weather index relating to the grass crop when there is no shortage of soil moisture. Grass growth is normally checked when the soil moisture deficit exceeds 50 mm and, if these periods are excluded, then the cumulative PE for the remainder of the summer, termed 'effective transpiration', should be a direct measure of the energy received by the growing crop. For annual hay yields in England, Smith produced the following relationship:

$$Y = 7{\cdot}07+1{\cdot}47E+0{\cdot}32N$$

where $Y$ = hay yield in cwt/acre, $E$ = effective transpiration in inches, $N$ = nitrogen application in cwt/100 acres.

Over the 1939–56 period, the above formula led to a correlation better than 0·9 between actual and estimated hay yields despite certain inaccuracies in the basic data and the assumptions of a uniform soil type. Smith (1967b) employed a similar approach to investigate how closely the incidence of grassland farming in Britain is tied to the distribution of a favourable grass-growing climate. The results indicated a close general relationship, since areal averages of effective transpiration, corrected for soil type, correlated highly with the proportion of land put down to grass and with the number of sheep and cattle carried relative to the area of available land.

More recently, Smith (1972) has used effective transpiration to suggest that climate is the dominant factor determining the type of farming and farm size in Britain. However, even in the most suitable areas of Britain, the grass-growing season rarely extends much beyond six months and, for the remainder of the year, livestock have to be fed on hay or silage. As shown by Warboys (1967), climate has an important role to play in the conservation of the grass crop as a winter feed, especially as regards the drying of hay in the field.

The New Zealand climate is rather more favourable for continuous grass production and Curry (1962) has suggested that the standard dairy cow requires about 500 mm of evapotranspiration per year in order to meet full forage requirements with a stock density of 1 cow per acre, whereas a flock of fat-lambs at 7 ewes per acre needs only 470 mm. On the other hand, potential evapotranspiration, grass growth, and livestock needs are still highly seasonal as shown in Fig. 4.5, which depicts the monthly dry-matter production for Rukuhia in relation to dairy cow needs. It can be seen that the main shortfall comes with the low rate of grass growth in winter and, although the winter deficit is balanced by the summer surplus, calving dates are arranged for July and August so that maintenance rations are required only during the winter months.

This seasonality affects the whole agricultural system and the fertilizer industry, for example, has its peak period of production and delivery in the autumn, as farmers topdress the land at this time to boost winter pasture production (Curry, 1963). In Alberta, livestock farming is subject to a rather different seasonal cycle as calving is planned for April or May when the last winter storms have passed and, as winter returns, the cattle are moved progressively closer to the ranch house (Brierley, 1967).

The climatic factors which help to determine the productivity of livestock vary greatly over the world. As shown by Maunder (1967), meat production in New Zealand is affected principally by adverse changes to the grass-growing season, with the reduction in livestock numbers following drought conditions as a particular

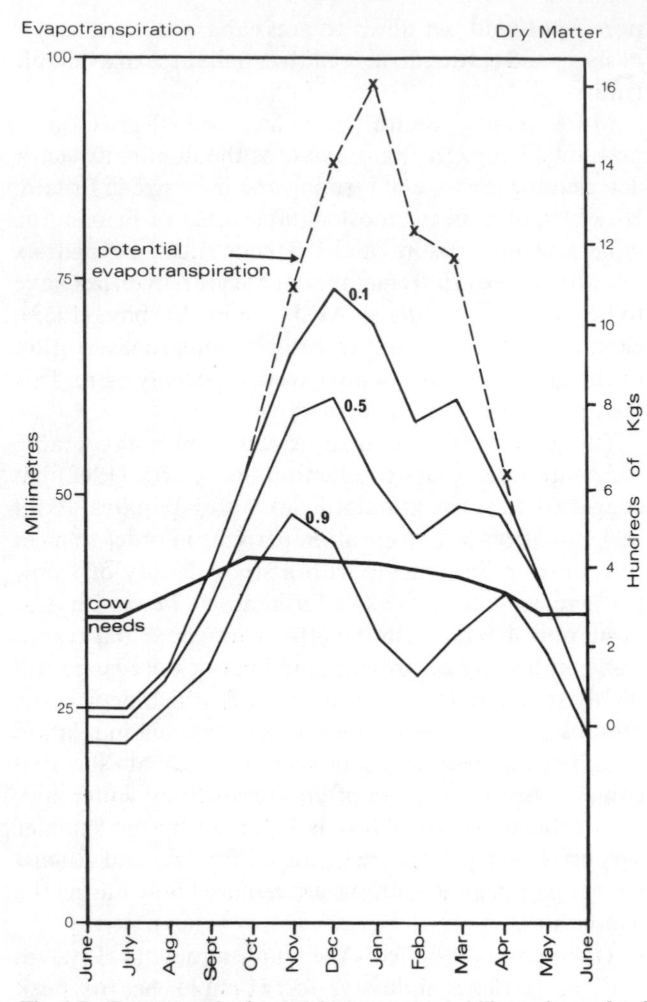


**Fig. 4.5 Evapotranspiration and dry-matter probability regimes related to cow needs for Rukuhia, New Zealand. After Curry (1962). Copyrighted by the American Geographical Society of New York.**

hazard. In the sheep-raising areas of Australia, rainfall deficiency is certainly the major risk and Carney (1950) has detailed how flock numbers are drastically reduced by the resultant deaths, by the sale of surplus stock, and by the decrease in lambings. Such specific hazards are less severe in Britain, where Blaxter (1965) has maintained that the livestock productivity of farm animals kept entirely out-of-doors is largely reduced by the limitations of the thermal environment. The low winter temperatures affect livestock both indirectly through the cessation of pasture growth and more directly through the physiological influence of cold stress and wind chill. Grass conservation for winter use combats the indirect effects and the direct climatic pressures are relieved by the provision of artificial shelters and the selective breeding of stock capable of standing up to the poor conditions.

On the other hand, in a different climatic environment, high temperatures can restrict yields. Thus, Hoxmark (1928) demonstrated that, over a five-year period, wool yields in Argentina varied inversely with mean annual temperatures, showing a decrease from 4446 g per fleece in Rio Negro (mean annual temperature 12·7 °C) to only 1625 g per fleece in Catamarca where the mean annual temperature is 20·6 °C.

Under tropical conditions, cattle may reverse their temperate grazing pattern and feed at night in order to take advantage of the low nocturnal temperatures. This feature is especially marked in the semi-arid areas and, in South Africa, Bonsma and Louw (1967) have stressed the importance of providing night pasture, which resulted in an improved growth rate even among adapted breeds of cattle. In some tropical countries, like Nigeria, there is a shortage of food from animal sources but stock are traditionally limited because of climatic stress. At present, most of the cattle in Nigeria consist of tropical breeds concentrated in the north of the country but Ojo (1971) has claimed that, provided artificial ventilation can be made available, the south as well as the north could support more cattle. The best prospects appear to be for the Shorthorn variety, which is a tropically evolved animal with wide tolerance, and only a narrow belt of country in the north has a potential probability of less than 95 per cent for this breed.

Milk production is known to depend on a number of climatic variables. In a review article, Hancock (1954) confined himself to temperature conditions, and concluded that temperatures between 4·4 °C and 21·1 °C had no influence on milk yield. At temperatures below this range, decreasing milk yields can be largely compensated by extra feed down to −15·0 °C, whereas above 26·7 °C there is a sharp drop in production. Despite the general significance of temperature, Chambers (1967) has stressed the complexity of the weather-milk reaction in Britain, where compensation through extra feeding and artificial housing can easily obscure the impact of severe atmospheric conditions. Short-term fluctuations tend to affect the cow directly, compared with longer changes which act through the grass crop, and Smith (1968) has used this seasonal factor to forecast annual milk yields. According to Smith, the main meteorological factors are:

(1) The relative earliness of spring. This parameter is reflected in the April milk yields.
(2) The summer rainfall which is largely unpredictable.
(3) The rainfall during the early haymaking season which controls the quality of the winter feed.

These factors have been translated into available weather data and incorporated into equations, which provide annual (April–March) yield forecasts at different periods of the year. For example, a forecast twelve months ahead, at the end of the previous March (Fig. 4.6a), can be made with a mean percentage error of 0·56 using November–March milk-production figures and March soil temperatures as an index of early spring conditions. As shown in Fig. 4.6b, a more accurate forecast can be

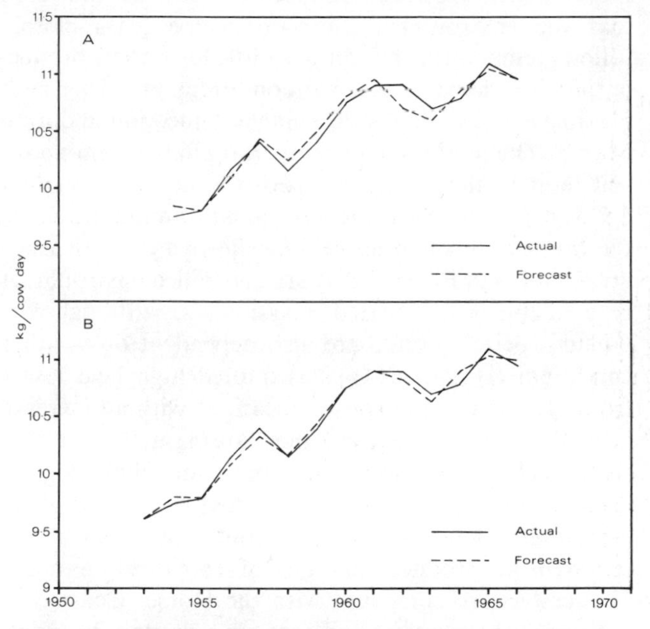


**Fig. 4.6 A comparison between actual and forecast milk production in Britain.**
**(A) Forecast attempted 12 months ahead at the end of March.**
**(B) Forecast attempted 9 months ahead at the end of June.**
**Based on data quoted by Smith (1968).**

attempted nine months ahead at the end of June by incorporating April–June production data and the June rainfall to produce a mean percentage error of 0·31. The marked upward trend in milk production indicated in Fig. 4.6 is due to improvements in dairy husbandry together with the introduction of higher yielding breeds of cows.

The production of honey is somewhat analagous to that of milk although dependent on different weather conditions. Hurst (1967) has commented on the general fall in honey production per colony from approximately 1950 onwards and has reported a highly significant positive correlation between annual honey yields and departures from the mean August temperature. This is a statistical confirmation of the fact that honey-bee activity increases during fine warm summers.

## 4.5 Horticulture

Horticultural production constitutes a highly specialized branch of agriculture and the intensive, market-oriented nature of the industry is well expressed both by the British term *market-gardening* and the North American equivalent *truck-farming*. Such a distinct identity has led to the emergence of a recognizable sub-disciplinc of horticultural climatology, the study of which has been furthered by various institutions, as shown by Winter and Stanhill (1957) in their account of work undertaken at the National Vegetable Research Station in central England. With horticulture there is a particular incentive to make the optimum use of the climatic environment in order to produce crops of high yield and high quality. In areas which have a short growing season, great emphasis is also attached to early cropping in order to avoid the seasonal glut when supply may be in excess of demand. This requirement means that horticulture is often concentrated where early growth is associated with a milder winter climate, such as in the truck-farming zone fringing the American Great Lakes; at the same time, the high value of the produce also makes economic the provision of artificial crop climates in many situations.

As with other crops, the weather is an important factor in production. The vinc, for example, is dependent on certain minimum temperature and sunshine criteria during the summer season and in California it has been claimed by Lave (1963) that 3000 degree-days above the threshold temperature of 10 °C are necessary in the main growing season before raisin grapes are of drying quality. Each degree-day beyond this minimum will result in a further 87 kg of raisins from the lower San Joaquin Valley, which is the dominant producing area of the US. Conversely, for a proper development of the flower buds, almonds need 200–500 'chilling hours' which may be defined as hours with temperatures less than 7·2 °C. There is some evidence that a further southward shift of almond production in California is precluded by insufficient chilling and Aron (1971) has presented a method for the prediction of total chilling hours from maximum and minimum temperatures.

In Europe, Wright (1968) has examined continuous records of wine production for Luxembourg dating back to 1626 and has been able to demonstrate a close statistical relationship between good wine harvests there and warm summers in England. Summer temperatures are critical for the quality of the Luxembourg vintage, since it lies near the northern limits of the European wine-producing area. The thermal environment is important for other tree crops as well as fruits, vegetables, and flowers. Thus, Shoemaker and Teskey (1959) have emphasized the importance of adequate summer temperatures without which ripening is delayed for apples and other orchard crops.

The siting of horticultural activity in order to receive the most favourable topoclimatic conditions is important both for early cropping and to minimize atmospheric hazards. Hogg (1965 and 1967a) has noted that the early production areas must have a supply of radiant energy sufficient to ensure rapid soil warming in spring, a soil texture light enough to be physically capable of early working, together with an above-average immunity to adverse weather factors. In Britain, the overall climatic advantage lies heavily with the south and south-west coastal areas of England and Wales including south Hampshire and the Isle of Wight, the Scilly Isles, south-west Cornwall, and Pembrokeshire. Apart from south Hampshire and the Isle of Wight, all these early production areas have a growing season which begins in the first half of February and, as shown in Fig. 4.2, the mean monthly temperatures are above 5·6 °C, thus indicating that growth will be interrupted by low temperatures for only short spells in winter.

Within the early production areas, every topographic opportunity must be utilized which means that, wherever possible, a sheltered south-facing site will be preferred. It is also more convenient and economic to avoid local climatic hazards such as frost and wind exposure than to protect crops artificially, but the profitability of horticulture does allow for considerable modification of the environment on a small-scale. Frost is probably the main risk factor and low-lying sites will rarely be adopted, although they may appear suitable from the point of view of shelter. Thus, Winter (1958) has described how a peach orchard in south-western Germany, located along a slope 72 m long with a height differential of only 3 m, suffered almost complete damage at the bottom of the slope whereas some trees near the top of the slope were not damaged at all.

The usual protection adopted against both frost and wind exposure is achieved by growing crops under glass either in heated cloches or the more common unheated Dutch lights. However, Jackson (1959a) has indicated that the environment under unheated glass closely follows temperature conditions outside, thereby producing a wide range of thermal conditions in spring with maxima in excess of 32·2 °C being not uncommon during March. The minimum air temperature at 76 mm above soil level in the protected crop was never more than 2·5 degC greater than the equivalent minimum outside the frame and was often less as shown by the thermograph traces in Fig. 4.7. A more controlled environment is available with heated glasshouses, although the greater operating costs are also dependent on weather conditions. Hogg (1964a) has quoted high heat losses from glasshouses in exposed locations with an increase in windspeed from 0 to 6·7 m/s causing a 100 per cent rise in fuel consumption. Since approximately 10 degree-days are equivalent to 1 tonne of coal per acre of glass, it has been calculated that 40–45 tonnes of extra coal are required to produce an acre of tomatoes in north Gloucestershire compared with the Dorset Coast.

Despite the physical and economic limitations involved, the improvement of horticultural climates is capable of producing significant returns. Unheated Dutch lights are widely used for the summer production of tender crops like melons and tomatoes in addition to out-of-season vegetables such as lettuce, which may be planted in December and cut during the first half of April some six weeks earlier than an outside crop, according to Jackson (1959b). Similarly, Hogg (1967a) has noted that although there are only 70–90 acres of cloched strawberries in Britain, the gross return per acre is probably

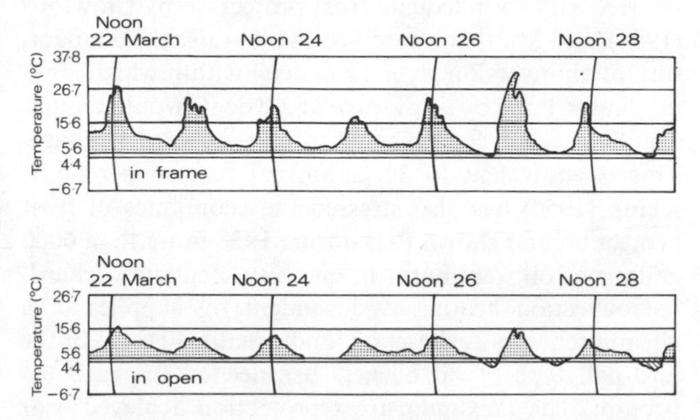


**Fig. 4.7 Air temperatures recorded 75 mm above soil level inside and outside cold frames during March in England. After Jackson (1959a).**

around £3000 because the berries are marketed a few days earlier than the main crop. Precise water control can also increase production values per unit acre on most market gardens and Hogg (1964a) instanced improvements from irrigation of £300 per acre for self-blanching celery, £150–160 for summer cauliflowers and lettuce, and £100 for dessert apples on shallow soil.

## 4.6 Weather hazards

So far in this chapter, weather hazards have only been introduced peripherally to the main climatic factors underlying agricultural production, but it is now time to consider some of the short-term risks and the methods available to combat them. Broad generalizations are difficult in view of the contrasting responses of different crops and animals to specific atmospheric events and this account will be limited to four common hazards—frost, drought, wind exposure, and severe storms.

*Frost.* Extremely low temperatures which occur during certain critical stages of the vegetative life cycle can adversely influence a wide variety of produce. In tropical and sub-tropical areas, crops such as rice may be at risk with temperatures well above the freezing point and Arakawa (1957) has described how 1782–87, 1833–39, and 1866–69 were all groups of years with disastrous rice harvests in Japan because of unseasonably cold summers. The development of temperate crops is often unhindered until the air temperature falls to near 0 °C. Winter freezing is expected in the mid-latitudes and is not the major problem, since many crops are immune to frost at that season. For example, fruit trees complete their fruiting in summer and are resistant to very low winter temperatures, and many root crops and tubers die back above ground in winter in order to retain a hardy existence below ground. On the other hand, the occurrence of late spring frost in April or May can be extremely damaging, especially to high-value crops aimed at an early market, and autumn frosts can also be a problem if they arrive before plants have made a physiological adjustment to winter conditions.

Several types of frost may be recognized, depending on the degree of severity and the prevailing meteorological process. An *air frost* occurs with a screen temperature below 0°C; a *ground frost* is observed when a similar temperature is recorded by a thermometer lying horizontally with its bulb just in contact with a short grass surface. These temperature conditions can be created either by a *radiation frost*, where the low values come from the air being cooled through contact with the ground under a temperature inversion, or by a *wind frost*, in which the cold air has been formed elsewhere and is then advected across the area concerned. Wind frosts tend to be less frequent but are more widespread when they do occur, unlike radiation frosts which may be entirely confined to, and are certainly most severe within, topographic hollows. Radiation frost is, therefore, more avoidable by careful site selection, although Matthews (1967a) has reported that some Tasmanian frosts in areas adjacent to the Central Plateau have certain characteristics of both types of frost.

The nature of frost risk is specific to individual crops and their stage of development. According to Critchfield (1966), the flowering period is a vulnerable growth point for most crops, although, in the spring-wheat zones of North America and the USSR, late summer frosts can be harmful even after the formation of the grain kernels. At temperatures in the −2·2 to 0 °C range an important limitation arises by the freezing of water solutions within the plant cells. Soft fruits are susceptible to this and Hogg (1970a) has stated that during April and May minimum temperatures $\leqslant -2{\cdot}2$ °C usually mean a complete loss of blackcurrants compared with only a partial crop loss with temperatures $\leqslant -1{\cdot}1$ °C. Severe frosts can bring about heavy financial losses within a few hours. Taylor (1970) has drawn attention to a late frost on the night of 22 June 1957, which damaged the mossland potato crop in Lancashire with farmers estimating individual losses varying from £1000 to £2000.

The first step towards frost protection is often achieved through a realistic assessment of the risk involved. The liability of the site to katabatic drainage can be estimated from large-scale contour maps on the assumption that slopes of at least 2° are necessary to support cold-air movement. Lawrence (1952) has described how the weekly radiation frost risk in spring may be estimated from records of minimum temperatures observed throughout at least one April–May period. Such short-term observations can then be related to longer climatic records for the area for the computation of frost probabilities. If the farmer knows the vulnerability of his crop and the gross return expected per acre, he is then in a position either to estimate probable losses due to the frost hazard or assess the cost/benefit situation of providing equipment to reduce the hazard.

In addition to the long-term frost risk, various methods are available for forecasting minimum temperatures a few hours ahead on individual nights. One of the most common techniques is that of Saunders (1949 and 1952), which depends on the identification of a discontinuity in the thermogram cooling trace followed by the use of regression equations to predict screen temperatures later in the night. This method was originally intended for radiation nights only but it has been applied more generally by Tinney and Menmuir (1968). Other methods of forecasting minimum temperature have been reviewed by Gordon and Virgo (1968).

If frost is expected, the risk may be limited by a variety of special actions including convection heating, wind machines, smoke generation, overhead sprinkling, and correct soil cultivation. Convection heating is one of the oldest methods and, like wind machines and smoke generation, is used mainly for orchard crops. Matthews (1967b) has shown that the efficiency of night air heating depends on several factors such as the height and strength of the temperature inversion, the speed of incoming cold air, the size of the area to be heated, the heat output, and the length of time that freezing temperatures persist. Large heat outputs have been shown to be necessary for adequate frost protection by Crawford (1964), and Matthews specified that to raise the temperature of an inversion layer 12 m deep within which air is moving at 1 m/s by an average of 2·5 degC would require 5 million Btu/h for every 30 m of windward frontage. This is equivalent to 30 gallons of fuel-oil per hour. Kemp (1956) has also stressed the economics of frost protection and shown that during 1953 more than 6000 gallons of oil were burnt in one New Zealand orchard.

Convection heating is dependent on supplies of a cheap fuel, usually diesel oil, and usually 40 to 75 of the lard-pail type of oil burners are needed for each 0·4 hectare. The maximum frost protection achieved near the burners is rarely much above 1–2 °C and may be less than half this over the area as a whole. However, Valli (1970) has reported the use of a natural gas heating system in the Appalachian fruit region of the US, where the minimum temperature at a height of 1·5 m within the test area was $-0{\cdot}6$°C compared with $-5$°C in the unprotected area.

Wind machines are found mainly in the US. They are normally horizontally or vertically bladed fans which either blow warm air from heaters or simply mix the cold air near the ground with the warmer air above. There is some evidence to suggest that only one machine per 2–6 hectares is needed if supplemented by heaters, but the fans need a lot of power and are often uneconomic. Furthermore, the warmer air has a natural tendency to rise and will not stay near the surface; protection is thereby reduced. Nevertheless, Bates (1972) has used a tower-mounted wind machine with two fans for frost protection in an Oregon cherry orchard with some success. On ten nights during April 1970, when the control temperature outside the orchard fell as low as $-7{\cdot}8$ °C, the temperature within the 3 hectare orchard was kept within the fruit safety zone.

Typical results are shown in Fig. 4.8 which indicates that temperatures within a radius of 80 m of the machine were maintained at −1·1 °C or higher and that the lowest orchard temperature was −2·8 °C. Although the method operates efficiently only in marked inversions with a source of warm air aloft, it was claimed that, under such conditions, the technique offers an economical, pollution-free method of frost protection in orchards up to 4 hectares in extent and that temperatures may be influenced over an area as wide as 10 hectares. In some cases, helicopters have been flown over orchards in an attempt to destroy the inversion but Small (1949) found that, although they could raise temperatures by some 1·7–4·5 degC, they were effective only if the frost danger lasted less than 4 or 5 hours.

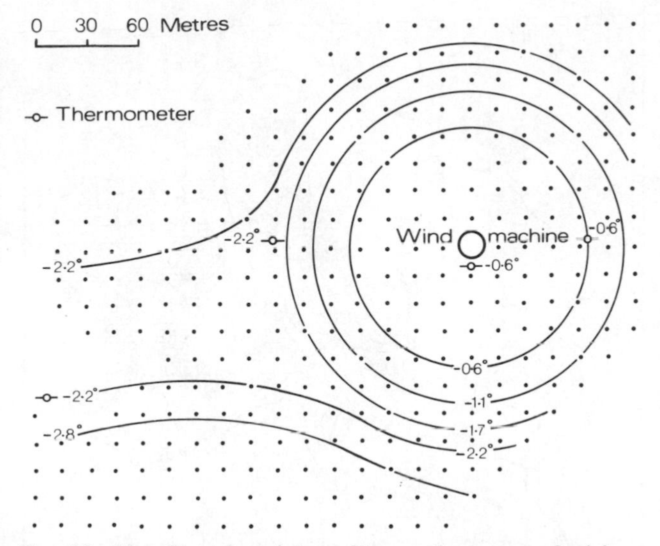


**Fig. 4.8 The effect of a wind machine on the pattern of minimum temperatures in an Oregon cherry orchard during radiation frost conditions on 12 April 1970. After Bates (1972).**

The generation of smoke or aerosol clouds may be used to create a reflective blanket above the crop in order to reduce the radiation losses from the ground. This method is only rarely employed as it is less efficient than either heaters or fans and the resultant smoke pall can be a source of public complaint in populated areas.

By comparison, overhead sprinkling techniques are one of the most efficient and versatile frost protection methods, being suitable for intensive vegetable growing as well as orchard trees. Although large quantities of water may be required, these will normally be cheaper than fuel supplies, and further economies can be achieved by the use of the spray lines for summer irrigation. The method depends on the continuous application of a fine water spray to the surface of the exposed plant throughout the period when the air temperature is at or below 0 °C. This produces a mixture of ice and water on the crop surface, the temperature of which then does not fall below freezing point. Sprinkling has the further advantage of working on both wind and radiation frosts, but the application of a continuous and even spray is essential to avoid damage.

Rather limited, but extremely cheap, frost protection can be achieved for low crops such as new potatoes or strawberries by correct soil tillage. It is well known that higher minimum temperatures occur over bare soil rather than grass and it is also warmer over moist compact soil than above a loose dry surface, so that spring-ploughed fields are more liable to spring frosts than autumn-ploughed fields. Therefore, the best soil condition for frost avoidance is the maintenance of a weed-free, consolidated, moist surface (Anon., 1964).

Ramdas (1957) has demonstrated how the thermal regime of surface soil in India may be partially controlled by manipulations of the heat and moisture balances, and Taylor (1967c) has described how marling can bring about favourable modifications to the temperature behaviour of peat soils in Lancashire. The insulating of soil with straw or similar material may be undertaken to protect the upper layers from low temperatures, and Desjardins and Siminovitch (1968) have reported the use of a protein-based foam as an insulating cover on grassed plots in Canada. The foam appeared most effective with radiation frosts when the 25 mm soil temperature was up to 3·6 degC higher in the treated plot, but low temperatures near the foam surface due to lack of heat conduction from the soil and evaporative cooling could be damaging to the upper parts of plants.

*Drought.* Unlike frost, drought or water deficiency is a difficult hazard to measure in the field. Heathcote (1969) has shown that drought assessment is often a complex three-fold problem of perception involving the identification of a water deficiency, the recognition of its effects, and the subsequent appraisal of its impact. From the viewpoint of crop growth, however, a drought may be said to occur when the soil moisture content within the plant root range falls below the permanent wilting point. Under these conditions, the soil moisture tension becomes too large for the rooting system to take up enough water to maintain the plant in a state of turgor, and a growth check occurs. The wilting point is reached by a progressive build-up of soil moisture deficit as evapotranspiration exceeds precipitation. It is the summer water balance, therefore, which is critical for agricultural production, although Smith (1971) has detailed the importance of winter rainfall over England and Wales in connection with problems of drainage, fertilizer practice, and autumn and spring cultivations.

The drought hazard can be eliminated by irrigation, which is no longer restricted to the arid countries but is now used in humid areas such as Britain, in order to increase crop yields and decrease the amount of year-to-year variability in agricultural production. This so-called *supplemental irrigation* has expanded rapidly over the past two decades as a result of a better understanding of moisture balance principles and a more precise knowledge of the benefits to be derived from soil-water management. Spray irrigation is widely used on a field scale for a wide variety of crops in addition to its traditional application to high-value horticultural produce. If irrigation is to be successful, the soil moisture content within the crop root range must be maintained between field capacity and the wilting point and, in practice, it has been found that water deficits begin to limit growth when about 50 per cent of the available water in the root zone has been used (Natural Resources (Technical) Committee, 1962). In turn, this depends on a fairly accurate assessment of the existing soil moisture deficit, which is usually based on the calculation of potential evapotranspiration after the method originally devised by Penman (1948). A detailed account of soil moisture deficit estimation and its relevance in the planning of irrigation requirements has been given (Pearl, 1954; Anon., 1967).

A general indication of irrigation need may be gained from a consideration of the frequency with which potential evapotranspiration exceeds precipitation by more than 76 mm, although the effective requirement is likely to be greater than this for shallow-rooted crops such as grass, vegetables, and potatoes and rather less for crops like sugar beet and cereals (Anon., 1962). Figure 4.9 indicates the annual frequency of irrigation need over England and Wales and it can be seen that the frequency is greater than 5 years out of 10 to the south-east of a line from about Hull to Torquay and increases to 9 years out of 10 around the Thames estuary and along part of the south coast.

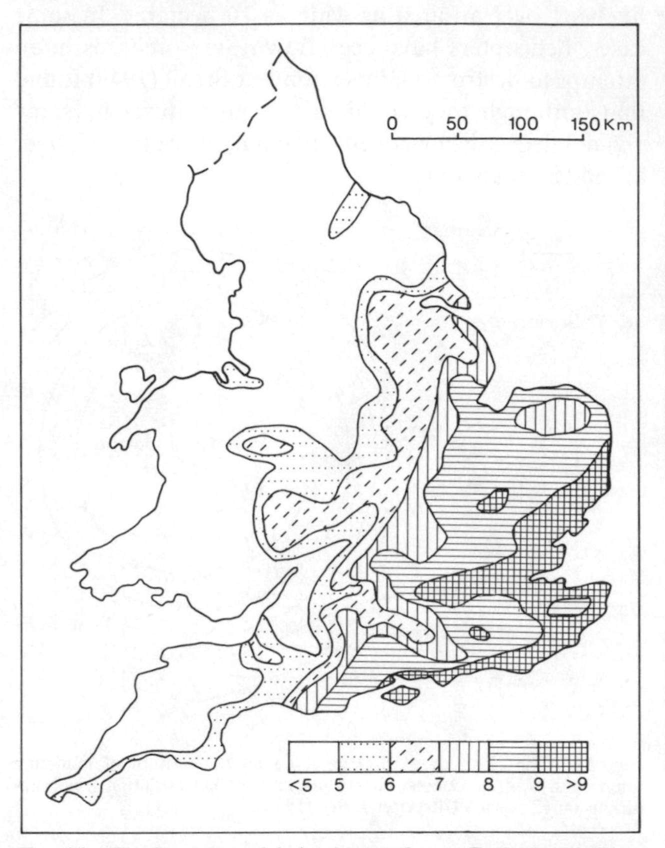


**Fig. 4.9 The frequency of irrigation need over England and Wales expressed as number of years in ten. After Pearl (1954). Reproduced by permission of the Controller of Her Majesty's Stationery Office.**

If irrigation is carefully applied, it acts as a fertilizer and, although the main effect is to improve yields, there may well be secondary benefits such as making the crop ready for market at an earlier date than a non-irrigated crop. The actual yield increases are dependent on a number of ancillary factors and the following are typical examples resulting from irrigation applications both of an experimental nature and on a farm scale (Anon., 1962).

Potatoes respond well to irrigation, especially if the soil moisture deficit is not allowed to exceed 25 mm, and total yield increases of about 50 per cent have been recorded over a wide range of conditions, with a doubling of yields in dry years. Irrigated potatoes are also of a more uniform size and of better quality than un-irrigated samples, but the timing of irrigation is extremely important in order to receive the best results. The yield of sugar beet has been increased by 30 per cent by irrigation, which is likely to be required between planting time until late August, and cereals are known to respond markedly to irrigation especially during a dry spring.

Grassland is sensitive to growth checks caused by water deficiency in summer and early work by Munro (1958) demonstrated that there is a progressive reduction in growth with an increasing soil moisture deficit and that the optimum yields are achieved where the deficit is kept below 25 mm. Irrigation may be required at any time between early April and mid-September over large areas of England with an average annual water need of 100–125 mm, rising to 250 mm in some areas in the driest years. The effect of grassland irrigation varies, with yield increases of more than 100 per cent on well-managed areas in dry years, although more typical results have been reported for experiments on grazing land in the Thames valley by Bone and Tayler (1963a, 1963b). Here, irrigation increased the amount of utilized herbage by between 30 and 40 per cent and produced a more uniform level of grass production throughout the growing season. Over a two-year period, an average annual water application of 140 mm increased the number of grazing days by 35 per cent and it was concluded that irrigation for milk production can be highly profitable in the right circumstances, especially if full exploitation is obtained by high stocking rates.

As might be expected, there is an extensive literature relating to the response of vegetable and fruit crops to irrigation. For example, early summer cauliflowers have been shown to be sensitive to water shortage at all stages of growth (Salter, 1961) and, like many other vegetables, can experience considerable yield increases (Anon., 1962). Goode and Hyrycz (1964) have shown important yield improvements on apple trees growing on a soil of low water-holding capacity, and more recently Goode (1970) has reported cumulative increases of yield for both soft fruits and orchard crops as a result of irrigation over a number of cropping seasons.

The extent and future growth of irrigation practice will be determined by the net economic return from the increased yields and the continued availability of water supplies. In 1962, it was estimated that there was enough equipment in Britain to irrigate 52 600 hectares of land and that the average rate of increase was 6050 hectares per year. It was expected that, given the continued availability of low-cost water, some 202 350 hectares would be irrigated by 1980. However, Prickett (1970) has claimed that in 1967 only about half of this area was being irrigated in England and Wales and there is some evidence of declining irrigation for the lower-value crops such as grass, sugar beet, and cereals. Attention is still being paid to the estimation of long-term irrigation needs based on water-balance principles (Hogg, 1970b) but the decreasing availability of cheap water supplies has led to some emphasis on methods of reducing the quantities of water required for irrigation.

Winter *et al.* (1970) have explored the possibility of increasing yields by the application of irrigation at rates below those determined by the full climatic moisture deficits, and Gangopadyaya and Venkataraman (1969) have conducted small-scale field experiments to see whether chemical sprays are capable of reducing water losses from cropped areas; they claim a 20–40 per cent reduction in evaporation after spraying an emulsion of alkoxy ethanol on to wet soil. In the southern US, flue-cured tobacco is normally irrigated when the soil moisture capacity has been depleted by 50 per cent, but Allen and Lambert (1971) have tested a more sophisticated irrigation-scheduling decision model and found that the new method led to comparable yield increases at a lower cost and with the use of less irrigation water.

Further economies in water use can be achieved through the careful timing of irrigation operations and Waggoner *et al.* (1969), among others, have produced results which suggest that, because of evaporation losses, spray irrigation is inefficient when applied in the daytime. Finally, it should be mentioned that, in certain cases, the quality of irrigation water may be as significant as the quantity and the cost. Thus, although Sale (1965) has concluded that low-temperature irrigation water applied by spray does not have an adverse effect on growth since the sprayed droplets soon reach thermal equilibrium with the air, Nishizawa and Yamabe (1970) have commented on damage to the rice crop in the northern and upland parts of Japan due to the application by gravity flow of cold irrigation water, often derived from snow-melt, which has been stored at depth in surface reservoirs.

*Exposure*. Agricultural hazards arising from overexposure to the atmosphere may take a variety of forms and, in the low latitudes, dessicating winds and high radiation inputs may be the most dangerous climatic elements. Thus, Ripley (1967) has examined the microclimatic effects of sheltering the tea crop and has concluded that the only real benefit would be to restrict the advection of dry air. In the mid-latitudes, however, exposure is usually associated with strong, physically damaging winds in the context of general farm productivity in windswept upland and coastal locations or with the risk to more vulnerable horticultural crops in the lowlands. Since a broad distinction can be maintained between these two categories, they will be treated separately in this account.

Some attention has been given to the exposure problem near the windward coast of upland Wales and Thomas (1959) has used tatter flags to estimate the need for shelter at two farm sites. Local topography rules out any possibility of a direct relationship between altitude and exposure in this area and Rutter (1968a and b) had indicated the shelter effect of local geomorphic features and established tree plantations. For example, it appears that lee slope eddies are an invariable feature of geomorphic shelter when slopes approach gradients of 1 in 4. For livestock protection, Rutter believes that shelter belts will be more beneficial on windward hillsides, but it is difficult to apply the results of exposure studies, since any potential benefits derived from sheltering on the most exposed parts of the farm have to be assessed in relation to possibly greater returns from a similar capital investment on another part of the farm. Apart from strong winds, west Wales also suffers from the deposition of marine salt, which is a function of rainfall conditions and local exposure as well as the strength of the on-shore wind (Rutter and Edwards, 1968). A similar hazard has been recognized by Lomas and Gat (1967) in an area of coastal citrus production in Israel, where the concentrations of airborne salt are so high within 2 km of the sea that total protection by windbreaks would not be economic so near to the coast.

Many studies have been made of the efficiency of shelter belts and Fig. 4.10, generalized by Food and Agriculture Organisation (1962) from Swiss data, depicts the well-known relationship between screen penetrability and downwind protection. If the shelter is very dense, there is a large decrease in windspeed immediately in the lee, but this is complemented by an almost equally rapid increase in strength at a distance only about five times the height of the barrier where the airflow is highly turbulent and gusty. Much better results are achieved with more open barriers which filter the wind rather than stop it abruptly. With windbreaks of about 40–50 per cent penetrability (i.e., about half the frontal area open) significant downwind reductions in windspeed can be found up to twenty times the barrier height and some slight effects may be noted at twice this distance.

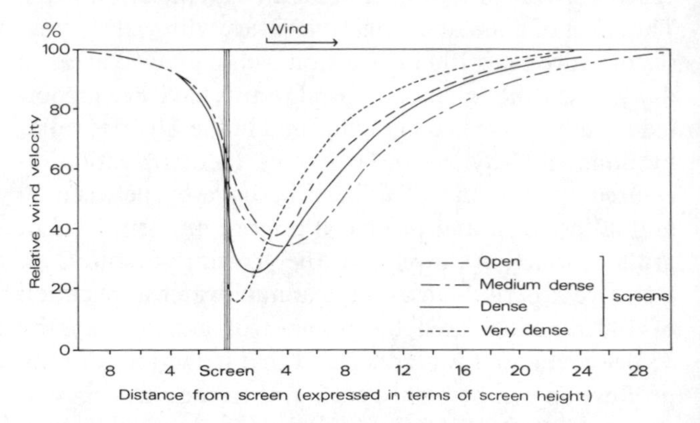


**Fig. 4.10 The influence of different shelterbelt densities on the velocity of airflow measured at 1·4 m above the ground surface. After Food and Agriculture Organisation (1962).**

Caborn (1955) has stressed the effect of shelter belts on local climatic factors other than windspeed. The relatively still air behind the barrier tends to increase the diurnal range of temperature with a slightly greater frost risk, while also raising the absolute humidity by some 2 to 3 per cent. Evaporation rates may be influenced for some considerable distance downwind, and recent experiments in Israel by Lomas and Schlesinger (1971) have indicated that, where the contribution of advective energy is dominant, the evaporation is markedly reduced but that, in the absence of advection, windbreaks have little influence on evaporation. Similarly, Skidmore and Hagen (1970) found that windbreaks reduced evaporation in proportion to the wind-speed reduction and the lowest relative evaporation was achieved with a 40 per cent porous barrier.

The broad agricultural implications of shelter belts have been discussed by Gloyne (1954 and 1970), who pointed out that protection may not always be required against the prevailing wind, as in Britain, where the shepherd and early vegetable grower find that the easterly winds of spring bring more danger than the stronger but warmer winds from the west and south. Another practical difficulty may arise out of the direct shading effects, especially by tree shelters, which may be beneficial to livestock in high summer but less favourable in winter and spring. In general, however, it does appear that dense barriers are best for livestock, especially if they can use the restricted protected area efficiently, whereas open belts are more suitable for sheltering crops. The results of measured yields of various horticultural crops grown experimentally behind a 50 per cent lath fence (Anon., 1964) showed that only cauliflowers failed to profit from the shelter but that the yield of anemones, beans, lettuce, and early potatoes increased by 37, 30, 27, and 21 per cent respectively. In Nebraska, Rosenberg (1966a and b) has demonstrated the advantages of shelter on an irrigated beanfield where, despite the evaporation reduction, the protected plants remained more turgid during the daylight hours and, for an area of irrigated sugar beets, germination was improved together with an increased root and total weight over beets grown in the open.

*Severe storms.* Some of the most dramatic and expensive agricultural losses occur as the result of severe storms, which bring a variety of hazards including strong winds, heavy rain, lightning, and hail. Even in so-called temperate areas like Britain, they are a definite threat and Smith (1964) has drawn attention to three separate tornadoes which developed along the Chiltern scarp overlooking the Vale of Aylesbury in May 1950, and caused some £50 000 damage to property including crops. Hogg (1967b) has described the impact of a deep depression on a bean field in Hampshire where the recently planted crop was reduced by about 80 per cent. Some parts of the crop were sheltered as part of a routine experiment on wind screens and it was concluded that very severe weather can increase the relative benefit of such shelter by about five times compared with normal growing weather. Storm damage occurs very quickly and Feteris (1955) reported how £330 000 of damage was accomplished in just fifteen minutes by a storm which struck the rich farming province of Zeeland in the Netherlands in June 1954. In this case, strong winds and hail were the destructive agents, but lightning can also be a problem. In Britain, ten people die every year from lightning with about one stroke of lightning falling on every 40 hectares of farmland annually, and Roberts (1968) has suggested methods of protecting farm buildings and livestock against lightning damage.

On the other hand, there is little doubt that hail constitutes the major storm threat, especially in North America. In an early review of American agriculture, Lemons (1942) found that between 1909 and 1925 seven leading crops suffered hail damage accumulating to $46 million, which represented more than 1 per cent of the average annual national agricultural income at that time. With the inclusion of losses to minor crops, livestock, and property, it was estimated that the total loss to farmers probably exceeded $100 million yearly. In 1928, a single hailstorm caused more than $2 million damage to fruit and sugar beets in eastern Colorado, but the most vulnerable area is the middle and upper Great Plains where large standing crop acreages, especially of wheat and corn, exist during the spring/summer hail season.

Hail is also common in the grain-farming region of southern and central Alberta where Paul (1966) has claimed that hail losses amount to more than $1 million annually. More precise information on the hail hazard in the US has emerged from studies combining meteorological data on hail with economic statistics on storm losses obtained from hail insurance bodies. Thus, Changnon (1967) has shown that in Illinois the monthly insurance indices reach a maximum in July, whereas the frequency of intense hail was found to increase progressively during the May–October crop season to reach a peak in September. This discrepancy is due to the fact that corn is more susceptible to physical damage in July than later in the growing season. Local variations in hail intensity were found, with a six-fold increase in northwestern Illinois compared with the central and southern parts of the State. In general, the occurrence of crop-hail damage was related to expected variables such as storm duration and hailstone size, but the most severe damage was caused in association with strong surface winds. The importance of this factor was confirmed in a wider regional study of the central and northwest US based on indirect measures of hail intensity developed from insurance records by Changnon and Stout (1967). Summer hail-fall intensities were found to be 5–15 per cent greater in the more mountainous States, such as Wyoming and Montana, than in adjacent States further east and it was suggested that this may be explained as a result of the hail-producing storms in the lee of the Rockies producing stronger surface gusts of wind.

Regional variations in hail losses may be partly attributed to areal differences in farming practice, as shown by the fact that thickly planted row crops in the Middle West can withstand wind-blown hail more easily than the thinner plantings in the Great Plains, but there are still genuine spatial patterns of hail risk with a rapidly decreasing intensity gradient eastward from the Rocky Mountain states. Changnon and Stout further concluded that there is less seasonal difference in the corn crop's susceptibility to damage than is found in the more widespread wheat States. For example, in Kansas the average June hailstorm day causes 15 times more wheat damage than the average April or July hailstorm day, but for corn in Illinois the average July hailstorm day is only about 3 times more destructive than the normal storm in June or August.

Apart from crop insurance, there are few protective measures that the farmer can take against the hail hazard. In some areas, such as over the Po valley vineyards in northern Italy, explosive rockets are released in an attempt to suppress hail. Sansom (1966) noted the lack of a fully accepted scientific basis for such methods but, nevertheless, organized a rocket experiment in the main tea-growing district of western Kenya where, in an area of 388 km$^2$, hail is reported on an average of 54 days per year and causes a mean annual loss of 675 000 kg of made tea. The results indicated a slight suppression of hail damage with a loss of 5·6 kg of tea per hectare over the experimental period compared with 5·8 kg per hectare during the unmodified control period. Although this difference could possibly be due to a changed storm distribution between the periods, the rockets were fired when the hail was falling and eyewitness evidence suggested a transition from hard hail to soft hail or rain.

## 4.7 Pests and diseases

In addition to the direct atmospheric hazards, agricultural production also faces restrictions from periodic outbreaks of pests and diseases. These epidemics are often weather-dependent either in terms of favourable local weather factors or the importation of airborne germs or spores by the prevailing winds. Whatever the immediate source of the outbreak, a complex interaction may well be established between the host—plant or animal—and the attacking organism, and Grainger (1967) has stressed the importance for plant pathology of studying both host and parasite in the case of potato blight forecasting. Some of the complex relationships between atmospheric conditions and the incidence of plant virus diseases have been summarized by Broadbent (1967). Most of the widely distributed viruses are spread by insects such as aphids and leafhoppers, and viruses are transmitted most efficiently under conditions suitable for these principal vectors. Thus, in the humid tropics and temperate latitudes, temperature is the main controlling factor, whereas rainfall is dominant in arid areas, as when locust swarms move with the wind towards a convergence zone and an area of recent rainfall and vegetation growth. Crop microclimate is of fundamental significance in virus epidemiology, but, since it is known that the optimum temperature for aphid reproduction is around 26° C, it is possible to generalize that they will be more active in warm continental climates than in cooler maritime areas and will flourish in warm dry summers rather than in cool wet ones.

Potato blight outbreaks occur every year in certain areas of west and south-west Britain, and in western Scotland the average loss of an untreated crop amounts to some 20 per cent (Grainger, 1955a). Early work on the weather-disease relationship in the Netherlands was clarified by Beaumont (1947), who showed that, when the air temperature remained above 10 °C and the relative humidity above 75 per cent for a period of 48 h, then potato blight was likely to occur within the following 14 days. This relationship illustrates the typical importance of temperature and humidity conditions, the rate of growth of all fungi being dependent on temperature, with the actual infection by the germination of fungal spores requiring a thin film of moisture on the plant leaves.

With potato blight, there is a 14-day period necessary for the parasite to become fully established after suitable weather conditions for germination, and this allows time for forecasting and subsequent control by fungicide sprays. In Britain so-called 'Beaumont periods' have been employed in a warning system since 1950, but local variations limit the uniform applicability of these data over the whole country. For example, Grainger (1955a) has claimed that meteorological stations located only 5 km away from the crop area in Scotland may provide an insufficiently accurate impression of crop risk, and Brenchley (1962) has drawn attention to the localized nature of blight-prone areas in the English Fens. As might be expected, the more humid parts of the West Country have a very high frequency of blight outbreaks, leading to routine spraying, although such protective action is not always economic, especially on main-crop potatoes. Potatoes may also be subject to losses from the potato-root eelworm. In parts of south-west Scotland, early potatoes make a lot of tuber growth before the soil temperature reaches 6·1 °C which is the minimum temperature for eelworm activity according to Grainger (1955b). By the time the earliest crop is lifted at the beginning of June, the eelworm cysts have been able to make little contribution to permanent soil infestation and consequently the loss in yield of first-earlies is restricted to 30–40 per cent as opposed to the 80–100 per cent loss which would probably result if main-crop varieties were grown.

The study of apple scab is similar to potato blight in that the link between weather and disease has been traced from laboratory research through field experiment to a practical warning system (Anon., 1964). The initial relationship of the disease with temperature and the duration of surface wetness has led to the adoption of simplified weather criteria which can be employed on a local basis. This is important, since remedial spraying action is necessary within 48 hours of the infection weather.

The micro-meteorological aspects of plant disease have been emphasized by Woodhead (1968 and 1969) in relation to coffee berry disease which is widespread in Africa; in the 1966–67 cropping season, this disease destroyed Arabica coffee in Kenya valued between £3 and 4 million. The germination of the causal fungus, which colonizes the maturing twigs, is dependent on high temperature and humidity conditions and there were indications that east-facing branches might be slightly more disease-prone than west-facing ones, because of the more rapid warming of the surface water films by morning insolation. The optimum germination temperature is about 22 °C and it was concluded that the plantation climate, as measured by a conventional screen between rows of coffee trees, under-estimated daytime twig surface temperatures by an average of 1 and 2·5 degC under wet and dry conditions respectively.

This disease may also be related to much wider atmospheric conditions and Nutman and Roberts (1969) have claimed that the rapid spread of coffee berry disease from higher to lower altitudes in Kenya since 1961 can be attributed to an increase in precipitation during the 'long rains'. As a result, prolonged periods of wetness combined with lower daytime temperatures are now experienced on the lower slopes where the spread of pathogens has been encouraged.

Several livestock diseases may be related to climatic variables. Liver fluke disease is a parasitic disease of sheep and cattle, which has been associated with wet summers in Britain since the classic year of 1879 when about 3 million sheep died. Ollerenshaw (1967) has described the life cycle of the parasite which, as shown in Fig. 4.11, is highly seasonal and develops within a particular variety of snail before emerging on to the herbage from which livestock are infected. Temperature factors are significant since there is no development of the parasite either in the egg or in the intermediate snail host below 10 °C but moisture is critical both for the hatching of fluke eggs and for the production of large numbers of host snails.

Livestock begin to show the effects of the disease in

late autumn, but an outbreak is dependent on at least three summer months with adequate moisture and the disease is most prevalent on lowland farms in the west, where the summers are usually sufficiently warm and wet to permit the completion of the parasite's life cycle.

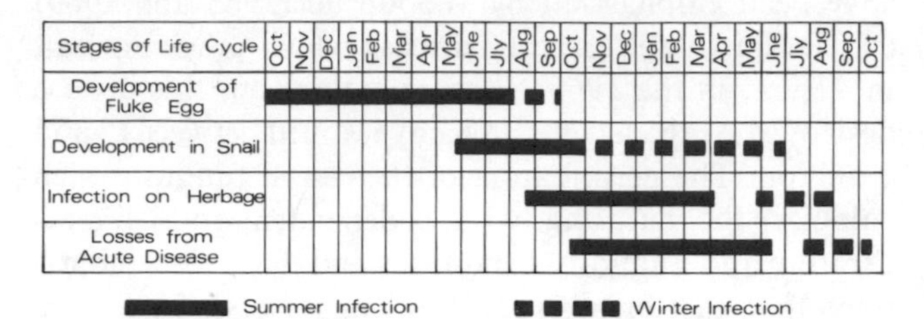


**Fig. 4.11 The life cycle of the liver fluke in relation to the seasonal periodicity of the disease. After Ollerenshaw (1967).**

Disease forecasting has been attempted since 1958 on the basis of monthly relationships between rainfall and evapotranspiration from May to October (Ollerenshaw, 1966). Epidemics are found to occur when summer rainfall exceeds evapotranspiration in all months but there will be little disease in the years with evapotranspiration exceeding rainfall in four or more summer months.

Smith and Ollerenshaw (1967) have illustrated some of the progress already made with respect to other livestock diseases such as nematodiriasis which is a worm disease of young lambs. It has been suggested that cold winters delay the hatching of the eggs, thus leading to a peak herbage infection just when the lambs are turned out to pasture. The onset of spring can be deduced from a comparative assessment of earth temperatures during March and these values have been used for forecasting purposes. In New Zealand, soil temperatures have also been employed in an understanding of the field ecology of a fungus responsible for epidemics of facial eczema amongst sheep. Mitchell *et al.* (1959) have reported that conditions are potentially dangerous if either the 0900 h soil temperature at 200 mm depth reaches an average of 17 °C in November or if the soil moisture deficit has reached 38 mm by the end of the same month.

The spread of pests and diseases by air currents has resulted in progressively greater co-operation between plant pathologists, veterinarians, and meteorologists. Hogg (1967c and 1970c) has concerned himself with outbreaks of black stem rust of wheat, which are largely controlled in the major wheat-farming areas by breeding resistant varieties of the crop. British wheat is more susceptible to damage and, when a severe epidemic occurred in 1955, there were yield reductions up to 75 per cent. Hogg has shown that such black rust epidemics originate in spore clouds, which form over parts of continental Europe and North Africa and are then carried to Britain in late spring and early summer. The actual trajectory of the spore clouds was determined from estimations of the geostrophic wind in conjunction with data from spore traps.

Tropical latitudes are particularly subject to swarms of insect pests including the tse-tse fly and the desert locust. The tse-tse fly exists over most of the non-desert areas of tropical Africa between latitudes 15° north and 20° south and carries human and animal *trypanosomiasis* or 'sleeping sickness'. According to Thompson (1955), the savana tse-tse can exist without food for more than one week in the high humidities found during the rainy season but can die within three days in the dry season. For the most part, it tends to exist in tree canopies where transpiration and shade maintain a combination of high humidity and moderate temperature.

Desert locusts also require adequate moisture both to enable the female to lay her eggs in wet ground and to provide vegetation growth to feed the immature, wingless locusts after they have hatched. Once they have become adult, winged locusts migrate between areas of rainfall and can consume vast quantities of crops and vegetation (Pedgley and Symmons, 1968). Rainey (1963 and 1969) has shown that locusts migrate with the prevailing wind and that the movement usually occurs during the day at temperatures between 20 and 40 °C. In this way, a swarm can migrate over thousands of kilometres in only a few weeks.

A major problem in locust control is that, given suitable weather, locusts can multiply at a spectacular rate. The most favourable weather appears to be a sequence of rains which fall at both sufficiently close intervals for successive generations to take advantage of the moisture for breeding and at locations sufficiently near to each other to be reached by wind-based migrations. Characteristically, therefore, major swarm movements occur down-wind, both towards and within areas of low-level wind convergence. According to Rainey (1973), the first conclusive recognition of an accumulation of locust swarms lying within a semi-permanent zone of convergence took place in July 1950 and Fig. 4.12 shows the distribution of these swarms in relation to the position of the Inter-Tropical Convergence Zone (ITCZ), which remained more or less stationary during the second half of the month. Some two months later, in September 1950, a southward movement of locusts, travelling in

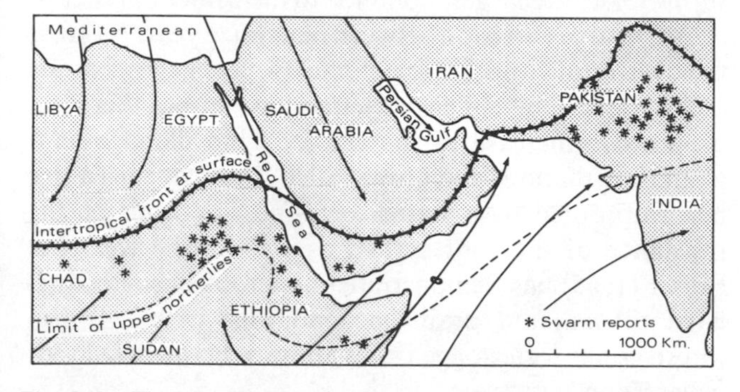


Fig. 4.12 The Inter-Tropical Convergence Zone and the distribution of desert locust swarm reports 12–31 July 1950. After Rainey (1973).

direct association with the seasonal migration of the Inter-Tropical Convergence Zone, was noted across the Horn of Africa in the Somali Republic. By using meteorological forecasts of the progress of the ITCZ, it was found possible to provide the appropriate authorities in the Northern Frontier Province of Kenya with 5 days advance warning of the arrival of the locusts, although the nearest observed swarms were still 500 km away to the north. Radar has been used to track migrations (Rainey, 1955) but it is desirable to know of locust upsurges before the insect is fully grown, and the detection of potential breeding sites by satellite offers one of the most promising methods likely to be adopted in locust control.

The back-tracking of airborne disease with the aid of synoptic meteorological information has also been attempted by Smith (1964) for the spread of fowl pest in England during the spring of 1960 and 1962. The initial outbreaks occurred in broiler houses, which were assumed to be continuous point sources of infectious dust particles, and the downwind dispersal of these particles through secondary and tertiary sources could be estimated up to 100 km. Similar evidence exists for the airborne spread of the highly infectious foot-and-mouth disease which affects cattle, sheep, and pigs. Hurst (1968) has attributed several outbreaks of this disease in Britain to windborne transmission over distances up to 200 km from the Continent, and Wright (1969) has demonstrated how the spread during the 1967–68 epidemic was largely due to the wind. It appears that, during anticyclonic weather, airborne transmission is limited to a distance of about 4 km from a source and that precipitation is involved in downwind dispersal beyond about 10 km.

In some circumstances, the use of artificial tracers such as smoke may be necessary. Smith (1967) has commented on the movement of the agrometeorologist into the field of the heating and ventilation engineer in order to ensure that intensively housed farm animals enjoy a maximum degree of comfort. Such comfort is related to high levels of productivity, which are found on large poultry stations and piggeries only if the ventilation system is capable of carrying off the waste heat and moisture generated by the animals. On the other hand, the successful provision of animal shelters must be based on sound economic considerations, as illustrated by Hahn and McQuigg (1970) for a number of selected stations in the US.

Local air currents and other climatic variables can be as important in controlling disease as in tracing its dispersal. Thus, Naya (1967) has made the point that the efficiency of fungicide applications, crop dustings, or fumigations is often influenced by the prevailing atmospheric conditions. Crop dusting or spraying from light aircraft or helicopters, for example, is most efficiently undertaken with very light winds which allow the pilot to fly close to the plants and do not lead to an excessive dispersal of the plant protection agent. In many tropical areas, high instability and atmospheric turbulence may limit the most suitable conditions to near sunrise or sunset.

## 4.8 Forestry

Like other agricultural crops, trees depend on adequate temperature and moisture conditions for growth, but, because of the long growth-period necessary before forests provide an economic timber harvest, their climatic dependency is somewhat distinctive. For example, trees probably have to withstand the same seasonal hazard many times compared with other agricultural crops which mature within a few months. The effect of a single adverse season can have a marked effect on growth: Rouse (1961) has indicated how the annual growth of Corsican pines in the relatively dry sands of the Breckland area of East Anglia is dependent on moisture conditions during the preceding winter. On the other hand, evidence from the New Forest in southern England suggested that the growth of Douglas fir is rather more sensitive to summer temperatures. Certainly the low tree-line in Britain relative to the limits of forest growth found in continental Europe is a direct reflection of the rapid deterioration of climate which takes place with altitude. As Millar (1964) has revealed for northern England, this fairly rigid ceiling on growth is due to a complex of temperature and exposure factors.

In the long term, reductions in forest yield are likely to result from permanent damage to the trees rather than growth restrictions even over several years. Forests tend to be exposed to a range of atmospheric hazards comparable to those affecting other crops plus some more specialized risks like windblow and fire. Thus, frost can be a major problem for young trees, especially in Britain where most of the exotic conifers grown are partially frost-susceptible (Day, 1939).

In certain climates, low-temperature conditions give rise to ice-storms which create widespread physical damage, as branches are snapped off by the weight of the ice deposits. Goebel and Deitschman (1967) working in Iowa have claimed that such damage is concentrated on the more mature trees and on specific species such as the eastern white pine and the Scots pine. In all areas, trees on the windward edge of plantations and in windbreaks suffered the greatest losses. This is a measure of the general exposure risk found in the uplands and also along windswept coasts, where deposits of windborne salt are an intermittent hazard causing salt-burning or browning of affected trees. Edwards (1968) has investigated salt deposition in some North Wales forests and found that the accumulations decreased rapidly with distance inland but increased with altitude. There was a highly significant negative correlation between point measurements of salt deposition and the height of trees growing at that site. The extent of salt-burn arising from a single gale in southern England has been reported by Edlin (1957). Only a few days after the gale, damage could be easily traced up to 16 km inland with extensive examples of affected trees up to 80 km from the coast. A large variation existed in the relative resistance of different species, with beech and elder very susceptible compared with the holm oak which was virtually immune.

The ultimate consequence of over-exposure is windthrow which is an important hazard in Britain. Windthrow losses are largely confined to the winter but there is a great deal of variation between individual years owing to the over-riding importance of a few severe gales. Forests themselves change in character and extent but there is a progressive increase in the risk of wind damage as a plantation matures and grows taller. The chance of a tree being uprooted by a stated wind-force depends on the resistance of the root system as well as the tree size. Fraser (1965) has demonstrated how root strength is related to soil and drainage conditions within British forests, since waterlogged soils allow only restricted root penetration.

The forester can adopt various measures, such as soil drainage or thinning, designed to delay windthrow and some of the economic implications of this are depicted in Fig. 4.13 for an average yielding crop of Sitka spruce trees grown in Britain. As shown by the horizontal scale, these trees would normally achieve a height of about 10 m at age 23 years, which is the usual time of first thinning, and the graph illustrates the likely financial returns

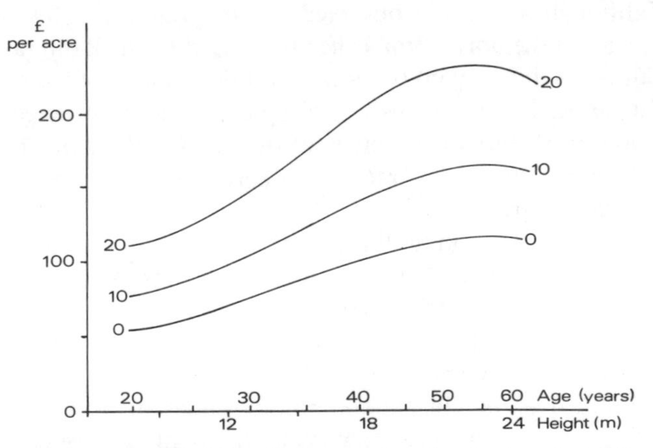


**Fig. 4.13 A typical relationship between the value of returns from Sitka spruce trees grown to various ages and measures taken to delay windthrow at 0, 10, or 20 years. (After Wardle, 1970.)**

in relation to treatments taken to delay windthrow at 0, 10, or 20 years in the subsequent life of the crop. Valuation is achieved by discounting net returns at 5 per cent to the time of valuation and it can be seen that there is a progressive increase in the value of the crop with time after treatment. In particular, the graph indicates the specially favourable financial returns obtained during the latter years of the crop's life, and Wardle (1970) has stressed the high increments of timber volume and value which are linked with quite small increases of height in this most vulnerable stage of the forest's development.

These general considerations can be exemplified with respect to the main British windblow losses in recent years, which were most pronounced in Scotland as a result of January gales in 1953 and 1968. The 1953 event, which blew down about 5 per cent of the total standing volume of coniferous timber in the UK, has been documented by Steven (1953) and Andersen (1954). In northeast Scotland, the varying degree of damage was related to the local distribution of extreme gusts as modified by topography, and there was a clear correlation between tree height and windthrow with few stands under 12 m in height experiencing damage. Rather less important variables were differences in wind tolerance of species, with larch resistant to windblow and Scots pine fairly easily thrown, together with edaphic factors such as compacted soil horizons which limited rooting depth.

A more detailed picture is available of Scottish forestry losses in the 1968 gale from the comprehensive report edited by Holtam (1971). This showed that, like the 1953 event, the gale of January 1968 had a return period of 75 years but destroyed 1·6 million cubic metres of timber which represented twice the annual cut for the whole of Scotland. Spruce plantations accounted for 51 per cent of the reported windblow damage, although this was a reflection of the dominance of this species rather than its vulnerability. In the west of Scotland there was a direct relationship between tree height and the incidence of windthrow, with only 0·04 per cent of the area with trees 0–6 m high suffering damage compared with over 20 per cent of the area with trees more than 15 m tall. In the more mountainous forested areas windthrow was closely related to local, topographically-induced accelerations of airflow on windward slopes, in funnels and above ridges. However, in areas of smoother relief, the damage was more extensively distributed according to increases in both tree height and altitude, as shown in Table 4.2.

The report details the ancillary problems of clearing, transporting, and marketing such a large volume of damaged timber, which must be removed quickly in order to prevent insect pests and fungal growths from inhabiting the fallen timber and infecting the surrounding stands. As the dead trees dry out, there is also an increasing risk of fire and in this particular case a target date for timber clearance was set eighteen months after the gale. The harvesting, sawmilling, and marketing involved was dependent on the co-operation of the entire timber industry in Scotland.

On a world scale, fire is the most significant hazard for forestry. According to Mercer (1971), the greatest risk exists in Australia where 2000 bushfires occur annually engulfing up to 400 hectares of forest in 30 min compared with only 0·5 hectare over the same time in the slower burning coniferous forests of the northern hemisphere. High temperatures and drought following a period of accelerated vegetation growth provide the most dangerous combination of weather and fuel conditions, although human variables such as the recreational mobility of the population and the weatherboard construction of many houses are additional factors.

Fire damage is extensive and expensive. In 1952, a fire in southern New South Wales destroyed 110 000 sheep and did \$15 m worth of damage in $7\frac{1}{2}$ h, and the 1967 Hobart fire resulted in material damage estimated at \$40 million. The forestry industry itself suffers severe losses and the effects of a 1939 bushfire, estimated to cost the industry \$200 million, are still being felt. Fire control is conservatively reckoned to cost \$6 million annually, and the general unavailability of large water supplies in Australia means that many fires, such as those of January 1962 (Whittingham, 1964), are ended largely by the arrival of rainfall.

In North America, Barrows (1966) has drawn attention to the fire damage from lightning storms which may produce multiple fires as on 12 July 1940, when 335 lightning-caused fires occurred in the National Forests of Montana and Idaho. Taylor and Williams (1968) have provided a detailed account of the synoptic conditions associated with the Hellgate Fire in the George Washing-

| Tree Height Classes | 0–4·5 metres | 5·0–6·0 metres | 6·5–7·0 metres | 7·5–9·0 metres | 9·5–10·0 metres | 10·5–12·0 metres |
|---|---|---|---|---|---|---|
| Area of trees above 305 m in hectares | 596 | 63 | 50 | 98 | 105 | 30 |
| Percentage windthrown | 0·6 | nil | 32 | 63 | 56 | 72 |
| Area of trees below 305 m in hectares | 590 | 251 | 232 | 115 | 187 | 41 |
| Percentage windthrown | 0·7 | 0·3 | 5 | 39 | 45 | 57 |

**Table 4.2** Windthrow in the Carron valley spruce forest, central Scotland, during the gale of 15 January 1968, in relation to tree height and elevation. (After Holtam, 1971.)

ton National Forest of western Virginia on 18 April, 1965. This fire was unusually severe with rapid horizontal spreading associated with intense convection and smoke columns reaching maximum heights estimated at 3000 m. Some 865 hectares were destroyed, representing a loss of $120 000 in control costs, timber destroyed, and loss of growth. Given suitable meteorological conditions, the smoke pall from forest fires may be advected over large distances; indeed Peterson and Drury (1967) have claimed that such a smoke haze was responsible for reducing the amount of solar radiation incident on the ground by about 25 per cent over a wide area of the Canadian tundra.

In Britain, most fires result from human action, either deliberate or inadvertent, and the weather is merely an environmental factor rather than the direct causal agent (Wardle, 1970). Nevertheless, numerous trees are struck by lightning, and Shipley (1946) has suggested that the oak is the most common victim, although conifers also carry a reasonably high risk. In most years, the average forest fire covers less than 0·5 hectare but, on the other hand, a large proportion of loss is caused by a few large fires. There is a pronounced seasonal and diurnal incidence of fire, with most losses occurring in spring and a high concentration of fire starts during the afternoon. Rouse (1959) has isolated the five main atmospheric factors which influence the start of forest fires—rainfall, temperature, relative humidity, wind velocity, and season—and incorporated them in the computation of a fire-danger index suitable for use in southern England. The importance of such meteorological variables has been noted by Kayll (1966) in the deliberate firing of *calluna vulgaris* heather in north-east Scotland, where the heather is burned to maintain the heath in its most productive state for sheep and grouse. Thus, with dry, windy conditions in autumn, 93 per cent of available fuel was burnt whereas spring burns on cool, moist days consumed only 30 per cent of the fuel with a consequent poor regeneration of the moorland vegetation.

## References


ALLEN, W. H. and LAMBERT, J. R. (1971). Application of the principle of calculated risk to scheduling of supplemental irrigation. II Use on flue-cured tobacco. *Agric. Met.*, **8**:325–340.

ANDERSON, K. F. (1954). Gales and gale damage to forests with special reference to the effects of the storm of January 1953 in north-east Scotland. *Forestry*, **27**:97–121.

ANONYMOUS (1962). *Irrigation*. Bulletin No. 138 (3rd Ed.) Ministry of Agriculture, Fisheries and Food, HMSO, London.

ANONYMOUS (1964). *The farmer's weather*. Bulletin No. 165 Ministry of Agriculture, Fisheries and Food, HMSO, London.

ANONYMOUS (1967). *Potential transpiration*. Tech. Bull. No. 16, Ministry of Agriculture, Fisheries and Food, HMSO, London.

ARAKAWA, H. (1957). Three great famines in Japan. *Weather*, **12**:211–17.

ARKLEY, R. J. and ULRICH, R. (1962). The use of calculated actual and potential evapotranspiration for estimating potential plant growth. *Hilgardia*, **32**:443–62.

ARON, R. H. (1971). Climatic chilling and future almond growing in southern California. *Prof. Geogr.*, **23**:341–3.

ASANA, R. D. and WILLIAMS, R. F. (1965). The effect of temperature stress on grain development in wheat. *Aust. J. Agric. Res.*, **16**:1–13.

BAIER, W. (1972). An agroclimatic probability study of the economics of fallow-seeded and continuous spring wheat in southern Saskatchewan. *Agric. Met.*, **9**:305–21.

BARROWS, J. S. (1966). Weather modification and the prevention of lightning-caused forest fires. In Sewell, W. R. D. (ed.) *Human Dimensions of Weather Modification*, Dept. of

Geography, University of Chicago, Research Paper No. 105: 169–82.

BASILE, R. M. (1954). Drought in relation to corn yield in the north-western corner of the Corn Belt. *Agron. J.*, **46**:4–7.

BATES, E. M. (1972). Temperature inversion and freeze protection by the wind machine. *Agric. Met.*, **9**:335–46.

BEAUMONT, A. (1947). The dependence on the weather of the dates of outbreak of potato blight epidemics. *Trans. Brit. Mycol. Soc.*, **31**:45–53.

BLAXTER, K. L. (1965). Climatic factors and the productivity of different breeds of livestock. In Johnson, C. G. and Smith, L. P. (eds),. *The biological significance of climatic changes in Britain,* Academic Press, 157–68.

BODLAENDER, K. B. A. (1963). The influence of temperature, radiation and photo-period on development and yield. In Ivins, J. D. and Milnthorpe, F. L. (eds.) *The growth of the potato*, Butterworths, London.

BONE, J. S. and TAYLER, R. S. (1963a). The effect of irrigation and stocking rate on the output from a sward. 1. Methods and herbage results. *J. Brit. Grassl. Soc.*, **18**:190–96.

BONE, J. S. and TAYLER, R. S. (1963b). The effect of irrigation and stocking rate on the output of a sward. 2. Dairy-cow production. *J. Brit. Grassl. Soc.,* **18**:295–99.

BONSMA, J. C. and LOUW, G. N. (1967). Heat as a limiting factor in animal production. In Tromp, S. W. and Weihe, W. H. (eds) *Biometeorology. Vol. 2 Pt. 1*, Pergamon Press, Oxford, 371–82.

BRENCHLEY, G. H. (1962). The control of potato blight in the Fens. In Taylor, J. A. (ed.) *Climatic factors and diseases in plants and animals*. Aberystwyth Memorandum No. 5, 15–21. Duplicated typescript.

BRIERLEY, J. S. (1967). Climate and the seasonal cycle of ranching, Pincher Creek, Alberta. *The Albertan Geographer*, **4**:38–44.

BROADBENT, L. (1967). The influence of climate and weather on the incidence of plant virus diseases, In Taylor, J. A. (ed.) *Weather and agriculture,* Pergamon Press, 99–104.

CABORN, J. M. (1955). The influence of shelter-belts on microclimate. *Q. Jl. Roy. Met. Soc.,* **81**:112–15.

CARDER, A. C. and HENNIG, A. M. F. (1966). Soil moisture regimes under summerfallow, wheat and red fescue in the upper Peace River region. *Agric. Met.*, **3**:311–31.

CARNEY, J. P. (1950). Drought feeding of sheep. *Q. Rev. Agric. Econ.*, **4**:11–14.

CHAMBERS, R. E. (1967). Weather hazards and milk production. In Taylor, J. A. (ed.) *Weather and agriculture*, Pergamon Press, 81–9.

CHANGNON, S. A. (1967). Areal-temporal variations of hail intensity in Illinois. *J. Appl. Met.*, **6**:536–41.

CHANGNON, S. A. and NEILL, J. C. (1968). A meso-scale study of corn-weather response on cash-grain farms. *J. Appl. Met.*, **7**:94–104.

CHANGNON, S. A. and STOUT, G. E. (1967). Crop-hail intensities in central and north-west United States. *J. Appl. Met.*, **6**:542–8.

COOPER, J. P. (1965). Climatic adaptation of local varieties of forage grasses. In Johnson, C. G. and Smith, L. P. (eds) *The biological significance of climatic changes in Britain*. Academic Press, 169–79.

CRAWFORD, T. V. (1964). Computing the heating requirements for frost protection. *J. Appl. Met.*, **3**:750–60.

CRITCHFIELD, H. J. (1966). *General climatology* 2nd Ed., Prentice-Hall, New Jersey, 420 pp.

CROXALL, H. E. and SMITH, L. P. (1965). Sowing dates for winter wheat. *NAAS Quart. Rev.*, **68**:147–9.

CURRY, L. (1962). The climatic resources of intensive grassland farming: the Waikato, New Zealand. *Geogrl. Rev.*, **52**:174–194.

CURRY, L. (1963). Regional variation in the seasonal programming of livestock farms in New Zealand. *Econ. Geog.*, **39**:95–117.

DAY, W. R. (1939). Local climate and the growth of trees with special reference to frost. *Q. Jl. R. Met. Soc.*, **65**:195–207.

DESJARDINS, R. L. and SIMINOVITCH, D. (1968). Microclimatic study of the effectiveness of foam as protection against frost. *Agric. Met.*, **5**:291–6.

DONALDSON, G. F. (1968). Allowing for weather risk in assessing harvest machinery capacity. *Amer. J. Agric. Econ.*, **50**: 24–40.

DUCKHAM, A. N. (1963). *The farming year*. Chatto and Windus Ltd, 525 pp.

DUCKHAM, A. N. (1967). Weather and farm management decisions. In Taylor, J. A. (ed.) *Weather and agriculture*, Pergamon Press, 69–80.

DUCKHAM, A. N. (1972). Meteorological forecasting and agricultural management. In Taylor, J. A. (ed.) *Weather forecasting for agriculture and industry*, David and Charles, 56–68.

EDLIN, H. L. (1957). Saltburn following a summer gale in south-east England. *Q. Jl. Forestry*, **51**:46–50.

EDWARDS, R. S. (1968). Studies of airborne salt deposition in some North Wales forests. *Forestry,* **41**:155–74.

FETERIS, P. J. (1955). £330,000 hail damage in fifteen minutes: analysis of a devastating hailstorm. *Weather*, **10**:223–32.

FOOD and AGRICULTURE ORGANISATION (1962). *Forest Influences*, Forestry and Forest Products Studies, No. 15, United Nations, Rome, 307 pp.

FRASER, A. I. (1965). The uncertainties of wind damage in forest management. *Irish Forestry*, **22**:23–30.

FRISBY, E. M. (1951). Weather crop relationships: forecasting spring wheat yield in the northern Great Plains of the U.S.A. *Trans. Inst. Brit. Geogr.*, **17**:77–96.

GANGOPADHYAYA, M. and VENKATARAMAN, S. (1969). Evapotranspiration reduction. *Agric. Met.*, **6**:339–45.

GEDDES, A. E. M. (1922). Weather and the crop yield in the north-east counties of Scotland. *Q. Jl. R. Met. Soc.*, **48**: 251–68.

GLOYNE, R. W. (1954). Some effects of shelterbelts upon local and microclimate. *Forestry*, **27**:85–95. .

GLOYNE, R. W. (1970). Shelter and local climate differences. *Weather*, **25**:439–44.

GOEBEL, C. J. and DEITSCHMAN, G. H. (1967). Ice storm damage to planted conifers in Iowa. *J. Forestry*, **65**:496–7.

GOODE, J. E. (1970). The cumulative effects of irrigation on fruit crops. In Taylor, J. A. (ed.) *The role of water in agriculture*, Pergamon Press, Oxford, 161–70.

GOODE, J. E. and HYRYCZ, K. J. (1964). The response of Laxton's Superb apple trees to different soil moisture conditions. *J. Hort. Sci.*, **39**:254–76.

GORDON, J. and VIRGO, S. E. (1968). Comparison of methods of forecasting night minimum temperatures. *Met. Mag.*, **97**:161–4.

GRAINGER, J. (1955a). The 'Auchincruive' potato blight forecast recorder. *Weather*, **10**:213–22.

GRAINGER, J. (1955b). Climate, host and parasite in crop disease. *Q. Jl. R. Met. Soc.*, **81**:80–8.

GRAINGER, J. (1967). Meteorology and plant physiology in potato blight forecasting. In Taylor, J. A. (ed.) *Weather and agriculture*, Pergamon Press, 105–13.

GRAINGER, J., SNEDDON, J. L., CHISHOLM, E. DE C., and HASTIE, A. (1954). Climate and the yield of cereal crops. *Q. Jl. R. Met. Soc.*, **81**:108–11.

GREGORY, S. (1954). Accumulated temperature maps of the British Isles. *Trans. Inst. Brit. Geogr.*, **20**:59–73.

HANN, L. and MCQUIGG, J. D. (1970). Evaluation of climatological records for rational planning of livestock shelters. *Agric. Met.*, **7**:131–41.

HANCOCK, J. (1954). Direct influence of climate on milk production: a review. *Dairy Science Abstracts*, **16**:89–102.

HANNA, L. W. (1971). Climatic influence on yields of sugarcane in Uganda. *Trans. Inst. Brit. Geogr.*, **52**:41–60.

HEATHCOTE, D. L. (1969). Drought in Australia: a problem of perception. *Geogrl. Rev.*, **69**:175–94.

HEWES, L. (1958). Wheat failure in western Nebraska, 1931–54. *Ann. Ass. Amer. Geogr.*, **48**:375–97.

HOGG, W. H. (1964a). Meteorology and horticulture. *Weather*, **19**:234–41.

HOGG, W. H. (1964b). Meteorology and agriculture. *Weather*, **19**:34–43.

HOGG, W. H. (1965). Climatic factors and choice of site, with special reference to horticulture. In Johnson, C. G. and Smith, L. P. (eds) *The biological significance of climatic changes in Britain*, Academic Press, 141–55.

HOGG, W. H. (1967a). Meteorological factors in early crop production. *Weather*, **22**:84–94 and 115–18.

HOGG, W. H. (1967b). The effect of a severe storm on 17 May, 1955 on a bean crop in Hampshire. In Taylor, J. A. (ed.) *Weather and agriculture*, Pergamon Press, 91–8.

HOGG, W. H. (1967c). The use of upper air data in relation to plant disease. In Taylor, J. A. (ed.) *Weather and agriculture*, Pergamon Press, 115–27.

HOGG, W. H. (1970a). Basic frost, irrigation and degree-day data for planning purposes. In Taylor, J. A. (ed.) *Weather economics*, Pergamon Press, 27–43.

HOGG, W. H. (1970b). Estimates of long-term irrigation needs. In Taylor, J. A. (ed.) *The role of water in agriculture*, Pergamon Press, 171–84.

HOGG, W. H. (1970c). Weather, climate and plant disease. *Met. Mag.*, **99**:317–26.

HOLMES, R. M. and ROBERTSON, G. W. (1970). Heat units and crop growth. In Nelson, J. G., Chambers, M. J., and Chambers, R. E. (eds) *Weather and climate*, Methuen, Toronto, 89–135.

HOLTAM, B. W. (ed.) (1971). *Windblow of Scottish forests in January 1968*. Forestry Commission Bull. No. 45, HMSO, Edinburgh, 53 pp.

HOOKER, R. H. (1922). The weather and crops in eastern England 1885–1921. *Q. Jl. R. Met. Soc.*, **48**:115–38.

HOPKINS, J. W. (1968). Protein content of western Canadian hard red spring wheat in relation to some environmental factors. *Agric. Met.*, **5**:411–31.

HOXMARK, G. (1928). The influence of the climatic conditions on the yield of wool in Argentina. *Mon. Weath. Rev.*, **56**: 60–1.

HUDSON, J. P. (1972). Agronomic implications of long-term weather forecasting. In Taylor, J. A. (ed.) *Weather forecasting for agriculture and industry*, David and Charles, 44–55.

HURST, G. W. (1965). Services to the agricultural community. *Met. Mag.*, **94**:121–6.

HURST, G. W. (1967). Honey production and summer temperatures. *Met. Mag.*, **96**:116–20.

HURST, G. W. (1968). Foot-and-mouth disease. *Vet. Record:* 610–14.

HURST, G. W. (1969). The variability with time and location of spring and summer temperatures in the United Kingdom. *Met. Mag.*, **98**:78–87.

HURST, G. W. (1970). Agrometeorology in the Meteorological Office. *Met. Mag.*, **99**:170–7.

HURST, G. W. and SMITH, L. P. (1967). Grass growing days. In Taylor, J. A. (ed.) *Weather and agriculture*, Pergamon Press, 147–55.

JACKSON, A. A. (1959a). The eco-climatology of Dutch lights. Pt. 1 Meteorological assessment. *Weather*, **14**:1–7.

JACKSON, A. A. (1959b). The eco-climatology of Dutch lights. Pt. 2. The effect on growth. *Weather*, **14**:9–16.

JOHNSON, W. C. (1964). Some observations on the contribution of an inch of seeding-time soil moisture to wheat yield in the Great Plains. *Agron. J.*, **56**:29–35.

JONES, E. L. (1964). *Seasons and Prices: The role of the weather in English agricultural history*. George Allen and Unwin Ltd, 193 pp.

KAYLL, A. J. (1966). Some characteristics of heath fires in north-east Scotland. *J. Appl. Ecol.*, **3**:29–40.

KEMP, W. S. (1956). Central Otago pip and stone fruit orchard. *N. Zealand J. Agric.*, **93**:421–30.

LAMB, H. H. (1965). Britain's changing climate. In Johnson, C. G. and Smith, L. P. (eds) *The biological significance of climatic change in Britain*. Academic Press, 3–31.

LANDSBERG, H. E. (1968). A comment on land utilisation with reference to weather factors. *Agric. Met.*, **5**:135–37.

LAVE, L. B. (1963). The value of better weather information to the raisin industry. *Econometrica*, **31**:151–64.

LAWRENCE, E. N. (1952). Estimation of weekly frost risk using weekly minimum temperatures. *Met. Mag.*, **81**:137–41.

LEMONS, H. (1942). Hail in American agriculture. *Econ. Geog.*, **18**:363–78.

LOMAS, J. (1972). Economic significance of dry-land farming in the arid northern Negev of Israel. *Agric. Met.*, **10**:383–92.

LOMAS, J. and GAT, Z. (1967). The effect of windborne salt on citrus production near the sea in Israel. *Agric. Met.*, **4**:415–25.

LOMAS, J. and SHASHOVA, Y. (1970). The effect of low temperatures on banana yields. *Int. J. Biometeor.*, **14**:155–65.

LOMAS, J. and SCHLESINGER, E. (1971). The influence of a windbreak on evaporation. *Agric. Met.*, **8**:107–15.

MCQUIGG, J. D. and CALVERT, O. H. (1966). Influence of soil temperature on the emergence and initial growth of upland cotton. *Agric. Met.*, **3**:179–85.

MATTHEWS, C. D. (1967a). Frost damage of horticultural crops. Part 1. *Tasmanian J. Agric.*, **38**:33–40.

MATTHEWS, C. D. (1967b). Frost damage of horticultural crops. Part 2. *Tasmanian J. Agric.*, **38**:89–98.

MAUNDER, W. J. (1963). The climates of the pastoral production areas of the world. *Proc. New Zeal. Inst. Agric. Sci.*, **9**:25–40.

MAUNDER, W. J. (1966). Climatic variations in agricultural production in New Zealand. *New Zeal. Geogr.*, **22**:55–69.

MAUNDER, W. J. (1967). Climatic variations and meat production: a New Zealand review. *New Zealand Sci. Rev.*, **25**:9–12.

MAUNDER, W. J. (1968). Agroclimatological relationships: A review. *Can. Geogr.*, **12**:73–84.

MERCER, D. (1971). Scourge of an arid continent. *Geogrl. Mag.*, **43**:563–7.

MILLAR, A. (1964). Notes on the climate near the upper forest limit in the northern Pennines. *Q. Jl. Forestry*, **58**:239–46.

MITCHELL, K. J., WALSHE, T. O., and ROBERTSON, N. G. (1959). Weather conditions associated with outbreaks of facial eczema. *New Zeal. J. Agric. Res.*, **2**:584–604.

MONTEITH, J. L. (1965). Radiation and crops. *Expt. Agric. Rev.*, **1**:241–51.

MUNRO, I. A. (1958). Irrigation of grassland. *J. Brit. Grassl. Soc.*, **13**:213–21.

NATURAL RESOURCES (TECHNICAL) COMMITTEE (1962). *Irrigation in Great Britain*. Office of the Minister for Science, HMSO, London.

NAYA, A. (1967). Insects, insecticides and the weather. (Part 2). *Weather*, **22**:211–15.

NISHIZAWA, T. and YAMABE, K. (1970). Change in downstream temperature caused by the construction of reservoirs. *Sci. Rep. Tokyo Univ. Educ. Sect. C.*, **10**:237–52.

NUTMAN, F. J. and ROBERTS, F. M. (1969). Climatic conditions in relation to the spread of coffee berry disease since 1962 in the East Rift Districts of Kenya. *East Afr. Agric. Forestry J.*, **35**:118–27.

OGUNTOYINBO, J. S. (1965). Agroclimatic problems and commercial cane sugar industry in Nigeria. *Nigerian Geogrl. J.*, **8**:83–97.

OGUNTOYINBO, J. S. (1966). Evapotranspiration and sugar cane yields in Barbados. *J. Trop. Geog.*, **22**:38–48.

OJO, O. (1971). Bovine energy balance climatology and livestock potential in Nigeria. *Agric. Met.*, **8**:353–69.

OLLERENSHAW, C. B. (1966). The approach to forecasting the incidence of *fascioliasis* over England and Wales, 1958–62. *Agric. Met.*, **3**:35–53.

OLLERENSHAW, C. B. (1967). Climatic factors and liver fluke disease. In Taylor, J. A. (ed.) *Weather and agriculture*. Pergamon Press, 129–35.

PAUL, A. (1966). The development of investigation into hail occurrence in Alberta. *The Albertan Geogr.*, **3**:4–10.

PEARL, R. T. (ed.) (1954). *The calculation of irrigation need.* Tech. Bull. No. 4, Ministry of Agriculture, Fisheries and Food, HMSO, London.

PEDGLEY, D. E. and SYMMONS, P. M. (1968). Weather and the locust upsurge. *Weather*, **23**:484–92.

PENMAN, H. L. (1948). Natural evaporation from open water, bare soil and grass. *Proc. Roy. Soc. (Lond.)* A, **193**:120–45.

PENMAN, H. L. (1962). Weather and crops. *Q. Jl. R. Met. Soc.*, **88**:209–19.

PENMAN, H. L. (1967). Climate and crops. *Mem. and Proc. Manchester Lit. and Phil. Soc.*, **110**:5–17.

PETERSON, J. T. and DRURY, L. D. (1967). Reduced values of solar radiation with occurrence of dense smoke over the Canadian tundra. *Geogrl. Bull.*, **9**:269–71.

PRICKETT, C. N. (1970). The current trends in the use of water for agriculture. In Taylor, J. A. (ed.) *The role of water in agriculture*, Pergamon Press, Oxford, 101–19.

RAINEY, R. C. (1955). Observations of desert locust swarms by radar. *Nature*, **175**:77.

RAINEY, R. C. (1963). Meteorology and the migration of desert locusts. *W.M.O. Tech. Note No. 54,* Geneva, 115 pp.

RAINEY, R. C. (1969). Effects of atmospheric conditions on insect movement. *Q. Jl. R. Met. Soc.*, **95**:424–33.

RAINEY, R. C. (1973). Airborne pests and the atmospheric environment. *Weather,* **28**:224–39.

RAMDAS, L. A. (1957). Natural and artificial modification of microclimate. *Weather,* **12**:237–40.

RIPLEY, E. A. (1967). The effects of shade and shelter on the microclimate of tea. *East African Agric. and Forestry J.*, **33**:67–80.

ROBERTS, M. J. G. (1968). Protection against lightning damage. *Agriculture (Lond.),* **75**:350–1.

ROSE, J. K. (1936). Corn yield and climate in the Corn Belt. *Geogrl. Rev.*, **26**:88–102.

ROSENBERG, N. J. (1966a). Microclimate, air mixing and physiological regulation of transpiration as influenced by wind shelter in an irrigated bean field. *Agric. Met.,* **3**:197–224.

ROSENBERG, N. J. (1966b). Influence of snow fence and corn windbreaks on microclimate and growth of irrigated sugar beets. *Agron. J.*, **58**:469–75.

ROUSE, G. D. (1959). Forest fire danger tables for southern England. *Forestry*, **32**:117–23.

ROUSE, G. D. (1961). Some effects of rainfall on tree growth and forest fires. *Weather,* **16**:304–11.

RUSSELL, E. W. (1967). Climate and crop yields in the tropics. A review of progress in reducing some harmful effects of climate on crop production. *Cotton Growers Rev.*, **44**:87–99.

RUTTER, N. (1968a). Geomorphic and tree shelter in relation to surface wind conditions, weather, time of day and season. *Agric. Met.*, **5**:319–34.

RUTTER, N. (1968b). Shelter effect of an old-established shelter block of European larch on a slope of mean gradient 1 in 4. *Agric. Met.,* **5**:335–49.

RUTTER, N. and EDWARDS, R. S. (1968). Deposition of airborne marine salt at different sites over the College Farm, Aberystwyth (Wales), in relation to wind and weather. *Agric. Met.*, **5**:235–54.

SALE, P. J. M. (1965). Changes in water and soil temperature during overhead irrigation. *Weather*, **20**:242–5.

SALTER, P. J. (1961). The irrigation of early summer cauliflowers in relation to stage of growth, plant spacing and nitrogen level. *J. Hort. Sci.,* **36**:241–53.

SANSOM, H. W. (1966). The use of explosive rockets to suppress hail in Kenya. *Weather*, **21**:86–91.

SAUNDERS, W. E. (1949). Night cooling under clear skies. *Q. Jl. R. Met. Soc.* **75**:154–60.

SAUNDERS, W. E. (1952). Some further aspects of night cooling under clear skies. *Q. Jl. R. Met. Soc.*, **78**:603–12.

SEN, A. R., BISWAS, A. K. and SANYAL, D. K. (1966). The influence of climatic factors on the yield of tea in the Assam valley. *J. Appl. Met.*, **5**:789–800.

SHIPLEY, J. F. (1946). Lightning and trees. *Weather*, **1**:206–210.

SHOEMAKER, J. S. and TESKEY, B. J. E. (1959). *Tree fruit production.* J. Wiley and Sons, New York.

SKIDMORE, E. L. and HAGEN, L. J. (1970). Evaporation in sheltered areas as influenced by windbreak porosity. *Agric. Met.,* **7**:363–74.

SMALL, R. T. (1949). The use of wind machines and helicopter flights for frost protection. *Bull. Amer. Met. Soc.*, **30**:79–85.

SMITH, C. V. (1964). Some evidence for the windborne spread of fowl pest. *Met. Mag.*, **93**:257–63.

SMITH, C. V. (1967). Airborne tracers in agricultural meteorology. *Met. Mag.*, **96**:150–5.

SMITH, C. V. (1970). Weather and machinery work-days. In Taylor J. A. (ed.) *Weather economics,* Pergamon Press, 17–26.

SMITH, L. P. (1960). The relation between weather and meadow-hay yields in England 1939–1956. *J. Brit. Grassl. Soc.*, **15**:203–8.

SMITH, L. P. (1962). Meadow hay yields. *Outlook Agric.*, **3**:219–24.

SMITH, L. P. (1964). Weather hazards in agriculture: a survey. *Mem. No. 7, University College of Wales*, 1–9.

SMITH, L. P. (1965). Possible changes in seasonal weather. In Johnson, C. G. and Smith, L. P. (eds) *The biological significance of climatic changes in Britain,* Academic Press, 187–91.

SMITH, L. P. (1967a). Meteorology applied to agriculture. *World Met. Org. Bull.*, **16**:190–4.

SMITH, L. P. (1967b). Meteorology and the pattern of British grassland farming. *Agric. Met.*, **4**:321–38.

SMITH, L. P. (1968). Forecasting annual milk yields. *Agric. Met.*, **5**:209–14.

SMITH, L. P. (1971). *The significance of winter rainfall over farmland in England and Wales*. Tech. Bull. No. 24, Ministry of Agriculture, Fisheries and Food, HMSO, London.

SMITH, L. P. (1972). The effect of climate and size of farm on the type of farming. *Agric. Met.*, **9**:217–23.

SMITH, L. P. and OLLERENSHAW, C. B. (1967). Climate and disease. *Agriculture (Lond.)*, **74**:256–60.

STALLINGS, J. L. (1961). A measure of the influence of weather on crop production. *Journ. Farm Econ.*, **43**:1153–62.

STANSFIELD, J. M. (1970). The effect of the weather on farm organization and farm management. In Taylor, J. A. (ed.) *Weather economics*, Pergamon Press, 11–16.

STEVEN, H. M. (1953). Wind and the forest. *Weather*, **8**:169–74.

TAGGART, W. J. (1970). Variations in the marginal value of agricultural labour due to weather factors, In Taylor, J. A. (ed.) *Weather Economics,* Pergamon Press, 45–50.

TAYLOR, D. F. and WILLIAMS, D. T. (1968). Severe storm features of a wildfire. *Agric. Met.*, **5**:311–18.

TAYLOR, J. A. (1967a). Economic and ecological productivity under British conditions: an introduction. In Taylor, J. A. (ed.) *Weather and agriculture*, Pergamon Press, 137–45.

TAYLOR, J. A. (1967b). Growing season as affected by land aspect and soil texture. In Taylor, J. A. (ed.) *Weather and agriculture*, Pergamon Press, 15–36.

TAYLOR, J. A. (1967c). Marling experiments to measure the modification of soil temperature regimes and relative productivity of Lancashire mosslands. In Taylor, J. A. (ed.) *Weather and agriculture*, Pergamon Press, 213–25.

TAYLOR, J. A. (1970). The cost of British weather. In Taylor, J. A. (ed.) *Weather economics*, Pergamon Press, 5–9.

THOMAS, D. (1959). The assessment of shelter-need upon exposed farm sites. *Weather*, **14**:375–84.

THOMPSON, B. W. (1955). Meteorology and the tsetse-fly. *Weather*, **10**:249–55.

TINNEY, E. B. and MENMUIR, P. (1968). Results of an investigation into forecasting night minimum screen temperatures. *Met. Mag.*, **97**:165–71.

TIPPETT, L. H. C. (1926). On the effect of sunshine on wheat yield at Rothamsted. *J. Agric. Sci.*, **16**:159–65.

TRUMBLE, H. C. (1937). The climatic control of agriculture in South Australia. *Trans. Roy. Soc. S. Aust.*, **16**:41–62.

TYRRELL, J. G. (1970). A note on the areal patterns in the value of early potato production in south-west Wales, 1967. In Taylor, J. A. (ed.) *Weather economics*, Pergamon Press, 51–65.

VALLI, V. J. (1970). The use of natural gas heating to prevent spring freeze damage in the Appalachian fruit region of the United States. *Agric. Met.*, **7**:481–6.

WAGGONER, P. E., BEGG, J. E., and TURNER, N. C. (1969). Evaporation of dew. *Agric. Met.*, **6**:227–30.

WARBOYS, I. B. (1967). Climatic factors in the development of local grass conservation techniques. In Taylor, J. A. (ed.) *Weather and agriculture*, Pergamon Press, 157–72.

WARDLE, P. A. (1970). Weather and risk in forestry. In Taylor, J. A. (ed.) *Weather economics*, Pergamon Press, 67–82.

WATSON, D. J. (1963). Weather and plant yield. In Evans, L. T. (ed.) *Environmental control of plant growth*, Academic Press, New York, 337–49.

WEAVER, J. C. (1943). Climatic relations of American barley production. *Geogrl. Rev.*, **33**:569–88.

WHITMORE, J. S. (1957). The influence of climatic factors on the agricultural development of South Africa. *S. African Geogrl. J.*, **39**:5–25.

WHITTINGHAM, H. E. (1964). Meteorological conditions associated with the Dandenong bush fires of 14–16 January, 1962. *Austral. Met. Mag.*, **11**:10–37.

WILLIAMS, G. D. V. (1972). Geographical variations in yield-weather relationships over a large wheat growing region. *Agric. Met.*, **9**:265–83.

WINTER, E. J., SALTER, P. J., and COX, E. F. (1970). Limited irrigation in crop production. In Taylor, J. A. (ed.) *The role of water in agriculture*, Pergamon Press, Oxford, 147–60.

WINTER, E. J. and STANHILL, G. (1957). Horticultural climatology at the National Vegetable Research Station. *Weather*, **12**:218–22.

WINTER, F. (1958). Das Spätfrostproblem im Rahmen der Neuordnung des südwestdeutschen Obstbaus. *Gartenbauwissensch.*, **23**:342–62.

WISHART, J. and MACKENZIE, W. A. (1930). The influence of rainfall on the yield of barley at Rothamsted. *J. Agric. Sci.*, **20**:417–39.

WOODHEAD, T. (1968). Micro-meteorological studies of coffee berry disease. *Ann. App. Biol.*, **62**:451–63.

WOODHEAD, T. (1969). An investigation into some micro-meteorological aspects of coffee berry disease. *Agric. Met.*, **6**:195–210.

WRIGHT, P. B. (1968). Wine harvests in Luxembourg and the biennial oscillation in European summers. *Weather*, **23**:300–4.

WRIGHT, P. B. (1969). Effects of wind and precipitation on the spread of foot-and-mouth disease. *Weather*, **24**:204–13.

ZACHS, M. R. (1945). Oats and climate in Ontario. *Canad. J. Res.*, **23**:45–75.

# 5. Water and power resources

Apart from the air itself, water is the most fundamental, atmospherically dependent natural resource available to man and therefore merits consideration in any text on applied climatology. On the other hand, the comprehensive study of water belongs to the separate but related science of hydrology, and several writers, such as Linsley (1960), have commented on the overlap between the applied aspects of climatology and hydrology. Although it is probable that a wholly satisfactory demarcation line can never be drawn between these two disciplines, applied climatology will be assumed to include *hydrometeorology* which, for the purposes of this book, will be interpreted as the study of the atmospheric mechanisms in the hydrological cycle insofar as they affect water resources and impinge on the work of the water engineer. In detail, this means that the following section will be largely concerned with the interactive role of precipitation and evaporation in determining the distribution of water, the assessment of water resources and the problems created by the extremes of flood and drought. The interrelationships between the atmosphere and *fuel* and *power* can be considered in terms of the climatic provision of power itself, the maintenance of power supply systems in adverse environmental conditions, plus the effects of weather on the demand for energy supplies.

## Water resources

### 5.1 The water cycle and the water balance

The earth's supply of fresh water is ultimately dependent on the workings of the hydrological cycle, which controls the circulation of all moisture between the various storage units within the earth–atmosphere system. As shown in Fig. 5.1, the atmosphere represents only a very small storage unit containing about 0·035 per cent of all fresh water. On average, the water content of the atmosphere is equivalent to approximately 25 mm of precipitation distributed evenly over the globe and, with a mean precipitation rate approaching 1000 mm per annum, this indicates an average residence time of only 10 days for each water molecule (Barry, 1969). Clearly, therefore, the cycle is maintained by the rapid flux and re-flux of atmospheric water in the twin processes of evaporation and precipitation, which in turn regulate the terrestrial water balance. From a water resource viewpoint, it is the terrestrial fresh water stored overground in lakes and rivers together with the underground water in aquifers which is of most interest. On a global scale, aquifer storage is considerably more important than surface reserves and Nace (1969) has estimated that recoverable ground water amounts to $7000 \times 10^{-3}$ km$^3$ compared with $125 \times 10^{-3}$ km$^3$ held in freshwater lakes and $1{\cdot}7 \times 10^{-3}$ km$^3$ stored in river channels.

Figure 5.1 indicates that most moisture exchange in the

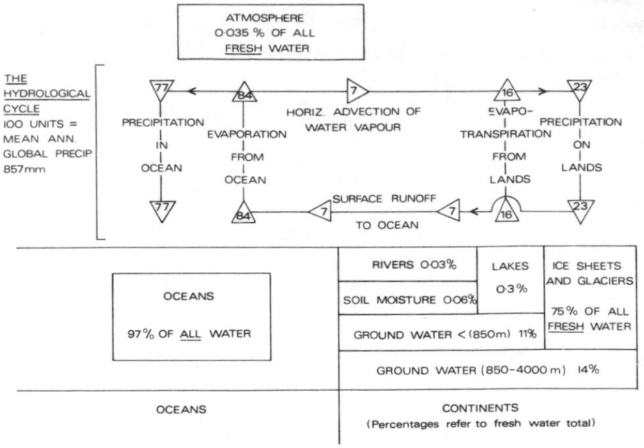

**Fig. 5.1 The global hydrological cycle and water storage. The oceanic percentage relates to all terrestrial water; the percentages for continental and atmospheric water to all fresh water. The units in the hydrological cycle are related to 100 units which represent the mean annual global precipitation of 857 mm. After More, R. J. Hydrological Models in Geography, in: Chorley, R. J. and Haggett, P. *Physical and information models in Geography*, Methuen, London, 1967.**

hydrosphere takes place between the atmosphere and the oceans rather than at the air-land interface. Generally, the oceans receive 77 per cent of all precipitation and supply 84 per cent of the world's evaporation, the deficit being made good by runoff from the continents over which precipitation exceeds evaporation. Sellers (1965) has shown that many geographical variations exist, however, and that the drier continents have a smaller fraction of the mean annual precipitation appearing as runoff. Thus, South America, which is the world's wettest continental area, loses less than two-thirds of its precipitation back to the atmosphere by evaporation compared with the dry continents of Australia and Africa, where more than three-quarters of the annual rainfall is lost to evaporation. In practical terms, this means that the gross availability of water resources is primarily determined by precipitation, and the latitudinal distribution of the water balance components, shown in Fig. 5.2, illustrates the broad relationship between rainfall and runoff amounts, with maximum runoff occurring where precipitation exceeds evaporation poleward of latitude 40° and between 10°N and 10°S.

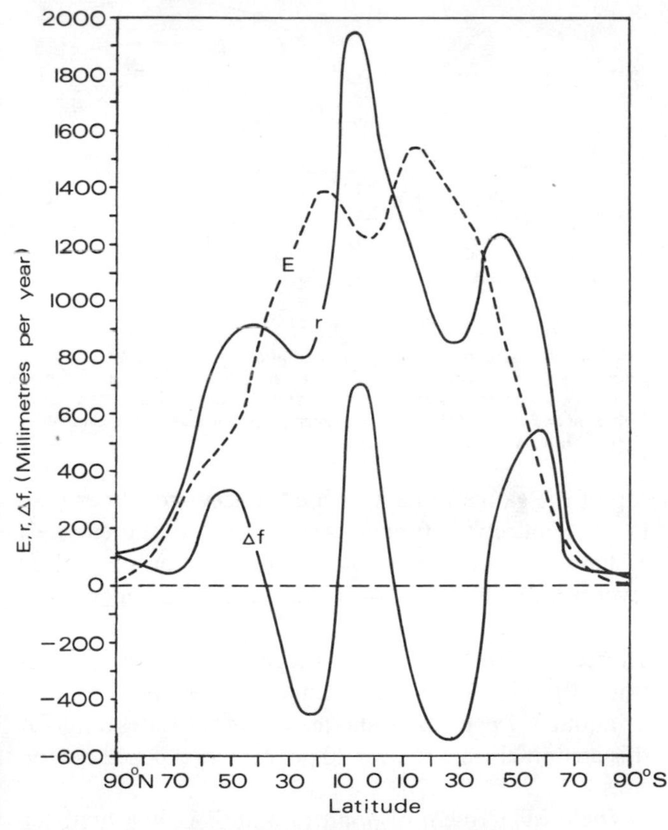


**Fig. 5.2 The average annual latitudinal distribution of evaporation (*E*), precipitation (*r*) and total runoff (*f*). After Sellers (1965). Copyright the University of Chicago. All rights reserved.**

On a more local scale, gross water resources are controlled by the seasonal water balance, since in the US, for example, a mean annual precipitation of 760 mm may produce a water surplus ranging from as little as 76 mm in Nebraska to 560 mm in the Rockies. Real understanding of the seasonal water balance stems from the work of Thornthwaite (1948) and the concept of *potential evapotranspiration* (PE) defined as the combined evaporation and transpiration that would occur from a vegetation-covered surface if soil moisture were sufficient to permit unrestricted transpiration. When the seasonal march of PE is compared with the precipitation and an allowance is made for soil moisture storage and its subsequent use, periods of water deficiency and surplus are clearly revealed as in Fig. 5.3 (Thornthwaite and Hare, 1955). The water surplus represents the total volume of water which is in excess of evaporative needs and therefore available for river discharge or the recharge of groundwater reserves. Not surprisingly, water balance techniques have been widely employed in water resource assessment.

## 5.2 Assessment of water resources

Ideally, water resources should be assessed directly by the field measurement of riverflow or groundwater fluctuations, and an account of such purely hydrological practices lies outside the scope of this book. On the other hand, in many circumstances a satisfactory indirect evaluation can be obtained from measurements of precipitation and evaporation and, although more specialized texts such as World Meteorological Organisation (1960 and 1965) or Chow (1964) should be consulted for a detailed account of hydrometeorological instruments and networks, a brief mention must be made here about the reliability of precipitation and evaporation measurement.

Routine measurements of precipitation are made in non-recording gauges, which are basically hollow cylindrical vessels, the height of which above ground level varies from 305 mm in Canada, Australia, and the UK to 787 mm in the US and 2007 mm in the USSR. According to Bleasdale (1959), however, it has been known for over 200 years that such gauges catch less precipitation than actually reaches the ground surface, owing to turbulence around the gauge aperture. Jevons (1861) was the first person to investigate the disturbed airflow caused by a gauge exposed above ground level and recent wind-tunnel trials of the standard British gauge by Robinson and Rodda (1969) have confirmed that even this small instrument produces an acceleration of air currents over the gauge by up to 37 per cent at a windspeed of 3·5 m/s (Fig. 5.4). Some precipitation is transported over the top of the gauge to be deposited to leeward rather than in the collecting funnel, and there is, therefore, a systematic underestimation of precipitation at ground level nearby (Rodda, 1967). The catch deficiency tends to increase with windspeed, and experimental evidence quoted by Rodda (1970a) suggests that the conventional British gauge under-estimates annual precipitation by about 3–7 per cent in sheltered, lowland sites and that this error increases to over 20 per cent at exposed upland sites.

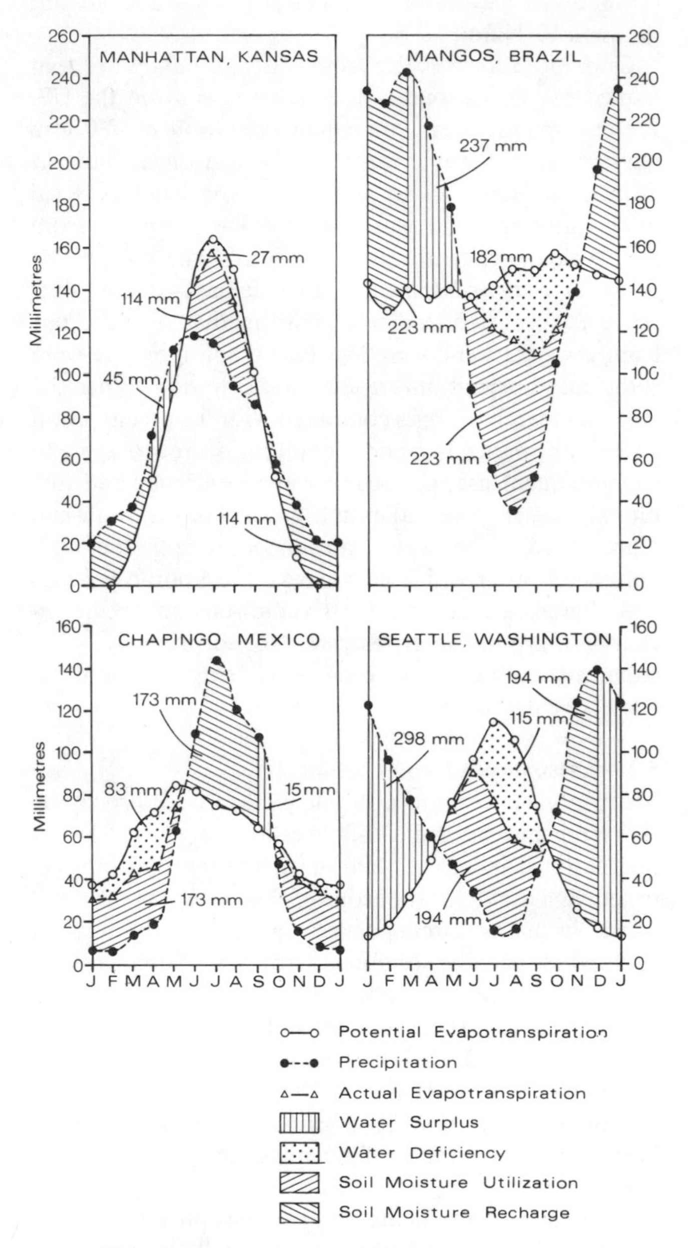


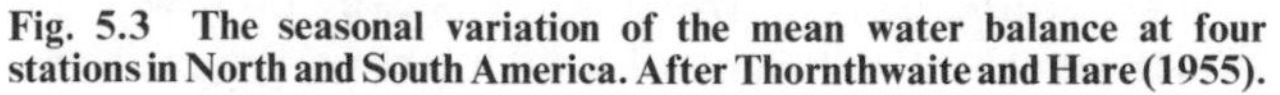

**Fig. 5.3 The seasonal variation of the mean water balance at four stations in North and South America. After Thornthwaite and Hare (1955).**

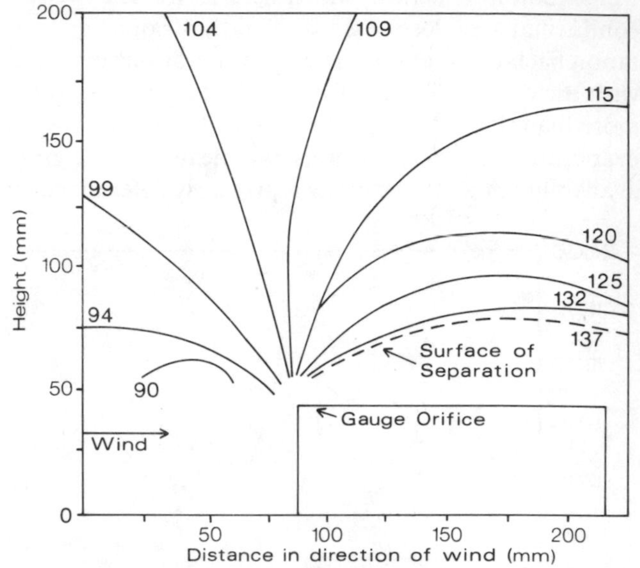


**Fig. 5.4 The structure of the windfield above a standard British rain-gauge based on wind-tunnel experiments. At a wind velocity of 3·5 m/s, air currents over the gauge are accelerated by up to 37 per cent. After Robinson and Rodda (1969). Reproduced by permission of the Controller of Her Majesty's Stationery Office.**

The measurement of point rainfall is only a first step in water resource studies, since areal estimates on a catchment scale are normally required. Thus, rain-gauging, like evaporation measurement, is basically a sampling process. Langbein (1960) has stated that a reasonable minimum density for a precipitation network would be more than 2 gauges per 1000 km$^2$, with further additions in more densely populated countries; even in Britain, however, with a gauge for about every 40 km$^2$, it is known that the sample is inadequate to support detailed analysis when compared with results from experimental networks (Holland, 1967a).

Most studies designed to calculate the number of raingauges necessary for areal measurements, such as those of Stephenson (1967) or Herbst and Shaw (1969), use statistical methods to determine the network density necessary to provide the required accuracy of estimate of the mean, although other workers, like McGuinness (1963), have simply established very dense networks and assumed that the results represent the true mean areal rainfall. The actual techniques for converting point to areal values have been summarized by Corbett (1967). As might be expected, the sampling requirements for monthly or annual precipitation are much lower than those for convective storms. Huff (1970) has obtained estimates of the sampling error of mean areal rainfall over areas ranging from 130 to 1430 km$^2$ in Illinois and found that, for a given sampling error, the gauge density for warm season storms was 2–3 times greater than that needed in winter.

Higher sampling requirements are also necessary as the period of rainfall decreases, and Huff and Shipp (1969) have shown that, for a minimum acceptance of 75 per cent explained variance between sampling points, a gauge spacing of only 0·5 km is needed for 1-min rainfall rates compared with 12 km for total precipitation in summer storms. Ideally, the design of a regional raingauge network should be tailored towards specific ends, such as water supply or flood forecasting, although in practice most have to serve for several water conservation purposes, as described by Shaw (1966) for Devon.

In this area of south-west England, as in parts of Scotland administered by the North of Scotland Hydro-Electric Board, it is apparent that automatic gauges, especially those capable of interrogation by telephone or radio, are destined to play an increasingly important role in precipitation measurement in remote areas (Reynolds, 1968). Other advances in recording rainfall include the use of radar, as discussed by Harrold (1966), and Wilson (1970) has shown from a study of Oklahoma thunderstorms how raingauge data may be employed to improve the accuracy of radar estimates.

On the other hand, the measurement of snowfall, which is subject to drifting and large gauge errors, remains a more formidable problem. For example, Findlay (1969) has claimed that annual precipitation assessment in northern Quebec and Labrador, where snowfall comprises an important component, may be subject to a systematic error of around 20 per cent and Uryvaev *et al.* (1965) have drawn attention to the deficiencies of snowfall measurement in the USSR, where gauges are maintained in sheltered local surroundings. However, Carlson and Marshall (1972) working in eastern Canada have compared radar records of falling snow with normal climatological observations and reported a degree of similarity comparable to that for summer rain within a distance of 160 km of the radar installation. In addition, experiments continue with specifically designed snow-measuring devices such as the pressure pillow (Penton and Robertson, 1967).

For a comprehensive review of the complexities surrounding evapotranspiration measurement, the reader should refer to summaries such as given by Veihmeyer (1964) or Ward (1971), but it can be stated that the most convenient records of open-water evaporation are obtained from tanks or pans. On a world scale, the US Weather Bureau Class A pan is the most common instrument, although it suffers from over-exposure and annual values have to be reduced by a coefficient of 0·70 to bring them into line with observed losses from lakes and reservoirs. A number of evaporation pans have been evaluated by Nordenson and Baker (1962), and work by Lapworth (1965) and Holland (1967b) has suggested that the standard British pan, which is sunk into the ground, suffers less from over-exposure than the Class A instrument and provides a more realistic indication of evaporation from a large water body.

Measurements of potential evapotranspiration may be made directly using specially constructed lysimeters, as described by Ward (1963), but few of these instruments have been installed and there are limited opportunities for making areal assessments of PE on this basis such as undertaken for the British Isles by Green (1970).

Normally, water resource studies rely on estimates of PE derived from formulae which, in turn, depend on meteorological observations available from the much denser network of climatological reporting stations. The formula first put forward by Penman (1948) is most widely used in this context and may well produce PE estimates which are as accurate as the areal assessments of rainfall employed in the water balance.

Mention has already been made in chapter 4 of the importance of the water balance concept for irrigation scheduling, but it will be apparent that many more applications exist on different time-scales for water resource assessment. On a monthly or seasonal time-scale, the water balance model has been used by Penman (1950) to estimate changes in ground-water storage over a chalk catchment, and Fig. 5.5 illustrates the close relationship

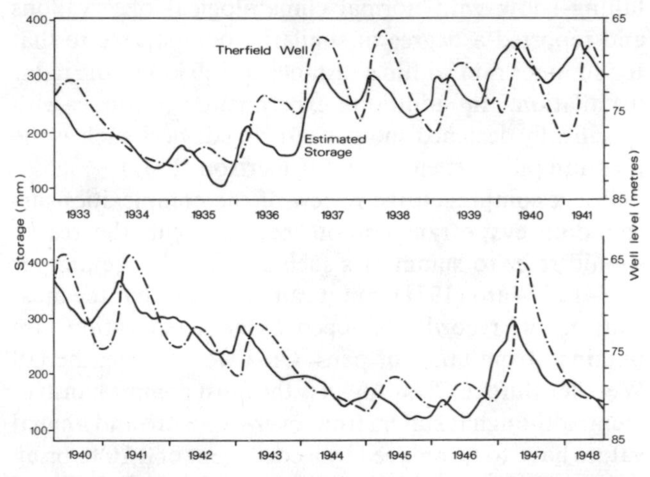


**Fig. 5.5 Estimated changes in monthly storage over a chalk catchment in eastern England compared with observed water-level changes in a well. The graph has been re-zeroed at the beginning of 1940 to reduce the significance of cumulative errors. Reproduced from the paper by H. L. Penman entitled 'The water balance of the Stour catchment area', *J. Instn. Wat. Engrs.*, Vol. 4, No. 6, p. 464.**

over a 16-year period between the estimated aquifer recharge and the actual fluctuation of the water-table measured in an observation well. More usually, however, the climatic water balance is employed with respect to surface resources and the work of Pardé (1955) and Beckinsale (1969), for example, confirms the extent to which seasonal river regimes are dependent on this mechanism.

A more important practical application lies in runoff forecasting which is the principal objective in hydrometeorology (Linsley, 1967). For example, Grindley (1960) used water balance techniques during the dry summer of 1959 to predict when the first appreciable increase in streamflow was likely to occur in the autumn over the UK, and routine calculations of soil moisture deficit are now supplied to most water authorities in Britain (Grindley, 1967). These official Meteorological Office calculations are based on the Penman formula but the Thornthwaite method has been used by Sanderson (1971) to estimate the variability of annual runoff in the Lake Ontario basin; Ward (1972) has shown that the Thornthwaite water balance can predict the runoff from a small clay catchment with an accuracy comparable with more sophisticated simulation techniques.

Theoretically, the safe yield of all water resource projects should be based on streamflow records obtained over a representative period of time but the frequent unavailability of such data means that a heavy, sometimes sole, reliance has to be placed on precipitation values in the planning and operation of water supply schemes. This has certainly been the case in Britain, where the early neglect of the river-gauging network led to virtually all surface-water supply schemes being established on the basis of mean rainfall and the size of the catchment area draining to the reservoir (Lapworth, 1930). An empirical rule was applied whereby the reservoir capacity was designed to cope with the three driest consecutive years and it was assumed that the annual rainfall during this period would not average less than 80 per cent of the long-term mean (Glasspoole, 1924).

More recently, Gregory (1956 and 1959) and Glasspoole (1955) have still stressed the importance of precipitation in understanding water supply problems in Britain but more emphasis is now placed on rainfall variability over periods shorter than three years. For example, Fig. 5.6, taken from Law (1955), illustrates the effect of different coefficients of variation of annual rainfall upon the yield of a typical reservoired catchment. It can be seen that, for the storage provided and assuming a probability of failure of, say, 1 per cent, the theoretical safe yield reduces from 108 million gallons per day with a coefficient of variation of 0·13 to 97 million gallons per day when the coefficient of variation increases to 0·17. Other yield estimates depend on statistical correlations of rainfall and runoff (Law, 1953), and Tucker (1960) has discussed some of the hydrometeorological factors which affect both the design and construction phases of reservoir projects.

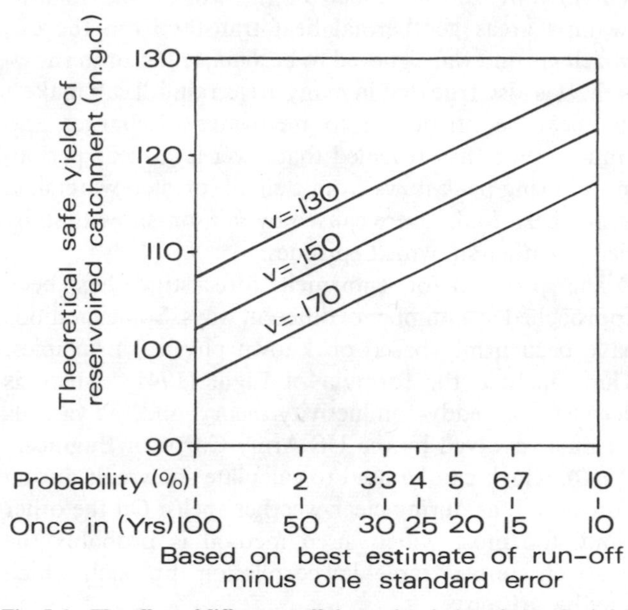


**Fig. 5.6 The effect of different coefficients of variation of annual rainfall on the yield of a typical reservoired catchment in upland Britain. Reproduced from the paper by F. Law entitled 'Estimation of the yield of reservoired catchments', *J. Instn. Wat. Engrs.*, Vol. 9, No. 6, p. 475.**

## 5.3 Hydrometeorology of snow

In many mid- and high-latitude areas, the availability of water resources is dependent on the hydrometeorology of snow rather than rainfall. Rooney (1969) has stated, for example, that not only is 51 per cent of California's streamflow derived from the snowmelt but this source also provides the most dependable supply and, in the southern Sierra Nevada, less than one year in ten experiences less than half of the mean annual precipitation in the snowpack zone compared with more than four years out of ten in the lower foothill zone. Pardé (1955), among others, has emphasized the regularity of snowmelt river regimes, although differences do occur such as between the strongly marked spring peak associated with some lowland rivers and the later, more protracted melt hydrograph of upland and glacial streams (Sommer and Spence, 1967).

Nevertheless, all these rivers are dominated by the annual temperature cycle which leads, as shown in Fig. 5.7, to a predictable flow maximum during the summer months. It will be apparent that the forecasting of such seasonal runoff is of fundamental importance for the efficient management of water resources, especially in areas like the western US where irrigators are largely dependent on meltwater and require advance information for planning crops (Fredericksen, 1958). In addition, Blanchard (1955) has described how forecasting is applied in the operation of hydroelectric schemes, and how multi-purpose reservoir schemes reap benefits, as when an accurate knowledge of snow storage to be released later in the year permits reservoir draw-down to meet urgent water demands or provide flood storage. Successful forecasting, therefore, depends both on an awareness of the water equivalent of lying snow and on an appreciation of the factors involved in snowmelt.

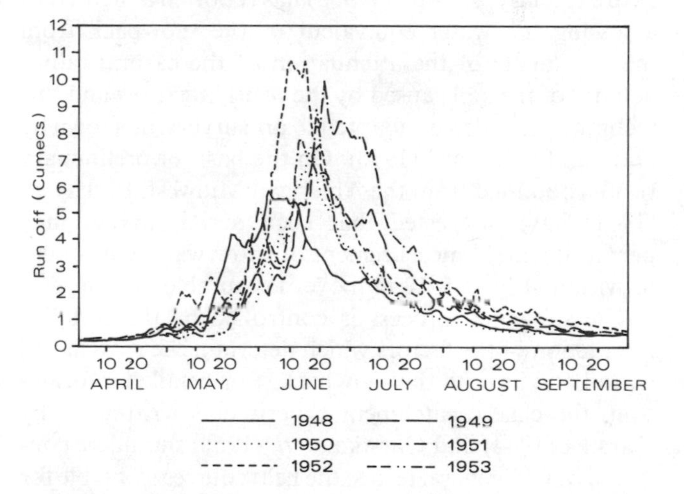


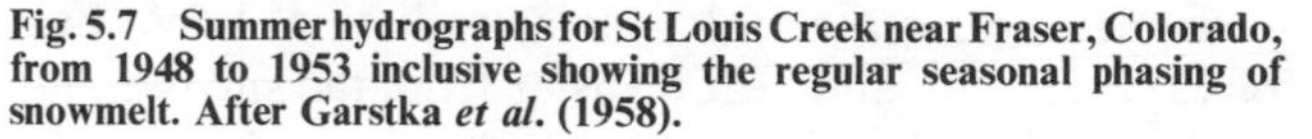


**Fig. 5.7 Summer hydrographs for St Louis Creek near Fraser, Colorado, from 1948 to 1953 inclusive showing the regular seasonal phasing of snowmelt. After Garstka *et al.* (1958).**

The traditional methods of surveying snow, as discussed by Church (1935 and 1942) and Garstka (1964), have been based on direct sampling of the snowpack at specific points along a snow course located, as far as possible, to be representative of watershed conditions. Depth samples are obtained by means of core-cutting devices, the actual water equivalent of the snow being determined by weighing at the time of sampling. Although access to the snow courses has been greatly improved by the development of special over-snow vehicles, many recent studies continue to document the sampling difficulties due to variations in the snow pack caused by altitude, aspect, season, and the occurrence of ice layers and high-density snow (Chinn, 1969 and Leaf, 1971).

The introduction of the radioisotope snow gauge linked to appropriate telemetering systems by Gerdel *et al.* (1950) in the US and Itagaki (1959) in Japan led to the provision of much-needed snowpack data for the remote and more inaccessible parts of river drainage basins. Aerial photographs have been used by Martinelli (1959 and 1965) to map the horizontal extent of snow in the alpine areas of Colorado and then, from computations of average snow density and vertical ablation, to estimate the water yield potential during July and August. More recently, Zotimov (1965) has reported a method of assessing the water equivalent of the snowpack from measurements of the attenuation of the natural radioactivity of the soil caused by the overlying snow and this technique can also be adapted to air surveys, as shown by Dahl and Odegaard (1970). On the basis of preliminary results obtained from the American Midwest, Peck *et al.* (1971) have suggested that such aerial surveys may permit the areal measurement of snow water equivalent to within at least 5–13 mm over favourable terrain.

The snowmelt process is controlled by the complex interaction of the factors which determine heat exchange within and around the snowpack. For detailed information, the classic catchment experiments written up by Garstka (1944) and Garstka *et al.* (1958) should be consulted but, in general terms, the heat sources available for snowmelting are threefold—solar radiation, latent heat of vaporization, and heat transfer from the surrounding environment such as the ground, the air, or falling rain. Most snowmelt models ascribe the dominant role to solar radiation and often regard environmental heat exchange as a negligible element. This interpretation may be largely attributed to the work of Wilson (1941), who concluded that, after the thirtieth day of snowcover, heat conduction from the ground was insignificant in terms of the melting process and that warm rain similarly represented such a small quantity of heat that, at a wet-bulb temperature of 10 °C, a 102 mm fall of rain would melt only some 12 mm of water from the snow. Recently, however, Santeford *et al.* (1972), experimenting near Lake Superior, Canada, have shown that the ground under the snowpack may remain unfrozen for much of the winter and in low, swampy areas geothermal heat transfer from the unfrozen ground was believed to be a major factor in snowmelt. It is also true that in many areas rainfall can make a significant contribution to meltwater discharges and Haupt (1968) has revealed that, over a 25-year period, more spring peak flows from a small forested watershed in northern Idaho were caused by rain-on-snow than by clear-weather snowmelt episodes.

The problem of snowmelt forecasting has been approached in a number of different ways. Some methods have been firmly based on known physical principles. These include the formula of Light (1941), which is derived from eddy-conductivity theory, and the various formulae evolved by the US Army Corps of Engineers (1960), which can be used to calculate snowmelt during rain as well as during clear-weather spells. On the other hand, the most widely used method is probably the relatively simple degree-day correlation approach, which may be written:

$$D = K(T-32)$$

where $D$ = average rate of snowmelt occurring over the basin in one day in in/day; $T$ = average daily air temperature in °F; $T-32$ = the daily number of degree-days above freezing, and $K$ = a melt coefficient. All the published work suggests that $K$ lies within 0·02 to 0·13 in/degree-day and, for a given catchment, a plot of cumulative degree-days against cumulative melt should approximate to a straight line (Johnson, 1966).

A reasonable degree of success has been achieved with this method. Pysklywec *et al.* (1968), for example, have reported results for spring snowmelt at New Brunswick, which are comparable with those obtained with more sophisticated techniques, and Johnson and Archer (1972) have suggested that the degree-day method can be employed for forecasting in Britain, where most of the snowmelt occurs with the influx of warm, moist and cloudy air from the Atlantic Ocean rather than as a result of large inputs of direct radiation. There is a particular need for such simple snowmelt models in the developing countries, where detailed heat budget data are unlikely to be available, and Hopkins (1972) has demonstrated how the upper limits of snowmelt flood discharges from a small mountainous catchment in Iran may be estimated on the basis of daily maximum temperatures observed in the lower part of the basin.

### 5.4 Water conservation

The world-wide trend of rising water consumption has brought, among other things, the demand in many countries for a more careful appraisal of the economic implications of water resources and Smith (1972) has shown that in Britain this has led to a call for higher water charges and for the introduction of universal domestic metering. So far, little evidence is available about the possible consequences of such changes, although data from the western US collected by Howe and Linaweaver (1967) suggest that, whereas only a small variation exists between metered and flat-rate areas in terms of average consumption, there is a reduction in some non-essential uses, such as garden sprinkling, in the metered areas, which show a greater efficiency of water use in relation to summer evapotranspiration. Similarly, Boyd (1971) has attempted to place an economic value on water quality and Toebes (1967) has drawn attention to the fact that, although over $800 million was spent on water development and control in New Zealand during 1966, only $0·5 million was spent on hydrological research. As these trends continue, it appears certain that more attention will be given to the further conservation of resources by conscious manipulation of the water balance for water supply purposes. The most far-reaching consequences may well stem from advances in the artificial stimulation of precipitation and Krick (1954) has commented on such water supply experiments in Colorado and Spain; however, rainfall modification is more conveniently discussed in chapter 8 and this section concentrates on more localized hydrometeorological manipulations.

Assuming that the precipitation increment remains unchanged, most modifications are designed to reduce the evaporative losses and ensure that a greater proportion of precipitation appears as streamflow. The direct suppression of open-water evaporation from lakes or reservoirs is an especially desirable aim in arid areas and may be achieved in a variety of ways including the construction of underground reservoirs, enclosing reservoirs under plastic covers or placing oil films on the water surface (Magin and Randall, 1960). The most usual method, however, is by the use of hexadeconal or similar monomolecular layers, pioneered in Australia and the southwestern US, and Mansfield (1959) has claimed a reduction of 20 per cent in evaporation losses from a 800 hectare reservoir near Broken Hill in western New South Wales as a result of this technique. Evaporation may also be reduced by altering the heat storage and net advected energy of a water body, as indicated by Koberg (1960) when cold water, drawn at depth from Lake Rooseveld in Arizona, was introduced into three downstream reservoirs.

As might be expected from the account of the water relations of different vegetation types in chapter 2, the management of watershed vegetation can provide an opportunity, through reduced interception and transpiration losses, to increase streamflow. In semi-arid areas, the clearance of phreatophytes from along stream courses is an important means of limiting transpiration, although recent work on salt cedar by van Hylckama (1970) has indicated that phreatophytes do not always transpire at the potential rate. On a catchment scale there have been many experiments which show that deforestation increases the water yield. Thus, Love (1955) claimed a streamflow increase of 22 per cent following the destruction of conifers in Colorado. Hardwood forest clearance has been reported by Hewlett and Helvey (1970) to produce 11 per cent higher runoff in North Carolina, and improvements in average annual water yield of 305 mm in New England and 114 mm in the Sierra Nevada foothills, California, have been documented by Hornbeck *et al.* (1970) and Lewis (1968). In Arizona, Hibbert (1971) has discussed the hydrological implications of converting chaparral to grass and shown that this may increase annual runoff by amounts varying from less than 50 mm to more than 300 mm depending on the seasonal distribution of rainfall. Conversely, a reduction in streamflow of 94 mm per year has been noted by Swank and Miner (1968) as a result of increased winter interception loss following the conversion of a hardwood cover to white pine, and McGuinness and Harrold (1971) have stated that the low flows from a catchment near Coshocton,

Ohio, have been further reduced after reforestation. The water yield from melting snow may also be improved by timber management, as shown by Goodell (1959) and Swanson (1970).

Despite the fact that ample evidence now exists for the increase of streamflow by timber harvesting in mid-latitude areas, few studies have considered the application of this knowledge to the water supply industry. However, Law (1956 and 1957), working in the English Pennines, estimated that the establishment of coniferous plantations over one water supply catchment represented a reduction in reservoir yield of more than one-third and claimed that, in terms of providing additional catchment and reservoir capacity, the capital cost of replacing the water lost by afforestation was at least £300 000 or some £500 per hectare. Hawkins (1969) has examined the possible effect of changes in streamflow on reservoir yield and stressed that, although an extra quantity of streamflow is best utilized with increasing storage availability downstream, a change in the timing of discharge shows a maximum increment at lower storage capacities.

Finally, it should be admitted that few of these studies have paid any attention to the wider ecological repercussions associated with deforestation, although a notable exception is that of Rich (1972) in Arizona, where most water is obtained from wildland watersheds. A conscious attempt is being made to manage the ecosystem for water, wood, wildlife, scenic value, and sediment control and, over a 5-year period, timber harvesting has produced an increase in annual runoff ranging from 10 to 35 mm without an attendant rise in sediment deposition.

Once artificial reservoir storage has been provided to smooth out the natural variations of the hydrological cycle, it is essential that optimum use is made of the water available. This is particularly true in the case of multi-purpose river-regulating reservoirs, since the often conflicting needs of the various river users require the evolution of precise but flexible operating rules for the control of reservoir releases. It has been suggested that for water supply purposes alone the wastage due to indifferent management techniques could amount to as much as 22 per cent of the total storage capacity of the reservoir. This wastage is largely the product of unnecessary discharges resulting in turn from an inability to forecast the behaviour of tributary streams in the catchment below the reservoir sluices (Jamieson, 1972). In order to optimize the operating procedures, increasing use is now being made of short-term rainfall forecasts, allied to the installation of radar and telemetering raingauges, as outlined by Goodhew and Jackson (1972).

## 5.5 The nature of floods

The broad economic and social importance of the flood hazard was outlined in chapter 1, and there can be little doubt of the real practical incentives which exist for a better understanding of all extreme hydrological events, including droughts. It is the civil engineer who is most immediately concerned with such problems and who has, therefore, a continuing interest in river behaviour, largely because riverflow offers the best opportunity for direct manipulations of the natural hydrological cycle. Wiesner (1964) has shown that, though the civil engineer employs hydrometeorology in both the design and operational phases of virtually all river-control projects, the chief application lies in flood estimation. Similarly, Wolf (1956) has noted that, in addition to assessing the physical risk, the hydraulic engineer must also consider the human implications of his structures and it is axiomatic that no works should ever endanger human life. On the other hand, inconvenience or damage to property often has to be tolerated if the average annual flood losses are estimated at less than the equivalent costs of the control project. For a comprehensive account of flood hydrology the reader is referred to works by Barrows (1948), Richards (1950), and Hoyt and Langbein (1955).

A flood may be defined as a discharge which exceeds the channel capacity of a river and then proceeds to spill onto the adjacent floodplain. Although many local hydrological factors relating to the drainage basin may influence floods, all the major flood-producing mechanisms have a hydrometeorological origin, which leads to an excess of precipitation over the prevailing infiltration capacity of the catchment surface. For our purposes, a three-fold classification of the atmospheric causes of floods can be recognized.

(1) *Rainfall excess.* Heavy rainfall is the most important cause of flooding on a world scale. This universal significance may be attributed mainly to the wide range

of both time and area over which rainfall excess operates, varying from the semi-predictable seasonal rains affecting areas of sub-continental extent to the highly episodic, intense storms which promote flash floods in small drainage basins.

Regular annual flooding from rainfall, such as occurs on the river Nile described by Hurst (1952), is principally a feature of low latitudes and results from extensive invasions of maritime tropical airmasses which provide much of the tropics and sub-tropics with a marked summer maximum of riverflow. Some rivers, such as the Congo which lies across the equator, may experience a double maxima of discharge, but most of these rivers, such as in monsoonal south-east Asia, have a characteristic cycle of alternate low water and flood seasons. Although the most widespread and prolonged rains tend to have a regular seasonal occurrence, they are also associated with shorter-term atmospheric events. For example, the unprecedentedly heavy rains which struck north Africa in September 1969 were due to a cyclonic storm originating over the central Mediterranean (Winstanley, 1970). Between 20 and 29 September, large areas recorded rainfall several times greater than the mean for the whole month, and in Tunisia and north-eastern Algeria almost 600 people were killed with a further 250 000 rendered homeless.

Prolonged heavy rains are frequently linked with the increasingly slow movement of hurricanes as they begin to stagnate on crossing land areas, and Pardé (1967) has described the intense rainfall lasting for more than three days when typhoon Gloria struck Taiwan in September 1963. During a 12-h period, the maximum riverflow was equivalent to between 91 and 95 per cent of the rainfall rate partly as a result of preceding rain saturating the soil.

The severity of flooding is not necessarily directly related either to the intensity or to the duration of the rainfall, and this is especially apparent in mid-latitude regions where the effect of isolated storms may depend a great deal on the antecedent weather conditions and the resulting infiltration conditions. If the infiltration capacity is low, then fairly modest storms may produce exceptional flooding. Thus, considerable damage was caused in the Canterbury area of New Zealand in July 1963 by relatively light but prolonged rainfall which thoroughly wetted the soil during a 6-day period (Rayner and Soons, 1965). More intense rainfall was responsible for heavy damage in a small area of Dallas, Texas, in April 1966 when a 5-h storm killed at least 7 people and created damage totalling \$2·5 million, of which \$1·3 million occurred in a drainage basin of 32 $km^2$. Although this storm had an average return period of less than 50 years, it generated the highest known floods in certain areas because antecedent rains had saturated the ground, according to Mills and Schroeder (1969).

Really exceptional flooding can occur when intense rainfall affects upland catchments with a low permeability. This happened on 15 August 1952, when a severe thunderstorm over Exmoor, England, produced a rainfall of 230 mm in 24 h (Bleasdale and Douglas, 1952). Rapid surface runoff ensued from a steep catchment already wetted by heavy rainfall during the preceding two weeks and Dobbie and Wolf (1953) estimated that, for periods, the rates of runoff were as high as, or even higher than, the rainfall intensity.

(2) *Snow and ice.* Winter accumulations of snow represent a potential flood hazard later in the year, when higher temperatures prevail, and violent spring floods are a recurring feature over large areas of eastern and northern Canada, northern Scandinavia, European Russia, and Siberia. The classic lowland snowmelt regime is found on the large Soviet rivers draining to the Arctic Ocean as a result of rapid simultaneous thawing across wide areas in May and June. Peak discharge occurs in June and on the lower Yenisey at Igarka, for example, the mean flow in June is 78 000 $m^3/s$, which is exceeded only on the Amazon (Beckinsale, 1969).

Damaging floods due to thawing conditions also occur much further south within the mid-latitudes, and Paulhus (1971) has recorded such an event during early 1969 in the Red River of the North and Mississippi-Missouri basins, when 9 lives were lost and losses of \$151 million sustained despite the operation of existing flood-control projects, flood forecasting, and federally sponsored aid, which prevented additional losses estimated at \$197 million. Under such conditions, severe land erosion can occur and Waananen *et al.* (1971) have indicated that a ground thaw which preceded high runoff in the states of Idaho, Washington, and Oregon during the 1964–65 winter produced suspended sediment concentrations ranging between 220 000 and 360 000 ppm in some streams.

Where the winters are cold enough to freeze the rivers to appreciable depths, as in the northern parts of Eurasia and North America, the spring snowmelt floods may be increased by the presence of ice-jams. These occur when thick ice floes, formed by rapid thawing and the mechanical disruption of the ice sheet, are floated downstream and then become locked together at constrictions in the river channel to create an effective dam for the meltwaters. Flooding can result either upstream as the level of the backwater rises or downstream when the ice and impounded waters are suddenly released. An example of the latter type of flooding occurred in the spring of 1952 on the Missouri river in North and South Dakota. Warm weather upstream in Montana led to the early breakup of the tributaries, which moved ice downstream to jam against the solidly frozen Missouri. Eventually, the ice gorge gave way and at Bismark, North Dakota, the flow of the Missouri increased from 75 000 to 500 000 cusecs within a few hours on 7 April (Hoyt and Langbein, 1955).

On the other hand, the worst snowmelt flooding is often associated with multiple causes, usually involving rainfall, rather than with temperature increases solely due to local changes in the heat budget. Thus, Kupczyk (1968) has examined the synoptic situations leading to meltwater floods on mountain streams in southern Poland and found that most were due to advection linked with a south-westerly cyclonic circulation rather than to large radiation inputs. The Rumanian floods of May 1970, which devastated the Transylvanian Basin, provide a good example of rain-on-snow flooding (Doneaud, 1971 and Boucher, 1972). The immediate cause was heavy rainfall on 12 to 14 May from a depression which was further deepened by the sharp temperature contrast with polar air to the north-west and drew in a warm, humid airmass from the sub-tropical Atlantic. The resulting snowmelt from the mountains combined with the precipitation to produce the worst natural disaster in Rumania during recorded history.

In the spring of 1965, floods in the upper Mississippi basin claimed 15 lives with the loss of over 11 000 houses and damage estimated at $160 million (Anderson and Burmeister, 1970). In this instance, flood peaks exceeded existing maxima at many sites owing to a complex interaction of rapid winter snowmelt, heavy rain on the snowpack, and the presence of deeply frozen ground making the soil almost impervious. The behaviour of the snowpack may be critical, and Wolf (1952) has described how a snow blanket some 305 mm thick in Glen Cannich, Scotland, absorbed heavy rainfall for an initial period of three hours before collapsing to give a flood peak 35 per cent higher than expected from the accumulated water.

(3) *Coastal factors*. Apart from the special category of inland floods caused by earth movements such as earthquakes, with landslides into lakes or reservoirs together with the collapse of dam structures, the only major type of flooding which occurs in the absence of precipitation is restricted to low-lying coastal areas. Such flooding of coasts or estuaries is invariably associated with strong onshore winds which help to press the sea over or through any coastal defences. On a world scale, the most dangerous situation is undoubtedly a low, enclosed coastline within hurricane latitudes, and the Bay of Bengal and the Gulf Coast of the US represent especially vulnerable areas. In Bangladesh perhaps as many as 300 000 people died in a hurricane-induced tidal surge on 13 November 1970. Altogether, over 2 million people were affected over an area of almost 7500 km$^2$ and the financial losses approached £50 million.

Similar disasters can occur with deep depressions in the mid-latitudes, such as on 31 January 1953 when the strongest northerly gale on record was responsible for a tidal surge in excess of 2·5 m in some areas around the North Sea (Douglas, 1953). According to Lane (1966), 1835 people were killed in Holland with a further 72 000 evacuated from their houses. More than 3000 houses were destroyed and 9 per cent of the country's agricultural land was flooded with a total cost placed at £150 million. In neighbouring Belgium, 22 lives were lost and damage was estimated at £20 million; 300 deaths were also recorded in eastern England (Steers, 1953).

Occasionally, estuarine areas may suffer floods from the cumulative effect of both inland and coastal factors. For example, Brooks and Glasspoole (1928) have shown how the lower Thames valley flood of 7 January 1928 was caused by a remarkable combination of high river discharges inland, arising from both heavy rain and snowmelt, together with the impedance of the flood crest's passage downstream of London by a spring tide which was almost 1·8 m higher than expected owing to the presence of a northerly gale.

## 5.6 Flood prediction and forecasting

Many techniques are used to estimate floods and, for a comprehensive review, the reader is referred to Wolf (1966). From a hydrometeorological viewpoint, however, a distinction may be made between *flood prediction* methods, which rely on largely statistical analyses of long-period precipitation data, and *flood forecasting* operations which are more concerned with the much shorter-term interpretation of specific storms as they arise.

(1) *Flood prediction*. It will be apparent that exceptional rainfall is the immediate cause of most floods and Reich (1970), for example, has studied the relationship between the extreme value statistics of rainfall and floods. For areas with a reliable run of data extending over many years, statistical methods may be used to derive functional relationships between extreme values and the observed probabilities or return periods for such events. The concept of frequency occurrence is basic to all flood studies because, without some notion of the probability of a stated risk, the flood hazard cannot be fully appreciated and engineering structures can neither be designed nor operated safely and economically. Some of the special problems associated with the statistical analysis of floods and droughts have been presented by Gumbel (1958).

In detail, the characteristics of storm rainfall are expressed in terms of frequency, intensity and areal extent. Figure 5.8 shows a rainfall-frequency graph for various storm durations at Cleveland, Ohio, from 1902 to 1947 (Linsley and Franzini, 1955). The straight-line relationships permit some extrapolation and the graph indicates the average time-period within which a rainfall of specified amount can be expected to occur once. Rainfall intensity, which is simply the amount divided by the storm duration, can be similarly expressed and Fig. 5.9 illustrates the frequency analysis of observed rainfall-intensity data for specified durations at Toronto, Ontario (Bruce and Clark, 1966). Since precipitation intensity increases markedly in the short-duration storms, Hershfield and Wilson (1957) have developed methods for estimating short-duration rainfall from larger observations over the US, based on mean annual precipitation and the numbers of precipitation and thunderstorm days per year. Similarly, Reich (1963), starting from work in South Africa, has been able to derive information on short-duration rainfall intensity for developing areas lacking autographic rainfall records and has proposed a tentative isopluvial map of 2-year, 1-h maximum precipitation for the world. Huff (1967) has examined the

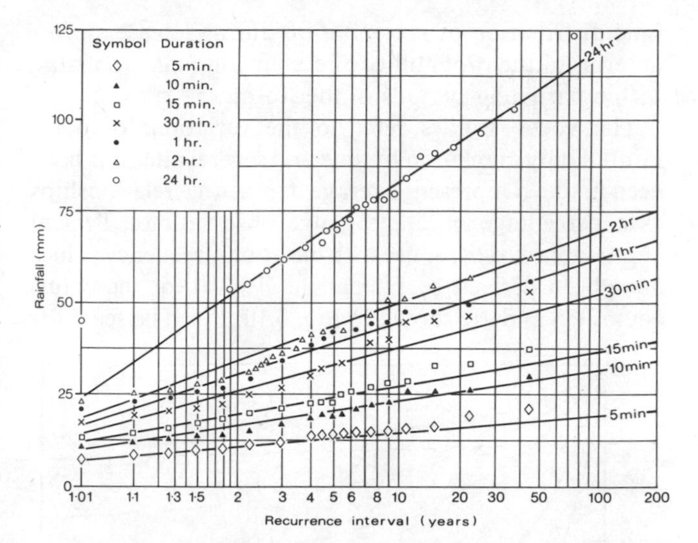


**Fig. 5.8 Rainfall-frequency graph for different storm durations at Cleveland, Ohio, 1902–47. From *Elements of hydraulic engineering* by Linsley, R. K. and Franzini, J. B., McGraw-Hill, New York, 1955. Used with permission of McGraw-Hill Book Company.**

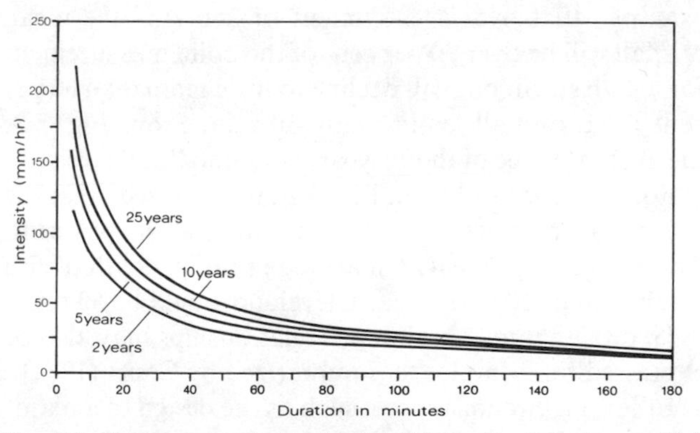


**Fig. 5.9 Frequency analysis of rainfall intensity data for specified durations at Toronto, Ontario. Reprinted with permission from Bruce, J. P. and Clark, R. H. *Introduction to hydrometeorology* (1966), Pergamon Press Ltd.**

time distribution of precipitation during intense storms in terms of the probability of certain amounts occurring within particular periods of the storm.

The above studies refer to the variations of point rainfall only in relation to time and several attempts have been made to represent average depth–area relationships as a percentage of single gauge observations. Typical depth–area curves for use with duration–frequency values have been derived by Hershfield (1961a) for the continental US and are shown in Fig. 5.10. It can be seen, for

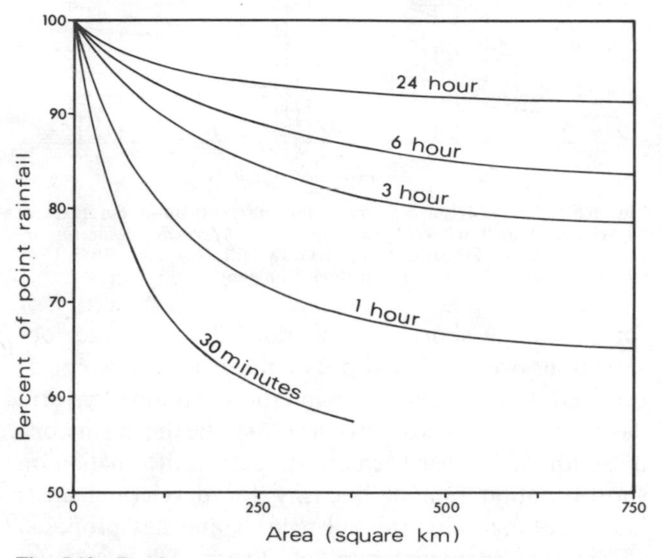


**Fig. 5.10 Depth–area curves for use with duration–frequency values for the continental US. After Hershfield (1961a).**

example, that over a catchment of 260 km$^2$ the mean rainfall will be over 90 per cent of the point measurement for a 24-h storm but will decline to little more than 60 per cent for a rainfall lasting only 30 min. From Fig. 5.8, the average value of the 20-year, 1-h rainfall at Cleveland, Ohio, is 50 mm, and from Fig. 5.10 it can be seen that the 1-h curve intersects the 200 km$^2$ area at 75 per cent. Therefore, the 20 year, 1-h average rainfall depth over a catchment of 200 km$^2$ near Cleveland would be 50 times 0·75, or 37·5 mm. Areal-depth relationships may also be expressed by formulae as summarized by Court (1961).

In certain circumstances, such as the design of a major dam, it may be desirable to know the magnitude of the largest flood that can physically occur. Such an event, known as the *probable maximum flood* (PMF) represents the upper limit of flooding for a stated catchment size which the existing climatic regime can produce and will result from the maximum possible combination of precipitation and snowmelt with minimum losses. Statistically, the probability of such a maximum is zero and the word 'probable' is used to convey some uncertainty of estimation rather than to indicate some finite frequency. Not surprisingly, the concept of the probable maximum flood has given rise to the idea of *probable maximum precipitation* (PMP). It is well known that precipitation intensity differs with synoptic type, as illustrated for two different areas of Britain by Lowndes (1968, 1969) and Matthews (1972), and Wiesner (1970) has accordingly made an important distinction between the PMP and the probable maximum storm. Thus, the PMP for areas of less than 260 km$^2$ and durations less than 6 h is likely to come from a thunderstorm, whereas frontal storms or tropical cyclones will usually be involved for larger areas and longer durations. Consequently, the PMP is the envelope to the depth–duration and depth–area curves of the probable maximum storms.

The hydrometeorological approaches to probable maximum flood estimation have been detailed by Wiesner (1970). One method involves the computation of PMP using storm models, which express rates of precipitation in terms of physical parameters such as surface dew point, height of convective cell and inflow and outflow of moist air. The storm model formulae are then tested against observed storm rainfalls and a statistical analysis of the meteorological factors within the equation enables the substitution of the extreme atmospheric values required to produce the PMP. In general, these models rely heavily on our existing knowledge of the meteorological processes leading to the uplift of moist airmasses and conform to either the upglide model, which describes forced ascent over a mountain barrier or frontal surface, or the convergence model, which relates to the dominantly convective mechanisms in thunderstorms and tropical hurricanes.

Such storm models appear to give reasonably appropriate results for long-duration storms operating over fairly wide areas, where the uplift processes are more easily simplified from both surface and upper-air data, although Pullen (1968), for example, has demonstrated a means of computing the depth of precipitable water during periods of storm rainfall in South Africa from surface-level climatological information. On the other hand, storm models are less suitable for short-period events and may produce rainfall estimates so far in excess of observed values that they are rejected as being unrealistic by some engineers (Wolf, 1966). As a consequence, more emphasis has been placed on the study of actual storms and the maximization of these observed events to their probable physical limits. This approach avoids many of the theoretical difficulties associated with storm models and depends largely for its success on the previous occurrence of enough major storms within the area in question, although relevant storms from adjacent areas, which are both climatically and topographically homogenous, can be maximized so that the precipitation results may be transposed to another drainage basin.

An interesting application of the storm-maximization technique has been reported by Binnie and Mansell-Moullin (1966) in connection with a PMF study of the Jhelum river in Pakistan, where the construction of a large dam at Mangla made it necessary to estimate the maximum spillway design flood to ensure that the impounding structure was not overtopped. In this area, severe flooding is most likely with storm rainfall during the summer monsoon season, together with a snowmelt contribution from the Himalayas. Storm transposition could not be employed because of the singular topographic characteristics of the Jhelum catchment but each major historic storm was maximized using a moisture-inflow index, which consisted of the product of two factors, namely a moisture factor representing the water vapour content of the airmass and a wind factor representing the rate at which moist air was transported into the storm area. Altogether, 14 observed storms were maximized to produce an estimate of the PMF from a 72-h storm. The resulting hydrograph, composed of runoff from the storm rainfall over both upper and lower parts of the main catchment, plus much smaller contributions from baseflow (including snowmelt) and the runoff from an antecedent major storm, is shown in Fig. 5.11. The flood of 4–7 August 1958 was selected to model the discharge from the antecedent storm, which was assessed to occur 72 h before the start of the probable maximum storm. It can be seen that the peak discharge is 73,620 cumecs and the 9-day flow volume equivalent to 7·65 million acre-feet.

The problem of PMP estimation has alternatively been approached through statistical analysis by workers such as Lockwood (1967) for Malaya. Not surprisingly, it is difficult to suggest meaningful return periods for PMP, although Lockwood (1968) has claimed that they probably vary between $10^4$ and $10^8$ years. Hershfield (1961b and 1965) has proposed a method based on the formula:

$$X_T = \overline{X}_n + KS_n$$

where $X_T$ = the rainfall for a return period $T$ years when a particular extreme value distribution is used, $\overline{X}_n$ and $S_n$ = the mean and standard deviation for a series of $N$ annual maxima and $K$ = the standardized variate. If, for a particular site, the observed maximum rainfall $X_m$ replaced $X_T$ and $K_m$ replaced $K$, then $K_m$ is the number of standard deviations which must be added to the mean to obtain $X_m$. From a study of 24-h maxima in the US, the enveloping value of $K_m$ was found to be 15. This method has been used in Alberta by McKay (1968) and in Iran by Gordon and Lockwood (1970). In the latter case, Hershfield's technique gave estimates of 170 mm for the 1-day PMP at Tehran compared with 75 mm and a return period of 1000 years for the Gumbel method.

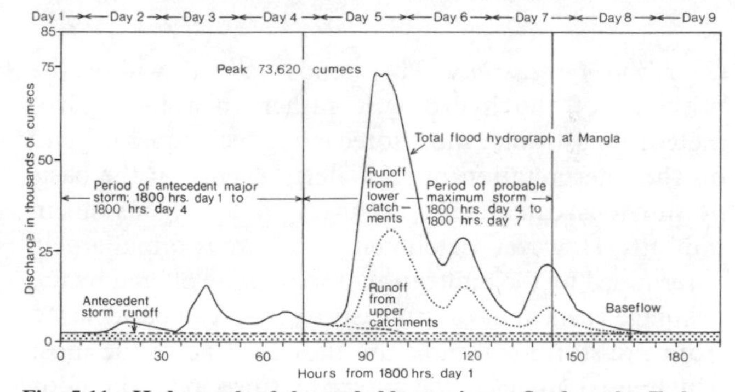


**Fig. 5.11 Hydrograph of the probable maximum flood on the Jhelum River at Mangla, Pakistan. The composite hydrograph includes contributions from baseflow, antecedent storm runoff, and runoff from both the upper and lower catchments. After Binnie and Mansell-Moullin (1966).**

It should be appreciated, of course, that all observed maximum falls become subject to revision as time progresses. Paulhus (1965) indicated how rainfall during March 1952 at Cilaos, La Reunion island, set new world records from 24 h up to 8 days with falls varying from

1870 mm to 4130 mm respectively. The magnitude-duration relationships for the world's greatest observed point rainfalls as at 1965 are shown in Fig. 5.12, the position of the envelopment line being expressed by the equation $P = 16{\cdot}6D^{0{\cdot}475}$, where $D$ is the duration in hours and $P$ the precipitation in inches over the duration $D$. According to Wiesner (1970), such values are unsuitable for maximization, as they frequently reflect highly localized conditions which do not permit easy extrapolation.

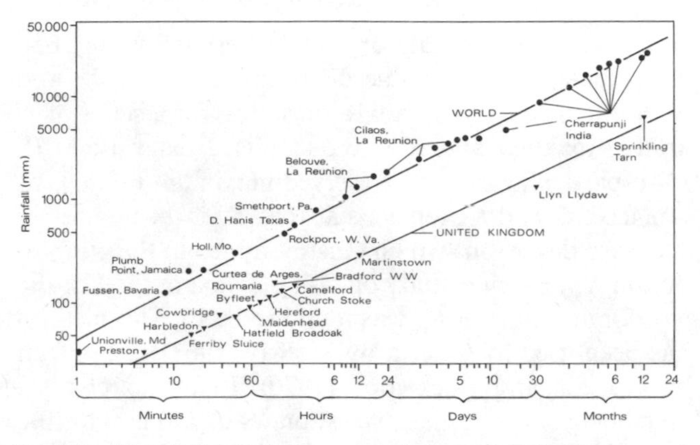


**Fig. 5.12 The relationship between magnitude and duration for observed world rainfall maxima compared with extreme rainfalls in the UK. After Rodda (1970b).**

(2) *Flood forecasting*. This tends to fall within the province of the hydrologist rather than the hydrometeorologist, since most forecasting techniques depend on the interpretation of short-term events on the basis of previous catchment responses to analogous storm rainfalls. However, catchment response is often largely determined by the infiltration opportunity offered by the drainage basin and some measure of this can be achieved from pre-storm soil-moisture indices. One of the most widely used guides is the antecedent precipitation index (API), which may be calculated for a number of days before the storm from the equation:

$$\text{API} = K_1P_1 + K_2P_2 + K_3P_3 + \ldots + K_nP_n$$

where $P_1$ is the precipitation the day before day 0, $P_2$ is precipitation two days before, $P_n$ is precipitation $n$ days before and $K$ is a constant $< 1$. The reduction constant is obtained through trial and error by relating various indices to a plot of rainfall minus runoff over the catchment.

The calculation of soil moisture deficit (SMD) has been described by Grindley (1967) and is based on the water balance method, whereby the difference between potential evapotranspiration and rainfall is accumulated progressively through the year. Goodhew (1970) considered that in Britain the API is the better indicator of potential flood-risk during winter because the SMD is always close to zero owing to low evapotranspiration rates. On the other hand, the SMD is more useful in summer as a rough approximation to the proportion of rainfall which will run off from a particular storm, provided that the rainfall intensity does not greatly exceed the infiltration capacity of the ground.

## 5.7 Human response to floods

Floods represent a social and economic problem simply because of man's persistent determination to settle in flood plain locations. This affinity for flood plains, justified by various factors such as the availability of level land for building and the presence of alluvium or water for agricultural or industrial purposes respectively, has been a feature throughout recorded history and, at the present time, it is estimated that between 10 and 15 per cent of the population of North America lives in areas subject to periodic flooding (White *et al.*, 1958 and Sewell, 1965). For many years, flooding has been recognized as being as much an economic problem as a hydrological one (White, 1939) and in the US, where basic flood-control legislation began in 1936, more than $7 billion have been spent since that date in an attempt to alleviate the hazard. Nevertheless, the cost of flood damage continues to increase. This rise in damage potential is partly due to purely economic processes, such as price inflation and higher living standards, but the continuing, and even accelerating, invasion of flood plains is an important explanation.

White *et al.* (1958) suggested that people occupy flood plains either because they are unaware of the real risks involved, perhaps as a result of an overestimation of the protection afforded by an existing flood-reduction programme, or because of their ability to transfer at least part of the incurred costs to other members of the community, such as the taxpayer or the subscriber to relief appeals. Certainly, Kates (1962) found that an accurate appreciation of the flood hazard existed only with the people who had experienced a direct and fairly recent flood event themselves, and there is also evidence that the problem of perception arises similarly in connection with the selection of the possible steps which may be taken either to alleviate floods or otherwise adjust to them.

For example, White (1964) indicated no less than eight possible adjustments to floods from which managers of flood-plain property are theoretically able to select an appropriate response. These alternatives included the modification of building structures to repel floodwaters, flood proofing, the regulation of flood plain land use, and flood insurance but it was shown that in the US, as elsewhere, the traditional choice is either simply to accept the loss when it occurs or to press the appropriate authorities to construct engineering works designed to eliminate the hazard. Engineering structures, however, often provide less than total protection, especially against the catastrophic flood, and Sewell (1969), in a case study of the Canadian Fraser, has drawn attention to the dangers of over-reliance on such measures. A further difficulty is that large-scale, integrated schemes for an entire drainage basin may take many years to complete and the implementation of the Fraser River Board's proposed scheme, planned in response to a disastrous flood in 1948, has been delayed by administrative problems, the pre-emption of the water-power market by other projects under construction in British Columbia and opposition from recreation interests.

Not surprisingly, much of the recent work on the human response to floods has concentrated on the adoption of some of the lesser-used alternatives. Krutilla (1966), for example, has emphasized that it is uneconomic to provide protection against the disastrous but infrequent flood, not least because the contribution of such a flood to the average annual damage is small, since the loss per occurrence is multiplied by the probability of occurrence in any one year, i.e., $P = 0{\cdot}001$ for the 1000-year flood. He therefore suggested a compulsory flood-loss insurance scheme as a means of achieving an efficient use of flood plains with premiums proportional to the risk involved and equal to both the private and social cost of flood-plain occupance. Alternatively, Day *et al.* (1969) have assumed that flood damage is a function of water depth (stage) and time and have attempted to evaluate the benefits to be derived from improved flood warning systems in Pennsylvania. They found that the reduction in damages due to adequate warning coupled with either evacuation or temporary flood proofing amounted to about one-third of the total residential loss, and Fig. 5.13 illustrates the various financial savings which can be achieved in a typical supermarket in relation to the length of warning given and the height of the flood. These estimates exclude the

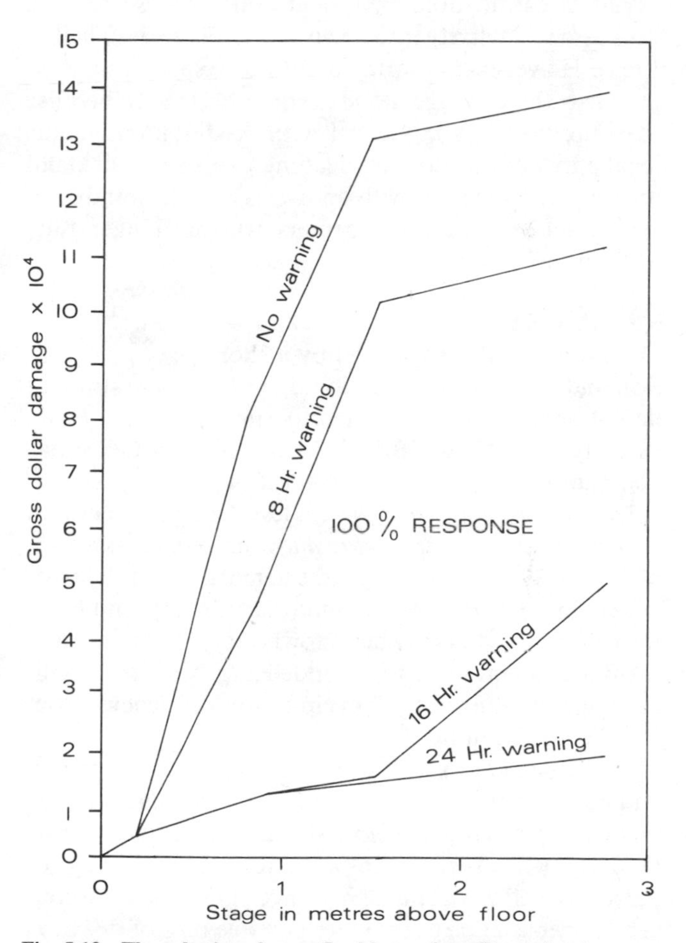


**Fig. 5.13 The reduction of gross flood losses for a Pennsylvania supermarket in relation to the height of the flood and the length of flood warning provided. It is assumed that 100 per cent efficiency is achieved in all necessary actions such as the raising of goods above the height of the flood crest. After Day *et al.* (1969). Copyright by American Geophysical Union.**

cost of providing flood-warning services, flood-plain evacuation, and re-occupation expenses and assume that a 100 per cent efficient response to the flood warning is made, whether the event occurs during the day or the night.

All cost-benefit analyses of adjustments to floods are beset by complications arising from the large number of peripheral influences and, in some cases, factors quite unrelated to flooding may become dominant. Thus, it was expected that, as a result of increased flood protection provided on Coon Creek, Wisconsin, there would be an intensification of agricultural land use and an expansion of activity into newly workable areas (Theiler, 1969). However, the farmers did not respond as anticipated to the changed flood frequency, largely because the improvements coincided with modifications to the local agricultural economy leading to a smaller demand for cropland together with an increase in the number of retired and semi-retired farmers who no longer fully utilized their land.

## 5.8 Droughts

Droughts are always started by a shortage of precipitation and many of the traditional approaches to drought definition have been restricted to rainfall analysis. From as early as 1887, the British Meteorological Office has distinguished between an *absolute drought*, or a period of at least 15 consecutive days with less than 0·2 mm of rain on any day, and a *partial drought* which is seen as a spell of at least 29 consecutive days, the mean daily rainfall of which does not exceed 0·2 mm (Skeat, 1961). Similarly, in the US, Hoyt (1942) has shown that the early assessment of drought severity depended largely on the magnitude and areal extent of precipitation deficiencies from mean annual conditions.

This type of approach will always retain some validity and recent improvements in statistical methods have even tended to place a new emphasis on rainfall studies, particularly with respect to a better understanding of persistence effects. Thus, Lawrence (1957) used the concept of persistence to estimate possible runs of dry days during the summer months in southern England, and Fitzpatrick and Krishnan (1967) and Watterson and Legg (1967) have employed Markov models for the investigation of the tendency for observed long dry periods to exceed the drought spells predicted by theory in Australia. On the other hand, Wallen (1967) has reviewed three basic methods of defining aridity and studied their applicability in relation to Middle Eastern agriculture, and concluded that the water balance approach was probably the most realistic. Certainly, Rodda (1965) has claimed that soil moisture content is the best hydrological criterion of drought, following an analysis of the incidence of absolute drought compared with the occurrence of soil moisture deficits in excess of 63 mm over a 21-year period at Oxford. It can be seen from Table 5.1 that these two methods have different seasonal phasings, with most rain-free days occurring in May and June as opposed to a late summer and early autumn incidence for the soil moisture deficits. Therefore, since soil moisture deficits have to be built up before rainfall deficiencies have any appreciable impact on water resource availability through the depletion of riverflow and the lowering of groundwater levels, it would appear that the water balance technique is more appropriate.

| | *J* | *F* | *M* | *A* | *M* | *J* | *J* | *A* | *S* | *O* | *N* | *D* |
|---|---|---|---|---|---|---|---|---|---|---|---|---|
| SMD | 0 | 0 | 0 | 1 | 5 | 16 | 22 | 21 | 22 | 15 | 4 | 0 |
| Abs. drought | 0 | 2 | 50 | 57 | 68 | 69 | 50 | 62 | 39 | 27 | 16 | 0 |

**Table 5.1** Average number of days with SMD > 63 mm compared with total number of days of absolute drought at Oxford, 1920–40. (After Rodda, 1965.)

According to Tannehill (1947), drought is unique among atmospheric and other environmental hazards in that it creeps up gradually on the afflicted area and consequently creates problems of recognition and perception. In Australia, the repercussions of drought tend to be most serious for the rural economy and here drought has to affect at least 10 per cent of the continent before it is generally reported. Heathcote (1969) has drawn attention to the difficulties involved in estimating economic losses due to drought, ranging from reduced crop and fodder growth, livestock deaths, losses in water transport and hydroelectricity production, to the cost of emergency relief measures and longer-term rehabilitation. Nevertheless, it has been calculated that the aggregate toll of drought on Australian primary production between 1900 and 1966 was about $1600 million,

| | County | Adams | Barber | Frontier | Finney | Cimarron | Kiowa |
|---|---|---|---|---|---|---|---|
| | Farmers' estimate of drought (yr/100) | 17 | 16 | 19·9 | 18·6 | 34·8 | 34·9 |
| Based on the Palmer Index | % Time drought | 42·4 | 46·9 | 41·6 | 47·2 | 48·7 | 47·2 |
| | % Mild drought and more severe | 32·8 | 39·6 | 32·0 | 37·0 | 39·8 | 34·8 |
| | % Moderate drought and more severe | 23·6 | 26·8 | 20·8 | 26·6 | 30·8 | 24·4 |
| | % Severe and extreme drought | 15·7 | 13·8 | 11·2 | 15·4 | 18·4 | 13·4 |

**Table 5.2** Comparison of farmers' estimate of drought frequency with actual drought conditions in the Great Plains. (After Saarinen, 1966.)

and that the drought of 1943–46 alone caused a loss of 26 million sheep (Anon., 1966). As in the case of the flood hazard, there is very often an imperfect awareness of the nature of the risk. Data from Saarinen (1966), presented in Table 5.2, indicate that in the Great Plains area of the US the farmer progressively underestimates the hazard as the risk becomes less probable. Thus, in the more humid areas, such as Adams and Barber Counties, the farmers' estimates of drought correlate best with extreme drought conditions on the Palmer Index, whereas in the drier areas (Frontier, Finney, and Cimarron Counties) the best relationship is with moderate and more severe droughts.

The Great Plains of North America provide a convenient example of an important and normally productive area which is subject to recurrent drought. These extensive grasslands stretch from Canada down to Mexico and are appreciably drier than their surroundings, owing to small and highly variable precipitation totals combined with the fact that most rainfall occurs in the summer months when evaporative demands are high. In a study of the Canadian dry belt of southern Alberta, Saskatchewan, and Manitoba, Villimow (1956) concluded that much of the total aridity was due to the area experiencing more anticyclones than regions to the north or south, and the position of the jet stream in summer nearly 1000 km to the south placed the dry belt in a vast area of subsidence which exists up to 3000 m in the atmosphere.

Laycock (1960) has applied water balance methods to the mapping of drought patterns in the Canadian prairies and, by assuming that crop yields became marginal with soil moisture deficits of 200 mm, found that all areas suffered some crop failure in at least one year during 1921–50. Figure 5.14 shows that the regional drought pattern varies considerably from year to year. In 1927, most of the southern areas, which are normally dry, were moist compared with the Peace River District to the north, whereas in 1936 a large part of the southern plains had severe drought compared with moist conditions further north.

Similar features exist in the US, where Thomas (1962) has identified a rhythmic sequence of four major drought periods. The midpoints of these severe dry spells occurred in 1892, 1912, 1934, and 1953, with a mean interval between the midpoints of 21 years and a deviation from the mean of 1–2 years. Borchert (1971) has drawn attention to important differences in the nature and incidence of these droughts. In the 1890s and the 1930s, virtually all of the Great Plains was affected and the droughts were accompanied by strong westerly circulations and a large number of frontal passages but little moist air. On the other hand, the droughts of the 1910s and 1950s were much more severe over the southern Plains and, although more moist air was available, there were few storms to release the moisture.

Through the work of Namias (1956 and 1960) and others, it is now generally accepted that severe droughts arise as a result of apparently chance variations of the atmospheric circulation. For example, Namias (1966) has shown that the rainfall deficiency, which affected the northeastern US mainly in spring and summer between 1962 and 1965, brought abundant precipitation to the far southwest and the northern Plains. The evidence suggests that this precipitation pattern, together with below-normal temperatures at all seasons, was associated with an anomalously deep mid-tropospheric trough just off the Atlantic seaboard, which gave a greater northerly component than usual and thus produced low temperatures in addition to the subsidence and dryness.

Borchert (1971) has claimed that results obtained by Dzerdzeevskii (1969) indicate that at least two of the major grassland droughts can be related to changes in the relative frequency of meridional and zonal air circulations in the northern hemisphere. As illustrated in Fig. 5.15, the coincidence appears closest with the persistence of each circulation type so that, when zonal patterns were at their highest frequency, as in the 1930s, the effect of dry continental air was especially great over the Grassland. However, with the swing back to more meridional circulations from 1935 to 1955, the persistence of this pattern produced the marked effect of anticyclonic conditions during the 1950s.

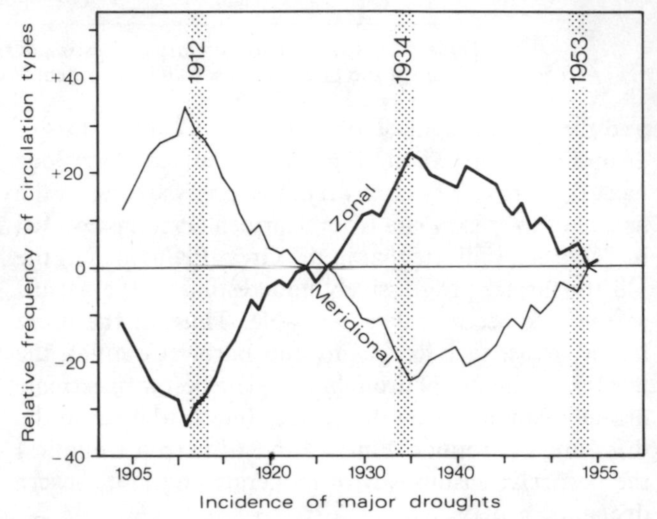


**Fig. 5.15 Relative dominance of meridional and zonal circulation types on the northern hemisphere daily weather maps from Dzerdzeevskii (1969) compared with major drought periods in the Great Plains and the southwestern US from Thomas (1962). After Borchert (1971). Reproduced by permission of the Association of American Geographers.**

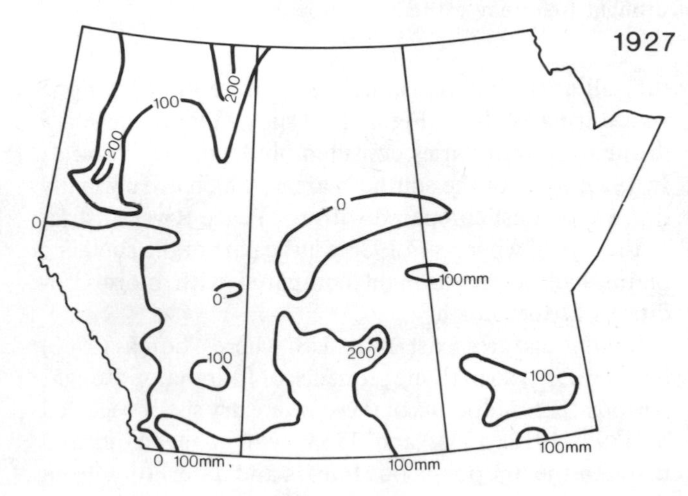

**Fig. 5.14 Regional patterns of drought in the Canadian Prairies in 1927 and 1936. The isolines represent soil moisture deficits in mm. After Laycock (1960).**

The output of American literature on drought can also be related to the incidence of rainfall deficiency over the Great Plains and there are numerous contemporary accounts of the 1930s drought (Henry, 1930; Chambers, 1935), including a description of the first great duststorm of the drought years which carried material as far east as New York (Miller, 1934). Much attention has also been devoted to the 1950s drought when sixteen States, comprising 18 per cent of the population living on 53 per cent of the US land area, were in receipt of drought assistance (Anon., 1958). In the three years 1954 to 1956, the Department of Agriculture spent $550 million on disaster relief plus a further $184 million to bolster livestock prices. In the State of Kansas, the average cost (at 1951 prices) of floods was $35 million per year, whereas the droughts of the 1930s and 1950s resulted in an average annual loss for the same period of $75 million also at 1951 prices, and excluding wind and soil erosion losses. Thus, over a period of about 28 years, drought losses were more than double those from floods. In the decade following the 1950s drought, farm mortgage and short-term debts rose by 130 per cent whereas gross farm income increased by only 33 per cent.

Borchert (1971) has claimed that, since the 1930s, severe droughts have accelerated basic contemporary changes taking place throughout the US with a more rapid trend to fewer, larger farms, more public controls and subsidies and greater consolidation of urban services and businesses within the Great Plains area. He further predicts that the next major drought could initiate further changes in the settlement pattern and provoke another urban water supply crisis.

## Fuel and power climatology

### 5.9 Climatic power sources

Atmospheric energy has been directly harnessed from the sun, winds, and moving water. Like the other sources, *solar energy* is ubiquitous, but is of low intensity and Tabor (1962) estimated that a central power station of about 50 000 kW installed capacity would require solar energy collectors of 2·5 km$^2$ covering a land area probably three times as large. This means that solar energy is most suitable for small-scale applications, such as domestic cooking, air conditioning, or refrigeration, where the energy can be utilized in the form of heat by absorption on black surfaces or through photonic utilization as light for photoelectric and photochemical conversion.

Nevertheless, Tabor (1958) has suggested that in areas of reliably high sunshine incidence not connected to existing electricity supply networks, such as the subtropical deserts, the prospects for small solar package developments are quite good. Probably the major difficulty associated with the widespread use of solar energy is that of heat storage, since short-term storage is necessary to cover both night-time periods and short spells of cloudy weather, whereas the entire outlook for solar heating would be transformed if long-term storage from summer to winter ever became practical.

*Wind energy* has a long history of use by sailing ships and windmills and, although less reliable than solar power, it is still more regular in many parts of the world than rainfall, on which all water power ultimately depends. Unfortunately, wind energy is comparable to solar power in that it is a random source of energy, which cannot be stored directly like water in a reservoir and, even when converted into electricity, conventional storage in batteries is very expensive (Golding, 1960). Another disadvantage, similar to the low intensity of solar energy, is the low density of air itself which means that large volumes are necessary for a stated output compared with the volume of water required for a hydroelectric turbine of the same power capacity. The icing of aerogenerators in cold climates and salt spray corrosion near coasts are further problems.

To ensure the most economical operation, cheap machines should be installed at the windiest possible sites where the cube law for wind power produces marked improvements in output. For example, an aerogenerator may have a specific output of 2700 kWh per annum per kilowatt with a mean annual windspeed of 13 kt, although on an adjacent hilltop the average windspeed could be 21 kt producing 5300 kWh per annum per kilowatt. According to Golding (1962), wind power has considerable potential in the arid areas of developing countries and, despite the fact that most large plants are designed only to supplement the main power network, there is some evidence that, even in a country like Britain, operating costs may be below those at large thermal power stations.

*Water power* was originally harnessed directly by means of water-wheels but is now most widely used for the generation of hydroelectric power. The relationship of weather and climate to hydroelectricity begins with the amount of precipitation on a watershed and the subsequent runoff regime, as stressed by Nye (1965) in terms of the location of undertakings in the State of Victoria, Australia. The supply of hydropower is sensitive to low riverflows caused either by droughts or snow and ice storage in the catchment, and most development has taken place in the temperate region experiencing rainfall throughout the year, although about 60 per cent of the potential water power of the world lies within the tropics. Maritime areas, such as north-west Europe,

Japan, and New Zealand provide especially suitable climatic regimes, and in the British uplands, for example, there is a winter to summer streamflow ratio which coincides with the winter to summer ratio in electricity demand of 1·7:1 (Aitken, 1963).

Nevertheless, even in the most climatically favourable areas, there are usually few rivers which are large enough to support base-load stations and artificial reservoir storage is necessary to augment low flows. Thus, Arakawa (1959) has demonstrated that, although many reservoirs are provided in Japan to conserve water from one period of the year to provide hydropower during the drier seasons, there are additional standby steam-generated installations and provision is also made for regional transfers of energy to combat rainfall deficiencies in localized catchment areas.

The correlation between rainfall and generated power has encouraged experiments designed to increase rainfall by cloud seeding for hydroelectric purposes by the Pacific Gas and Electric Company of San Francisco, as reported by Eberly (1966) and Sewell (1966). On the basis that precipitation increases of 10 per cent and perhaps more can be achieved in the mountainous areas of the western US, where the Company's 67 hydroelectric power plants are located, it was found that, on large watersheds of approximately 1300 km$^2$, an increase of less than 2 per cent in average runoff might cover the cost of the cloud-seeding project, although a 10 per cent improvement might be required over much smaller areas. The chief economic benefits arise from the savings in fuel costs when thermal generation does not have to be employed, since the operating and maintenance costs of a hydro scheme are only about one-half the equivalent charges for a thermal system, and it was estimated that in a dry year a 10 per cent rainfall increase over a large catchment could be worth as much as $210 000 to the Company. The cost-benefit reduces with increasing river discharge above the level where all the water cannot be fully utilized.

## 5.10 Maintenance of power systems

Overhead power lines for the distribution of electricity are vulnerable to a number of weather hazards, including strong winds, which may snap supporting poles, especially if additional strain is introduced by the accumulation of ice, snow, or even wind-blown trees on the high-voltage lines. High humidities affect the efficiency of electrical transmission and the performance of insulators but a large proportion of supply interruptions are due to lightning, which was responsible for 39 per cent of the disturbances to the New England grid system in 1948 (Corey, 1949). According to Forrest (1950a and b), lightning was responsible for 1157 breakdowns on the 9500 km of transmission line of the British grid system from 1934 to 1947 and, as illustrated in Fig. 5.16, there

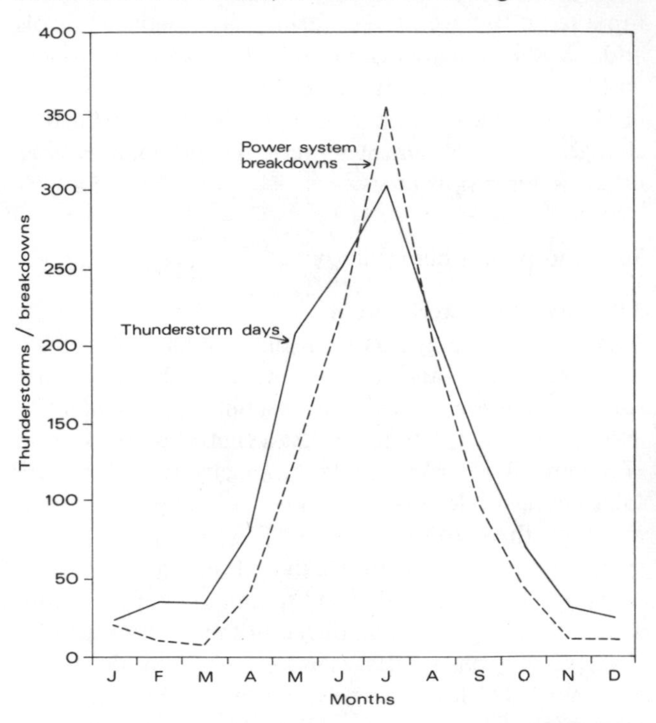


**Fig. 5.16 The relationship between power system breakdowns and thunderstorm days in Britain. Compiled from data in Forrest (1950a).**

is a marked seasonal correlation between the number of thunderstorm days and the power system breakdowns due to lightning. It can be seen that the breakdowns have a sharper peak in July than the thunderstorm-day incidence and this has been explained by Davis (1969) as the result of the ratio of thunderstorm-hours to thunderstorm-days being nearly 3 in July compared to only just over 2 in the remainder of the summer months. The detailed time-scale is important, as breakdowns have a daily maximum between 1600 and 1700 h GMT, which coincides exactly with the diurnal cycle of thunderstorm activity. During periods of risk, it is often necessary to reduce power transmissions to ensure that, if one line is tripped because of a lightning strike, another line cannot be overloaded and thereby initiate cascade tripping of a complete section of the grid (Davies, 1960).

When power is supplied from oil or natural gas wells, the supply system depends partly on the maintenance of the extraction operations in the areas where the deposits have been found. The world-wide extension of off-shore drilling operations during recent years has led to the need for highly specialized weather-forecasting services and Hibbert (1966) has described how oil companies in the Arabian Gulf and off the Nigerian coast have combined to provide an organized network of weather-reporting stations. Drilling for oil in the UK sector of the North Sea began in late 1964 and the high frequency of gales in the area has created special operational and forecasting problems. The rig operators are interested mainly in the accurate prediction of winds with the associated wave conditions, and since some procedures, such as the relocation of a drilling platform, can take a long time, the forecasters may have to make predictions for up to 72 h ahead. Some of the forecasting difficulties have been outlined by Ogden (1972) but it is encouraging to note that it has so far proved possible to extrapolate from known conditions to predict the most extreme local, wind-generated waves at latitude 57°N.

## 5.11 Weather and power consumption

The demand for fuel and power is very sensitive to changes in atmospheric conditions. One of the most clear-cut relationships is the inverse correlation which exists between space-heating requirements and air temperatures, as illustrated in Fig. 5.17 for weekdays during December 1964 at St Louis, Missouri (Turner, 1968). With the marked increase in gas space-heating appliances in Britain over recent years, the peak demands for this fuel have become increasingly temperature-dependent, although the approximate linear relationship is not close enough for forecasting purposes. Therefore, Berrisford (1965) has rejected the degree-day method and devised an exponential weighting model, which depends on the cumulative temperature experience of the immediate past and, because it allows time for human reaction to temperature changes, is a sensitive index of day-to-day fluctuations. On the annual time scale, Manley (1957a and b) explored the connection between coal consumption and temperature during the heating season in England and concluded that a drop of 30 per cent in the accumulated temperature deficit below normal in any one season, such as occurred several times between 1878 and 1895, would result in an additional requirement of 5–6 million tonnes of coal.

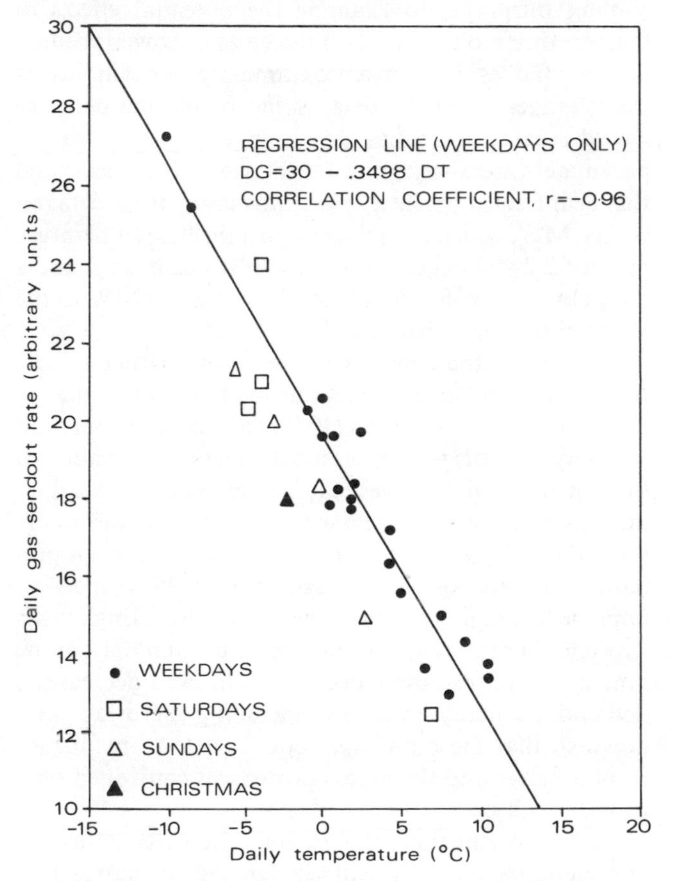


**Fig. 5.17 The correlation between the daily gas sendout rate and daily air temperature on weekdays during December 1964 at St Louis. After Turner (1968).**

Most investigations have concentrated on electrical power systems, because electricity cannot be stored on a large scale, and so accurate load-forecasting is essential for the most efficient operation. According to Dryar (1949), the weather component of the highest load demand in Philadelphia during December may vary from 40 000 kW to 140 000 kW and appears equally sensitive in other areas. Not surprisingly, air temperature is an important influence and Davies (1960) claimed that, around freezing point, a sustained fall of 0·5 degC can increase demand up to 160 MW in Britain.

Similarly, Johnson *et al.* (1969) have used data from 14 electricity generating companies in the American Mid-West, where peak loads occur in the summer for cooling purposes, to examine the potential effects of temperature modification on the costs of power production and the results from a programming model indicated that changes of 1·7–2·8 degC in mean daily temperature would create substantial economies. But other weather parameters are also significant. Davies (1960) has stated that near freezing point a 25 kt wind may increase demand by 700 MW, which is equivalent to a further temperature drop of 2·2–2·8 degC, compared with a calm day, and a dark cloud over the city of London adds 350 MW to the load for lighting purposes.

In practice, the complex influence of weather factors may be difficult to isolate from load variations due to other causes and Stephens (1951) has outlined a method whereby the effects of a seasonal trend on the habits of the consumers themselves may be eliminated. Sometimes the pattern of consumption within a community is totally unpredictable, as described by Corey (1949), when preparations were made during January 1925 to meet an unprecedented lighting load expected in New England as a result of a total eclipse of the sun. Unfortunately, at the time of the eclipse there occurred a marked decrease in demand as many premises, including factories, shut down so that the population could observe the phenomenon better and the excess power was controlled only with difficulty.

At the present time, the gas and electricity industries are placing increasingly stringent demands on the weather-forecasting services in Britain. For example, the gas industry requires regular 2-h forecasts of temperature for the first 38 h of each forecast period to facilitate accurate short-term estimates of demand and also needs a prediction of maximum and minimum temperatures for a further 2–3 days for an approximation of the longer-term situation. In order to be reasonably successful, the forecasts should be within ±2 degC according to Parrey (1972). For the electricity industry, the basic meteorological information is first processed into a number of working variables which express the 'effective' temperature, the cooling power of the wind, the amount of daylight illumination, and the rate of precipitation. The 'effective' temperature takes account of the time-lag of consumer response to temperature fluctuations due to thermal storage in the fabric of buildings and, for a detailed account of the demand-forecasting models in current use, the reader is referred to Barnett (1972).

## References


AITKEN, P. L. (1963). Hydro-electric power generation. *Conservation of water resources*. Inst. Civil Engrs., London, 34–42.

ANDERSON, D. B. and BURMEISTER, I. L. (1970). Floods of March–May 1965 in the Upper Mississippi River Basin. *U.S. Geol. Survey Water Supply Paper 1850-A*, 448 pp.

ANONYMOUS (1958). *A Report on Drought in the Great Plains and Southwest*. U.S. Govt. Print. Office, Washington, D.C., 45 pp.

ANONYMOUS (1966). Drought. *Current Affairs Bulletin*, **38**:51–64.

ARAKAWA, H. (1959). Hydroelectric power generation and the climate of Japan—a case of engineering meteorology. *Bull. Amer. Met. Soc.*, **40**:416–22.

BARNETT, C. V. (1972). Weather and the short-term forecasting of electricity demand. In Taylor, J. A. (ed.) *Weather forecasting for agriculture and industry*. David and Charles, Newton Abbot, 209–23.

BARROWS, H. K. (1948). *Floods: their hydrology and control*. McGraw-Hill, New York, 432 pp.

BARRY, R. G. (1969). The world hydrological cycle. In Chorley, R. J. (ed.) *Water, earth and man*. Methuen, London, 11–29.

BECKINSALE, R. P. (1969). River regimes. In Chorley, R. J. (ed.) *Water, earth and man*. Methuen, London, 455–71.

BERRISFORD, H. G. (1965). The relation between gas demand and temperature: a study in statistical demand forecasting. *Operational Res. Quart.*, **16**:229–46.

BINNIE, G. M. and MANSELL-MOULLIN, M. (1966). The estimated probable maximum storm and flood on the Jhelum river, a tributary of the Indus. *Paper 9, River Flood Hydrology*, Instn. Civil Engrs., London, 189–210.

BLANCHARD, F. B. (1955). Operational economy through applied hydrology. *Proc. Western Snow Conf.*, Colorado State University, Fort Collins, 35–48.

BLEASDALE, A. (1959). The measurement of rainfall. *Weather*, **14**:12–18.

BLEASDALE, A. and DOUGLAS, C. K. M. (1952). Storm over Exmoor on August 15, 1952. *Met. Mag.*, **81**:353–67.

Borchert, J. R. (1971). The Dust Bowl in the 1970's. *Ann. Ass. Amer. Geogrs.*, **61**: 1–22.

Boucher, K. (1972). The Rumanian flood disaster of May 1970. *Weather*, **27**: 55–62.

Boyd, J. H. (1971). Pollution charges income and the costs of water quality management. *Water Resour. Res.*, **7**: 759–69.

Brooks, C. E. P. and Glasspoole, J. (1928). *British floods and droughts*. E. Benn, London.

Bruce, J. P. and Clark, R. H. (1966). *Introduction to hydrometeorology*. Pergamon Press, Oxford, 319 pp.

Carlson, P. E. and Marshall, J. S. (1972). Measurement of snowfall by radar. *J. Appl. Met.*, **11**: 494–500.

Chambers, M. J. (1935). The drought of 1933–34 in New Mexico. *Mon. Weath. Rev.*, **63**: 14–15.

Chinn, T. J. H. (1969). Snow survey techniques in the Waitaki catchment, South Canterbury. *J. Hydrol.* (N.Z.), **8**: 68–76.

Chow, V. T. (1964). *Handbook of applied hydrology*. McGraw-Hill, New York.

Church, J. E. (1935). Principles of snow surveying as applied to forecasting streamflow. *J. Agric. Res.*, **51**: 97–130.

Church, J. E. (1942). Snow and snow surveying; ice. In Meinzer, O. E. (ed.) *Hydrology*. McGraw-Hill, New York, 83–148.

Corbett, E. S. (1967). Measurement and estimation of precipitation on experimental watersheds. In Sopper, W. E. and Lull, H. W. (eds), *Forest Hydrology*. Pergamon Press, Oxford, 107–29.

Corey, C. P. (1949). The effects of weather upon electric power systems. *Bull. Amer. Met. Soc.*, **30**: 159–67.

Court, A. (1961). Areal-depth rainfall formulae. *J. Geophys. Res.*, **66**: 1823–31.

Dahl, J. B. and Odegaard, H. (1970). Areal measurements of water equivalent of snow deposits by means of natural radioactivity in the ground. *Isotope Hydrology 1970*, I.A.E.A., Vienna, 191–210.

Davies, M. (1960). Grid system operation and the weather. *Weather*, **15**: 18–24.

Davis, N. E. (1969). Diurnal variation of thunder at Heathrow Airport. *Weather*, **24**: 166–72.

Day, H. J., Bugliarello, G., Ho, P. H. P. and Houghton, V. T. (1969). Evaluation of benefits of a flood warning system. *Water Resour. Res.*, **5**: 937–46.

Dobbie, C. H. and Wolf, P. O. (1953). The Lynmouth flood of August 1952. *Proc. Instn. Civ. Engrs.*, **2**: 522–88.

Doneaud, A. (1971). Meteorological factors causing Romanian floods in 1970. *W.M.O. Bulletin*, **20**: 28–31.

Douglas, C. K. M. (1953). Gale of January 31, 1953. *Met. Mag.*, **82**: 97–100.

Dryar, H. A. (1949). Load dispatching and Philadelphia weather. *Bull. Amer. Met. Soc.*, **30**: 159–67.

Dzerdzeevskii, B. L. (1969). Climatic epochs in the twentieth century and some comments on the analysis of past climates. In Wright, H. E. (ed.) *Quaternary Geology and Climate*. Proc. 7th Congress of Int. Ass. for Quaternary Res., National Academy of Sciences, Washington, 49–60.

Eberly, D. L. (1966). Weather modification and the operations of an electric power utility: the Pacific Gas and Electric Company's test program. In Sewell, W. R. D. (ed.) *Human dimensions of weather modification*. University of Chicago, Dept. of Geography Research Paper No. 105: 209–26.

Findlay, B. F. (1969). Precipitation in Northern Quebec and Labrador. *Arctic*, **22**: 140–50.

Fitzpatrick, E. A. and Krishnan, A. (1967). A first-order Markov model for assessing rainfall discontinuity in Central Australia. *Arch. Met., Wien. Ser. B*, **15**: 242–59.

Forrest, J. S. (1950a). Variations in thunderstorm severity in Great Britain. *Q. Jl. R. Met. Soc.*, **76**: 277–86.

Forrest, J. S. (1950b). The performance of the British grid system in thunderstorms. *Proc. Inst. Elec. Eng.*, **97**, Pt. 2: 345–64.

Fredericksen, D. G. (1958). Use of snow survey data by soil conservation districts. *Proc. Western Snow Conf.* Colorado State University, Fort Collins, 32–5.

Garstka, W. U. (1944). Hydrology of small watersheds under winter conditions of snow-cover and frozen soil. *Trans. Amer. Geophys. Un.*, **25**: 838–71.

Garstka, W. U. (1964). Snow and snow survey. In Chow, V. T. (ed.) *Handbook of applied hydrology*. McGraw-Hill, New York.

Garstka, W. U., Love, L. D., Goodell, B. C. and Bertle, F. A. (1958). *Factors affecting snowmelt and streamflow*. US Bureau of Reclamation and Forest Service, Fraser, Colorado. 189 pp.

Gerdel, R. W., Hansen, B. L. and Cassidy, W. C. (1950). The use of radioisotopes for the measurement of the water equivalent of a snow pack. *Trans. Amer. Geophys. Un.*, **31**: 449.

Glasspoole, J. (1924). Fluctuations of annual rainfall: three driest consecutive years. *Trans. Instn. Wat. Engrs.*, **29**: 83–101.

Glasspoole, J. (1955). Rainfall in relation to water supply. *Q. Jl. R. Met. Soc.*, **81**: 271–86.

Golding, E. W. (1960). Using the wind for power. *Weather*, **15**: 113–21.

Golding, E. W. (1962). Energy from wind and local fuels. *The problems of the arid zone. Arid Zone Research 18*. UNESCO, Paris: 249–58.

Goodell, B. C. (1959). Management of forest stands in western United States to influence the flow of snow-fed streams. *Int. Ass. Sci. Hydrol.*, **48**: 49–58.

Goodhew, R. C. (1970). Weather is my business. 1. The hydrologist. *Weather*, **25**: 33–9.

Goodhew, R. C. and Jackson, E. (1972). Weather forecasting and river management. In Taylor, J. A. (ed.) *Weather forecasting for agriculture and industry*. David and Charles, Newton Abbot, 166–83.

GORDON, A. H. and LOCKWOOD, J. G. (1970). Maximum one-day falls of precipitation in Tehran. *Weather*, **25**:2–8.

GREEN, F. H. W. (1970). Some isopleth maps based on lysimeter observations in the British Isles in 1965, 1966 and 1967. *J. Hydrol.*, **10**:127–40.

GREGORY, S. (1956). Rainfall studies and water supply problems in the British Isles. *Adv. of Science*, **13**:347–51.

GREGORY, S. (1959). Climate and water supply in Great Britain. *Weather*, **14**:227–32.

GRINDLEY, J. (1960). Calculated soil moisture deficits in the dry summer of 1959 and forecast dates of first appreciable runoff. *Int. Ass. Sci. Hydrol.*, **51** (Helsinki): 109–20.

GRINDLEY, J. (1967). The estimation of soil moisture deficits. *Met. Mag.*, **96**:97–108.

GUMBEL, E. J. (1958). Statistical theory of floods and droughts. *J. Instn. Wat. Engrs.*, **12**:157–84.

HARROLD, T. W. (1966). The measurement of rainfall using radar. *Weather*, **21**:247–9, 256–8.

HAUPT, H. F. (1968). The generation of spring peak flows by short-term meteorological events. *Bull. Int. Ass. Sci. Hydrol.*, **13**:65–76.

HAWKINS, R. H. (1969). Effect of changes of streamflow regimen on reservoir yield. *Water Resour. Res.*, **5**:1115–19.

HEATHCOTE, R. L. (1969). Drought in Australia: A problem of perception. *Geogrl. Rev.*, **59**:175–94.

HENRY, A. J. (1930). The great drought of 1930 in the United States. *Mon. Weath. Rev.*, **58**:351–4.

HERBST, P. H. and SHAW, E. M. (1969). Determining raingauge densities in England from limited data to give a required precision for monthly areal rainfall estimates. *J. Instn. Wat. Engrs.*, **23**:218–30.

HERSHFIELD, D. M. (1961a). *Rainfall frequency atlas of the United States, for durations from 30 minutes to 24 hours and return periods from 1 to 100 years*. US Weather Bureau, Hydrologic Services Division, Tech. Paper 40.

HERSHFIELD, D. M. (1961b). Estimating the probable maximum precipitation. *J. Hydraul. Div. Amer. Soc. Civ. Engrs.*, **87**:99–116.

HERSHFIELD, D. M. (1965). Method for estimating probable maximum rainfall. *J. Amer. Waterwks. Assoc.*, **57**:965–72.

HERSHFIELD, D. M. and WILSON, W. T. (1957). Generalising of rainfall–intensity–frequency data. *Int. Ass. Sci. Hydrol.* (Toronto), **1**:499–506.

HEWLETT, J. D. and HELVEY, J. D. (1970). Effects of forest clear-felling on the storm hydrograph. *Water Resour. Res.*, **6**:768–782.

HIBBERT, A. R. (1971). Increases in streamflow after converting chaparral to grass. *Water Resour. Res.*, **7**:71–80.

HIBBERT, D. (1966). Weather forecast services for off-shore oil operations. *Weather*, **21**:114–19.

HOLLAND, D. J. (1967a). The Cardington rainfall experiment. *Met. Mag.*, **96**:193–202.

HOLLAND, D. J. (1967b). Evaporation. *Br. Rainf. 1961*. HMSO, London, 3–34.

HOPKINS, J. S. (1972). A study of snowmelt floods in a mountainous catchment using limited meteorological data. *Met. Mag.*, **101**:221–8.

HORNBECK, J. W., PIERCE, R. S. and FEDERER, C. A. (1970). Streamflow changes after forest clearing in New England. *Water Resour. Res.*, **6**:1124–32.

HOWE, C. W. and LINAWEAVER, F. P. (1967). The impact of price on residential water demand and its relation to system design and price structure. *Water Resour. Res.*, **3**:13–32.

HOYT, W. G. (1942). Droughts. In Meinzer, O. E. (ed.) *Hydrology*. McGraw-Hill, New York, 579–91.

HOYT, W. G. and LANGBEIN, W. B. (1955). *Floods*. Princeton University Press, New Jersey, 469 pp.

HUFF, F. A. (1967). Time distribution of rainfall in heavy storms. *Water Resour. Res.*, **3**:1007–19.

HUFF, F. A. (1970). Sampling errors in measurement of mean precipitation. *J. Appl. Met.*, **9**:35–44.

HUFF, F. A. and SHIPP, W. L. (1969). Spatial correlations of storm, monthly and seasonal precipitation. *J. Appl. Met.*, **8**:542–50.

HURST, H. E. (1952). *The Nile*. Constable, London, 326 pp.

ITAGAKI, K. (1959). An improved radio snow-gage for practical use. *J. Geophys. Res.*, **64**:375–83.

JAMIESON, D. G. (1972). River Dee research program. 1—Operating multi-purpose reservoir systems for water supply and flood alleviation. *Water Resour. Res.*, **8**:899–903.

JEVONS, W. S. (1861). On the deficiency of rain in an elevated rain gauge as caused by wind. *Lond. Edinb. Dubl. Phil. Mag.*, **22**:421–33.

JOHNSON, P. (1966). Flooding from snowmelt. *Civil Eng. Public Wks. Rev.*, **61**:747–50.

JOHNSON, P. and ARCHER, D. R. (1972). Current research in British snowmelt river flooding. *Bull. Int. Ass. Hydrol. Sci.*, **17**:443–51.

JOHNSON, S. R., MCQUIGG, J. D. and ROTHROCK, T. P. (1969). Temperature modification and costs of electric power generation. *J. Appl. Met.*, **8**:919–26.

KATES, R. W. (1962). *Hazard and choice perception in flood plain management*. University of Chicago, Dept. of Geography Research Paper No. 78, 157 pp.

KOBERG, G. E. (1960). Effect on evaporation of releases from reservoirs on Salt River, Arizona. *Bull. Int. Ass. Sci. Hydrol.*, **17**:37–44.

KRICK, I. P. (1954). Technical and economic aspects of weather modification in relation to water supply. *J. Instn. Wat. Engrs.*, **8**:336–52.

KRUTILLA, J. V. (1966). An economic approach to coping with flood damage. *Water Resour. Res.*, **2**:183–90.

KUPCZYK, E. (1968). Warunki synoptyczne wystepowania wezlvan roztopowych w Polsce Poludniowej. *Przeglad Geofizyczny*, **13**:143–56.

LANE, F. W. (1966). *The elements rage* (Third Ed.). David and Charles, London.

LANGBEIN, W. B. (1960). Hydrologic data networks and methods of extrapolating or extending available hydrologic data. *Hydrologic Networks and Methods, U.N. Flood Control Series* No. 15, Bangkok, 13–41.

LAPWORTH, H. (1930). Meteorology and water supply. *Q. Jl. R. Met. Soc.*, **56**:271–86.

LAPWORTH, H. (1965). Evaporation from a reservoir near London. *J. Instn. Wat. Engrs.*, **19**:163–81.

LAW, F. (1953). The estimation of the reliable yield of a catchment by correlation of rainfall and runoff. *J. Instn. Wat. Engrs.*, **7**:273–92.

LAW, F. (1955). Estimation of the yield of reservoired catchments. *J. Instn. Wat. Engrs.*, **9**:467–87.

LAW, F. (1956). The effect of afforestation upon the yield of water catchment areas. *J. Brit. Watwks. Ass.*, **38**:489–94.

LAW, F. (1957). Measurement of rainfall, interception and evaporation losses in a plantation of Sitka spruce trees. *Int. Ass. Sci. Hydrol.* (Toronto), **2**:397–411.

LAWRENCE, E. N. (1957). Estimation of the frequency of 'runs of dry days'. *Met. Mag.*, **86**:257–69, 301–4.

LAYCOCK, A. H. (1960). Drought patterns in the Canadian prairies. *Int. Ass. Sci. Hydrol.* (Helsinki), **51**:34–47.

LEAF, C. F. (1971). Areal snow cover and disposition of snowmelt runoff in central Colorado. *Forest Service Research Paper FSRP–RM–66*, Rocky Mountain Forest and Range Experiment Station, 25 pp.

LEWIS, D. C. (1968). Annual hydrologic response to watershed conversion from oak woodland to annual grassland. *Water Resour. Res.*, **4**:59–72.

LIGHT, P. (1941). Analysis of high rates of snowmelting. *Trans. Amer. Geophys. Un.*, Pt. 1: 195–205.

LINSLEY, R. K. (1960). Common ground of meteorology and hydrology. *Bull. Amer. Met. Soc.*, **41**:423–8.

LINSLEY, R. K. (1967). The relation between rainfall and runoff. *J. Hydrol.*, **5**:297–311.

LINSLEY, R. K. and FRANZINI, J. B. (1955). *Elements of Hydraulic engineering*. McGraw-Hill, New York.

LOCKWOOD, J. G. (1967). Probable maximum 24-hour precipitation over Malaya by statistical methods. *Met. Mag.*, **96**:11–19.

LOCKWOOD, J. G. (1968). Extreme rainfalls. *Weather*, **23**:284–9.

LOVE, L. D. (1955). The effect on streamflow of the killing of spruce and pine by the Englemann spruce beetle. *Trans. Amer. Geophys. Un.*, **36**:113–18.

LOWNDES, C. A. S. (1968). Forecasting large 24–hour rainfall totals in the Dee and Clwyd River Authority area from September to February. *Met. Mag.*, **97**:226–35.

LOWNDES, C. A. S. (1969). Forecasting large 24-hour rainfall totals in the Dee and Clwyd River Authority area from March to August. *Met. Mag.*, **98**:325–40.

MCGUINNESS, J. L. (1963). Accuracy of estimating watershed mean rainfall. *J. Geophys. Res.*, **68**:4763–7.

MCGUINNESS, J. L. and HARROLD, L. L. (1971). Reforestation influences on small watershed streamflow. *Water Resour. Res.*, **7**:845–52.

MCKAY, G. A. (1968). Meteorological conditions leading to the project design and probable maximum flood on the Paddle River, Alberta. *Trans. Amer. Soc. Agric. Engrs.*, **21**:173–7.

MAGIN, G. B. and RANDALL, L. E. (1960). Review of literature on evaporation suppression. *U.S. Geol. Survey Professional Paper 272–C*. Govt. Printing Office, Washington, 53–69.

MANLEY, G. (1957a). Climatic fluctuations and fuel requirements. *Scott. Geogrl. Mag*, **73**:19–28.

MANLEY, G. (1957b). Climatic fluctuations and fuel requirements. *Adv. of Science*, **13**:324–6.

MANSFIELD, W. W. (1959). The influence of monolayers on evaporation from water storages. *Austral. J. Appl. Sci.*, **10**:65–72.

MARTINELLI, M. (1959). Some hydrologic aspects of alpine snow fields under summer conditions. *J. Geophys. Res.*, **64**:451–5.

MARTINELLI, M. (1965). An estimate of summer runoff from alpine snowfields. *J. Soil and Water Conserv.*, **20**:24–6.

MATTHEWS, R. P. (1972). Variation of precipitation intensity with synoptic type over the Midlands. *Weather*, **27**:63–72.

MILLER, E. R. (1934). The dustfall of November 12–13, 1933. *Mon. Weath. Rev.*, **62**:73–7.

MILLS, W. B. and SCHROEDER, E. E. (1969). Floods of April 28, 1966 in the northern part of Dallas, Texas. *U.S. Geol. Survey, Water Supply Paper* 1870-B., 37 pp.

NACE, R. L. (1969). World water inventory and control. In Chorley, R. J. (ed.) *Water, earth and man*. Methuen, London, 31–42.

NAMIAS, J. (1956). Some meteorological aspects of drought, with special reference to the summers of 1942–1954 over the United States. *Mon. Weath. Rev.*, **83**:191–205.

NAMIAS, J. (1960). Factors in the initiation, perpetuation and termination of drought. *Int. Ass. Sci. Hydrol.*, **51**:81–91.

NAMIAS, J. (1966). Nature and possible causes of the northeastern United States drought during 1962–65. *Mon. Weath. Rev.*, **94**:543–54.

NORDENSON, T. J. and BAKER, D. R. (1962). Comparative evaluation of evaporation instruments. *J. Geophys. Res.*, **67**:671–9.

NYE, R. H. (1965). The value of services provided by the Bureau of Meteorology in planning within the State Electricity Commission of Victoria. In *What is weather worth?* Bureau of Meteorology, Melbourne, 87–91.

OGDEN, R. J. (1972). Forecasting for North Sea oil and gas rigs. *Weather*, **27**:336–42.

PARDÉ, M. (1955). *Fleuves et rivières*. Armand Colin, Paris, 224 pp.

PARDÉ, M. (1967). La crue fantastique de Takekan Creek à Formose en Septembre 1963 par l'effet du cyclone 'Gloria'. *C.R. Acad. Sci. Paris Sér. D,* **264**:1592–6.

PARREY, G. E. (1972). Forecasting temperature for the gas and electricity industries. *Met. Mag.,* **101**:264–70.

PAULHUS, J. L. H. (1965). Indian Ocean and Taiwan rainfalls set new records. *Mon. Weath. Rev.,* **93**:331–5.

PAULHUS, J. L. H. (1971). The March–April 1969 snowmelt floods in the Red River of the north, upper Mississippi and Missouri basins. *National Weather Service, Office of Hydrol. Tech. Rev.* NOAA–TR–NWS–13, 101 pp.

PECK, E. L., BISSELL, V. C., JONES, E. B. and BURGE, D. L. (1971). Evaluation of snow water equivalent by airborne measurement of passive terrestrial gamma radiation. *Water Resour. Res.,* **7**:1151–9.

PENMAN, H. L. (1948). Natural evaporation from open water, bare soil and grass. *Proc. R. Soc. Lond. Ser. A.,* **193**:120–45.

PENMAN, H. L. (1950). The water balance of the Stour catchment area. *J. Instn. Wat. Engrs.,* **4**:457–69.

PENTON, V. E. and ROBERTSON, A. C. (1967). Experience with the pressure pillow as a snow measuring device. *Water Resour. Res.,* **3**:405–8.

PULLEN, R. A. (1968). Computation of the moisture content of the atmosphere using surface-level climatological data. *J. Hydrol.,* **6**:168–82.

PYSKLYWEC, D. W., DAVAR, K. S. and BRAY, D. I. (1968). Snowmelt at an index plot. *Water Resour. Res.,* **4**:937–46.

RAYNER, J. N. and SOONS, J. M. (1965). The storm in Canterbury of 12–17 July 1963. *New Zeal. Geogr.,* **21**:12–25.

REICH, B. M. (1963). Short-duration rainfall-intensity estimates and other design aids for regions of sparse data. *J. Hydrol.,* **1**:3–28.

REICH, B. M. (1970). Flood series compared to rainfall extremes. *Water Resour. Res.,* **6**:1655–67.

REYNOLDS, G. (1968). Automatic raingauges in North Scotland. *Weather,* **23**:88–93.

RICH, L. R. (1972). Managing a ponderosa pine forest to increase water yield. *Water Resour. Res.,* **8**:422–8.

RICHARDS, B. D. (1950). *Flood estimation and control.* Chapman and Hall, London, 173 pp.

ROBINSON, A. C. and RODDA, J. C. (1969). Rain, wind and the aerodynamic characteristics of raingauges. *Met. Mag.,* **98**:113–20.

RODDA, J. C. (1965). A drought study in South-east England. *Wat. and Wat. Engng.,* **69**:316–21.

RODDA, J. C. (1967). The systematic error in rainfall measurement. *J. Instn. Wat. Engrs.,* **21**:173–7.

RODDA, J. C. (1970a). Definite rainfall measurements and their significance for agriculture. In Taylor, J. A. (ed.) *The role of water in agriculture.* Pergamon Press, Oxford, 1–10.

RODDA, J. C. (1970b). Rainfall excesses in the United Kingdom. *Trans. Inst. Brit. Geogrs.,* **49**:49–60.

ROONEY, J. (1969). The economic and social implications of snow and ice. In Chorley, R. J. (ed.) *Water, earth and man.* Methuen, London, 389–401.

SAARINEN, T. F. (1966). *Perception of the drought hazard on the Great Plains.* University of Chicago, Dept. of Geography Research Paper No. 106:198 pp.

SANDERSON, M. (1971). Variability of annual runoff in the Lake Ontario basin. *Water Resour. Res.,* **7**:554–65.

SANTEFORD, H. S., ALGER, G. R. and MEIER, J. G. (1972). Snowmelt energy exchange in the Lake Superior region. *Water Resour. Res.,* **8**:390–7.

SELLERS, W. D. (1965). *Physical climatology.* University of Chicago Press, Chicago and London, 272 pp.

SEWELL, W. R. D. (1965). *Water management and floods in the Fraser River Basin.* University of Chicago, Dept. of Geography Research Paper No. 100.

SEWELL, W. R. D. (1966). Weather modification and hydroelectric power. *Water Power,* **18**:353–7.

SEWELL, W. R. D. (1969). Human responses to floods. In Chorley, R. J. (ed.) *Water, earth and man.* Methucn, London, 431–51.

SHAW, E. M. (1966). The Devon River Authority raingauge network. *Weather,* **21**:291–7.

SKEAT, W. O. (ed.) (1961). *Manual of British water engineering practice* (3rd Edn.). Heffer, Cambridge.

SMITH, K. (1972). *Water in Britain.* Macmillan, London, 241 pp.

SOMMER, A. and SPENCE, E. S. (1967). Some runoff patterns in a permafrost area of northern Canada. *The Albertan Geogr.,* **4**:60–4.

STEERS, J. A. (1953). The east coast floods January 31–February 1, 1953. *Geogrl. J.,* **119**:280–95.

STEPHENS, F. B. (1951). A method of analysing weather effects on electrical power consumption. *Bull. Amer. Met. Soc.,* **32**:16–20.

STEPHENSON, P. M. (1967). Objective assessment of adequate numbers of rainfall gauges for estimating areal rainfall depths. *Int. Ass. Sci. Hydrol.,* **78**:252–64.

SWANK, W. T. and MINER, N. H. (1968). Conversion of hardwood-covered watersheds to white pine reduces water yield. *Water Resour. Res.,* **4**:947–54.

SWANSON, R. H. (1970). Local snow distribution is not a function of local topography under continuous tree cover. *J. Hydrol. (N.Z.),* **9**:292–8.

TABOR, H. (1958). Industrial energy from sunshine. *New Sci.,* **4**:473–5.

TABOR, H. (1962). Solar energy. *The Problems of the Arid Zone. Arid Zone Research 18,* UNESCO, Paris, 259–70.

TANNEHILL, I. R. (1947). *Drought: its causes and effects.* Princeton University Press, Princeton, N.J., 264 pp.

THEILER, D. F. (1969). Effects of flood protection on land use in the Coon Creek, Wisconsin, watershed. *Water Resour. Res.,* **5**:1216–22.

THOMAS, H. E. (1962). The meteorologic phenomenon of drought in the Southwest. *U.S. Geol. Survey. Professional Paper 342A*, Washington.

THORNTHWAITE, C. W. (1948). An approach toward a rational classification of climate. *Geogrl. Rev.*, **38**:55–94.

THORNTHWAITE, C. W. and HARE, F. K. (1955). Climatic classification in forestry. *Unasylva*, **9**:51–9.

TOEBES, C. (1967). The place of hydrology in New Zealand's economy. *J. Hydrol. (N.Z.)*, **6**:57–8.

TUCKER, G. B. (1960). Some meteorological factors affecting dam design and construction. *Weather*, **15**:3–13.

TURNER, D. B. (1968). The diurnal and day-to-day variations of fuel usage for space heating in St. Louis, Missouri. *Atmos. Envir.*, **2**:339–51.

URYVAEV, V. A. *et al.* (1965). Principal shortcomings of methods of observing snow cover and precipitation and proposals of the State Hydrological Institute for their improvements. *Gos. Gidrol. Inst. Trudy vyp*, **175**:31–58.

U.S. ARMY CORPS OF ENGINEERS (1960). *Runoff from snowmelt.* Engineering and Design Manuals, EM1110–2–1406.

VAN HYLCKAMA, T. E. A. (1970). Water use by salt cedar. *Water Resour. Res.*, **6**:728–35.

VEIHMEYER, F. J. (1964). Evapotranspiration. In Chow, V. T. (ed.) *Handbook of applied hydrology*. McGraw-Hill, New York.

VILLIMOW, J. R. (1956). The nature and origin of the Canadian dry belt. *Ann. Ass. Amer. Geogrs.*, **46**:211–32.

WAANANEN, A. O., HARRIS, D. D. and WILLIAMS, R. C. (1971). Floods of December 1964 and January 1965 in the far western states. Part 1—Description. *U.S. Geol. Survey, Water Supply Paper* 1866–A, 265 pp.

WALLEN, C. C. (1967). Aridity definitions and their applicability. *Geograf. Ann.*, **49A**:367–84.

WARD, R. C. (1963). Measuring potential evapotranspiration. *Geography*, **48**:49–55.

WARD, R. C. (1971). Measuring evapotranspiration: a review. *J. Hydrol.*, **13**:1–21.

WARD, R. C. (1972). Estimating streamflow using Thornthwaite's climatic water-balance. *Weather*, **27**:73–84.

WATTERSON, G. A. and LEGG, M. P. C. (1967). Daily rainfall patterns at Melbourne. *Austral. Met. Mag.*, **15**:1–12.

WHITE, G. F. (1939). Economic aspects of flood forecasting. *Trans. Amer. Geophys. Un.*, **20**:218–33.

WHITE, G. F. (1964). *Choice of adjustment to floods.* University of Chicago, Dept of Geography Research Paper No. 93, 150 pp.

WHITE, G. F. *et al.* (1958). *Changes in urban occupance of flood plains in the United States.* University of Chicago, Dept. of Geography Research Paper No. 57.

WIESNER, C. J. (1964). Hydrometeorology and river flood estimation. *Proc. Instn. Civ. Engrs.*, **27**:153–67.

WIESNER, C. J. (1970). *Hydrometeorology*. Chapman and Hall, London, 232 pp.

WILSON, J. W. (1970). Integration of radar and raingauge data for improved rainfall measurement. *J. Appl. Met.*, **9**:489–97.

WILSON, W. T. (1941). An outline of the thermodynamics of snowmelt. *Trans. Amer. Geophys. Un.*, Pt. 1: 182–95.

WINSTANLEY, D. (1970). The north African flood disaster, September 1969. *Weather*, **25**:390·403.

WOLF, P. O. (1952). Forecast and records of floods in Glen Cannich in 1947. *J. Instn. Wat. Engrs.*, **6**:298–324.

WOLF, P. O. (1956). The civil engineer's interest in heavy rainfall. *Weather*, **11**:241–8.

WOLF, P. O. (1966). Comparison of methods of flood estimation. *Paper 1, River flood hydrology*. Institution of Civil Engineers, London, 1–23.

WORLD METEOROLOGICAL ORGANISATION (1960). *Hydrologic networks and methods.* United Nations Flood Control Series 15, Bangkok, 157 pp.

WORLD METEOROLOGICAL ORGANISATION (1965). Guide to Hydrometeorological Practices. *Tech. Paper 82, W.M.O.*, Geneva.

ZOTIMOV, N. V. (1965). A surface method of measuring the water equivalent of snow by means of soil radioactivity. *Gos. Gidrol. Inst. Trudy vyp.*, **130**:148–50.

# 6. Transport and the atmosphere

### 6.1 Introduction

Atmospheric conditions affect all forms of transportation and this section attempts a systematic coverage of the three major forms of movement of both people and goods: by air, by land (road and rail), and by water (ocean and inland waters).

Meteorology and climatology have been applied to the problems of *aviation* ever since man first started to fly, because the atmosphere is a three-dimensional medium for movement quite unlike that for surface-based travel systems. But, just as aviation has benefited from increased atmospheric knowledge, so the development of climatology itself owes a great deal to the information gained from flight and its requirements. The principal function of most national meteorological agencies is to provide a service for the state's military and civil aircraft, and without a global network of co-operating services modern aviation would be impossible. Weather knowledge helps both to increase safety in the air and to improve the operational efficiency of airlines. To a certain extent, the economic incentive is now more important than the safety factor, since flying hazards have been much reduced, although weather-induced crashes still occur and changing aviation technology, such as the advent of supersonic transport, continually introduces new problems as well as solutions. In any case, safety cannot be entirely divorced from economic considerations, because no commercial airline could remain profitable with a poor safety record.

The application of weather knowledge to air transport is highly efficient. According to Beckwith (1966), only about 1·3 per cent of the scheduled airline route mileage in the US has to be cancelled because of poor weather conditions, and BOAC (now part of British Airways) cancels about 0·5 per cent of its flights for the same reasons (Mason, 1966). Nevertheless, Barry (1965) has demonstrated that, although flight cancellations reduce airline revenue almost to zero, fixed costs totalling 68–80 per cent of operating expenses still have to be met with a consequent deterioration of profit margins. Mason (1966) has stated that estimated weather losses to scheduled airlines in the US during 1965 amounted to $97 million compared with a total loss of $211 million for general aviation including private and charter flights. The loss to the airlines was assessed at 3·5 per cent of total revenue with costs for carrying additional fuel and delays to passengers the major items. In the UK, equivalent weather losses suffered by BEA (also now part of British Airways) were estimated at about £1·5 million per year or some 2·3 per cent of revenue. These losses are due to a variety of factors including cancellations, diversions, in-flight delays, costs of carrying additional fuel, and provision of extra maintenance required plus the loss of traffic due to the irregularity of the service.

Many of the dangers of *water transport* are attributable to the atmosphere, as are many of the factors influencing the efficient operation of ships. Consequently, the maritime nations of the world have always taken a keen interest in the inter-relationships between climatology and sea transport and in the associated problem of inland water transport. Compared with transportation by air and water, travel by *land* is less vulnerable to weather influences, although this relative immunity may occasionally lead to a false sense of security and an underestimation of risk when a particular hazard does intrude. In broad terms, weather sensitivity is greater in the case of roads rather than railways.

## Aviation climatology

### 6.2 Airport weather

Critchfield (1966) has rightly claimed that aviation climatology starts with the selection of an airport site, which ideally should experience a minimum of adverse weather likely to hinder landing and take-off. The most important atmospheric hazards are those affecting visibility or runway conditions and include fog or low cloud, turbulent eddies and crosswinds, and the accumulation of water, snow, or ice on the landing strip. Although little can be done to prevent severe weather affecting the runway surface, particularly when heavy rain or snow are characteristic of the wider regional climate, airports show considerable adaptation to wind conditions. Thus, the main runway is normally aligned in the direction of the prevailing wind in order to minimize the effects of crosswinds on landings. Such crosswinds can be troublesome at certain airports and Gloyne (1966) has reported that at Turnhouse, Edinburgh, they were responsible for the diversion of 0·8 per cent of all scheduled landings. At the North Front airfield at Gibraltar, the southerly sea breeze blows almost at right-angles to the only runway (Ward, 1954). During the summer months, the breeze often attains a mean speed of over 7·5 m/s, with gusts up to 15 m/s, and under these conditions landing becomes a somewhat hazardous operation.

According to Roberts (1971), a runway is considered to be unsafe when the wind component at right-angles to it exceeds a certain critical value depending on aircraft type. In Fig. 6·1 A′OA is the alignment of the runway and, if PO represents direction and speed $V$ of the surface wind, then the cross-component is given by PL or $V \sin a$. If the critical value of the cross-component is $U$, then the critical speed for a wind inclined to the runway at an angle $a$ is represented by

$$V \sin a = U,$$

or

$$V = U \operatorname{cosec} a.$$

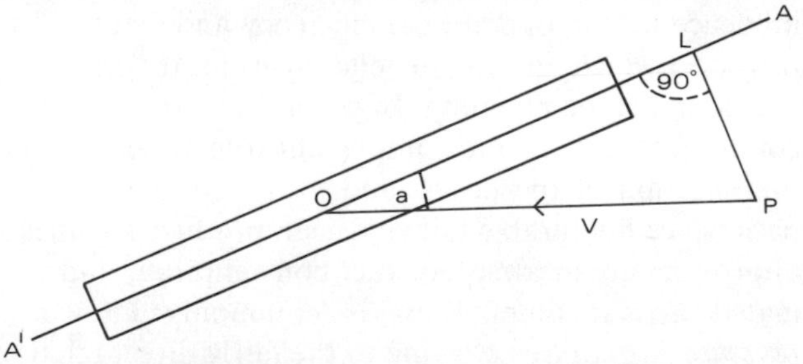

**Fig. 6.1 The location of airport runways in relation to the speed and direction of the wind. After Roberts (1971). Reproduced by permission of the Controller of Her Majesty's Stationery Office.**

The overall usability of a runway in terms of wind is the proportion of time at which the cross-component is equal to or less than the critical value and may be determined from a knowledge of surface wind frequencies at the airport. To ensure a successful landing, the pilot requires detailed information on wind behaviour in the last 30 m of descent. He is usually appraised of surface conditions by the controller about 10 min before touch-down, and data analysed by Keddie (1971) have shown that a mean wind speed measured over a $4\frac{1}{2}$ min period provides a reasonable estimate of speeds 10 min later.

Despite the increasing sophistication of automatic landing equipment, poor visibility from fog and low cloud ceilings is probably the major impediment to airport operations throughout the world and Fig. 6.2A illustrates the common situation, where a layer of radiation fog covers an airport although there is clear air

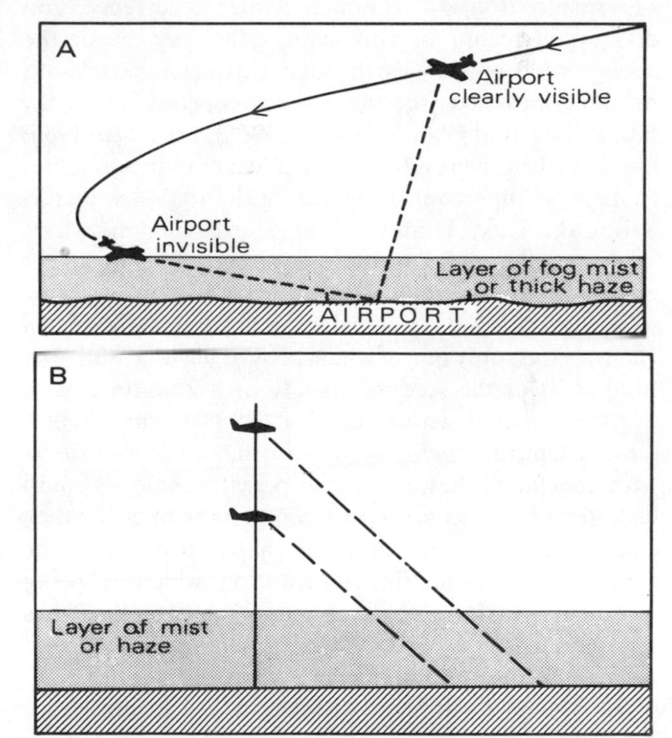


**Fig. 6.2 Some aspects of air-to-ground visibility.**
**(A) Airport obscured by surface fog as an aircraft descends to land.**
**(B) Horizontal range of visibility from an aircraft flying above a haze layer near the ground.**
**After Roberts (1971). Reproduced by permission of the Controller of Her Majesty's Stationery Office.**

above. At heights well above the fog, the airport may be visible through the small depth of the fog layer but, when the plane comes in to land, runway visibility will be much reduced because of the greater thickness of fog which has to be penetrated. Similarly, Fig. 6·2B shows that, where poor visibility is due to a fog layer below the aircraft, the greatest horizontal distance at which the ground is visible increases progressively with the height of the aircraft.

Much effort has been devoted to the study of the physics of artificial fog dispersal. Most progress has been achieved with *supercooled fogs*, in which fog particle temperatures lie below freezing point, and follows from the work of Schaefer (1946) and Vonnegut (1947) on the formation of ice crystals from supercooled water droplets using dry ice and silver iodide. Early experiments were conducted in the Pacific Northwest of the US where supercooled fog is a common winter occurrence, and dissipation techniques involving either dry ice or the release of propane gas through expansion nozzles to produce localized freezing are now operational in the US, Russia, and France (Wycoff, 1966). Beckwith (1966) has described the results and significance of programmes to disperse supercooled fog during the 1963–4 winter at Salt Lake City, Utah, and Medford, Oregon, where supercooled fogs comprised respectively 97 and 57 per cent of all fog with visibility less than 0·8 km. The fogs were seeded by dry ice pellets from light aircraft on 16 occasions, only one of which proved unsuccessful. The total costs of the seeding operations were estimated at $3825 compared with returns of $19 000, which represents a typical benefit/cost ratio of about 5, and Beckwith concluded that, if these dispersal techniques could be extended up to ambient temperatures of 5 °C, they could well provide a viable alternative to some of the expensive electronic instrumentation which is being installed at major airports. Similarly, Hicks (1967) has stressed that an outlay of $20 per hour may be sufficient to maintain a fog-free airport approach zone on the basis of tests with liquefied propane in the US and Greenland.

So far, however, the dissipation of so-called *warm fogs* above 0 °C has proved more difficult, although Osmun (1969) has indicated some of the strides recently made. Thermal techniques were used first and aimed at evaporating the constituent water droplets by the input of large amounts of heat energy as in the FIDO (Fog Investigation and Dispersal Operation) system used in Britain during World War II to clear landing strips for military aircraft. This method depended on open flares produced from oil burners along the edges of the runways but was only really effective in shallow radiation fogs, created a fire hazard for aircraft, and was too expensive for national commercial operation.

A comparable technique called Turboclair has been tried out in France, where the heat and kinetic energy of jet engines have been used to vaporize fog droplets and lift the fog near the end of the runway enough for the pilot to complete his landing procedures with safety (Dubois, 1965). Radiation fog may also be dispersed by a slightly different adaptation of the same principles, whereby warm, dry air above the thermal inversion is mixed with the fog particles below, and Plank (1969) has illustrated how such shallow fogs may be evaporated with the aid of helicopters to create the necessary air turbulence. Other warm fog-dispersal methods attempt to promote droplet coalescence in order to produce particles large enough to fall out as drizzle. In some cases, electrical charges have been injected into the fog to create attraction between the droplets, and finely ground deliquescent salt particles have also been used to induce droplet growth (Wycoff, 1966). But, like methods of driving the fog particles to the ground mechanically by a water spray released from an aircraft flying above the fog, none of these techniques is at present sufficiently effective to justify routine operation.

### 6.3 In-flight weather

During flight, the atmosphere exerts its most complete influence on the operational efficiency and safety of air transport which, in turn, depend on accurate forecasting and sound route planning. In general, the forecast wind conditions have the most important role in determining the selection of the fastest and most economical flight path, since favourable tailwinds can produce significant improvements in airspeed, fuel consumption, and estimated arrival times. This is commonly known as *pressure pattern flying* owing to the fairly direct relationship between the isobaric pattern and the upper winds. Critchfield (1966) has noted that long-range flights often follow different routes in summer and winter as a result of the latitudinal shift of pressure and wind systems, but forecast winds can usually provide more accurate information than climatological winds. Thus, Mason (1966) has estimated that, on average, forecast winds save about 3 per cent in flight time on a normal transatlantic crossing compared with climatological winds.

Where especially strong winds are anticipated, as in the jet streams, it may be necessary for the flight path to deviate considerably from the shortest great circle route in order to allow for head- or tailwind effects. Vertical variations in wind speed will help to influence the optimum cruising altitude for the flight, although air temperatures and the height of the tropopause are also important because the efficiency of jet engines decreases with increasing ambient temperatures. On the other hand, other operational considerations may preclude the choice of the fastest route, and the actual conditions encountered in flight may require continuous modifications to the planned route. In addition, the calibration of aneroid altimeters and certain airspeed indicators is based on assumed standard conditions of atmospheric pressure and temperature, so that indicated readings may have to be revised in accordance with existing conditions. For example, if a constant aneroid altimeter reading is maintained during a flight from a high-pressure area to a region of lower pressure, the aircraft will in fact be slowly losing height and this loss of altitude can clearly be critical in mountainous country or when landing procedures are attempted, as shown in Fig. 6.3.

The question of in-flight safety can be considered in relation to turbulence, icing, and the effects of storms. The implications of *turbulence* become more and more important with the rise in aircraft speeds. Small eddies produce a bumping sensation which is uncomfortable to passengers and may ultimately lead to structural fatigue, and large, isolated motions may impose such extreme loads that the aircraft becomes out of control or even suffers direct structural damage. Some low-level turbulence may occur as a result of thermals produced by intense solar heating in the tropics or as a result of mechanical disturbances of the airflow near the ground, but these difficulties may easily be avoided by flying either at night or at a sufficiently high altitude.

Much more serious, however, are the lee wave disturbances set up downwind of mountain chains, since there is always a danger of an aircraft being caught in one of the downdraughts with a loss of normal aerodynamic and handling qualities. Corby (1957) has prepared notes for forecasters and pilots on this phenomenon and Mason (1954) has described the effects of waves over northern Spain and warned that normal safety heights, calculated to give clearance of 305 m over all obstacles *en route*, may be insufficient for wave oscillations over such mountainous country.

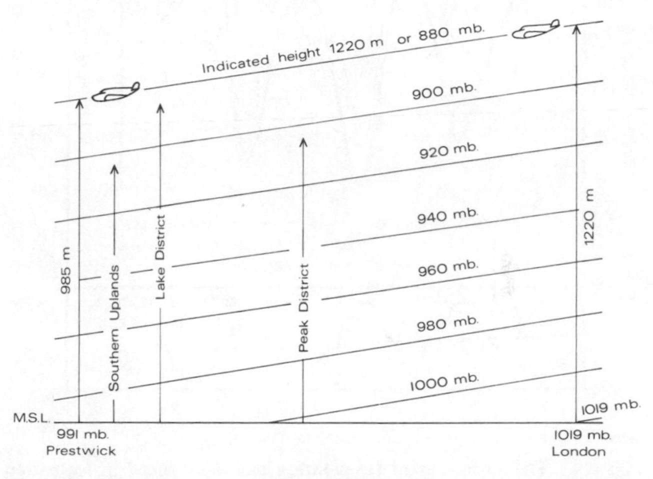


**Fig. 6.3 Schematic representation of the need for zero correction to an altimeter as a result of surface pressure variations. When the aircraft arrives at Prestwick from London the indicated height over-reads by 234 m. After Roberts (1971). Reproduced by permission of the Controller of Her Majesty's Stationery Office.**

More recently, Nicholls (1973) has identified a resurgence of interest in the airflow over mountains and has ascribed this as partly due to the fact that the advent of higher-flying aircraft has given a new importance to the propagation of lee waves and their associated turbulence in the stratosphere. Even so, the oscillations introduced into low-level tropospheric flows remain of high significance and Fig. 6.4 illustrates some measurements obtained on a flight by a Varsity aircraft into a moderate southwesterly airstream in the lee of the Black Mountains, Wales, on 14 December 1971. The aircraft was flown horizontally upwind through large waves in the inversion at an altitude of 1525 m, and it can be seen that considerable variations in temperature and wind speed occurred within the waves. These caused marked effects on the aircraft which was flown, as far as possible, under constant power, and the airspeed dropped to 120–125 kt in the downdraughts and rose to over 160 kt in the updraughts. Since the stalling speed of the Varsity at 1525 m is only 95 kt, this example shows the potential danger of flying through lee waves at a constant altitude without changing power.

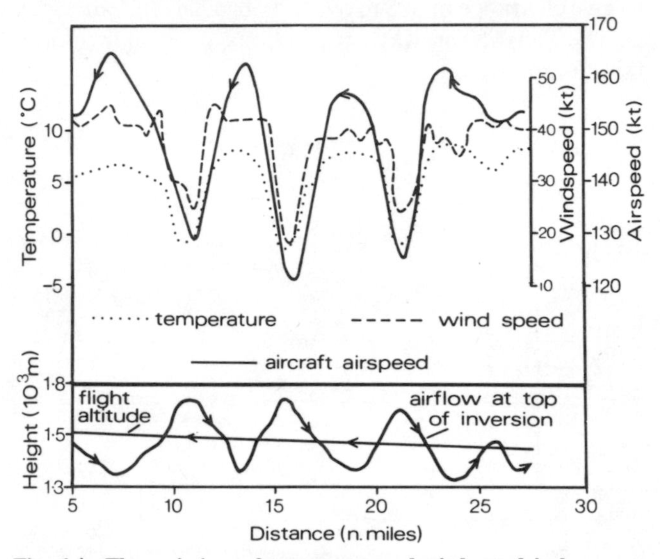


**Fig. 6.4 The variations of temperature and wind speed in lee waves downwind of the Black Mountains, Wales, on 14 December 1971 and the effects on the airspeed of an aircraft flown at approximately constant power and altitude. Arrows indicate the direction of the flight and of the airflow. After Nicholls (1973).**

On the other hand, the generation of lee waves can be forecast with reasonable reliability. They are, therefore, less of a hazard than clear-air turbulence (CAT), which occurs at high-levels in cloudless areas or zones of thin cirrus and consequently provides little visible warning of its onset. Mancuso and Endlich (1969) have defined CAT as random, three-dimensional eddies, which cause aircraft to experience high-frequency accelerations, and have shown that, over the US, the best meteorological indicators were the vertical vector wind shear and the product of wind shear and horizontal deformation. The causes of clear air turbulence are not fully understood but most cases are found near the tropopause with the most severe instances associated with the cold low-pressure side of the jet stream, since this is the most suitable location for rapid vertical change of the horizontal wind. Such wind shear turbulence is most dangerous when the related waves are in the initial rotor or billow phase, which, although short-lived, produces the greatest vertical velocities (Ludlam, 1967).

CAT is always more severe if it exists in combination with lee waves, as when a jet stream crosses a mountain region, and, according to Roach (1970), has never been reported as more than very uncomfortable away from upland areas. The turbulent layer is often only about 150 m deep, however, and can therefore be avoided by a slight alteration of aircraft altitude, although layers over ten times thicker than this have been observed. CAT appears to be widespread in the northern hemisphere with many isolated reports, such as that of McGinnigle (1970), and over a 5-year period the British Meteorological Office received 300 reports of severe clear-air turbulence from UK-based aircraft operating mainly on European and Atlantic routes (Roach, 1969). In the southern hemisphere, Spillane (1967) has commented on frequent occurrences over southern Australia in winter, where CAT seems to be related to the subtropical jet stream, which intensifies rapidly during May to reach maximum development in July. However, Woods (1972) has recently pointed out that satellite observations by means of infra-red spectrometers can be used to determine atmospheric radiance gradients, which represent zones of rather large-scale vertical wind shear. Since CAT is least likely to occur where the radiance gradient is small, such satellite data could be employed to design flight paths which have a low probability of encountering CAT.

*Ice formation* on aircraft may give rise to several problems resulting in a loss of performance, although most modern aircraft are fitted with anti-icing equipment. Alternatively, high-speed aircraft may readily move out of an icing zone, but this advantage is diminished by the fact that faster aircraft tend to accumulate ice more rapidly than slower ones as long as the kinetic heating produced by friction between the aircraft and the surrounding air does not raise the surface temperature of the plane above 0 °C. Since ice builds up mainly on the leading edges of the aeroplane, the main effect is to alter the aerodynamic characteristics during flight, causing an increase of drag and stalling speed plus a decrease of lift, air speed, and rate of climb (Meteorological Office, 1957).

In the most severe conditions, the weight of ice may displace the centre of gravity of an aircraft and cause the controls to become more unbalanced.

Jet engines themselves may be subject to impact icing, particularly at the mouth of the air intake with a resulting loss of power, and piston engines also suffer from induction or carburettor icing. This can occur at temperatures as high as 25 °C, largely because of the cooling effect of fuel evaporation, but is most severe in moist air with an inlet temperature of about 13 °C. Carburettor icing is most likely to occur in rain or clouds but may also appear in clear air if the relative humidity is greater than 60 per cent.

Icing is most likely with air temperatures between 0 °C and −20 °C but the rate of accretion varies with the meteorological situation and the type of ice deposit which occurs. Translucent rime or glaze ice is more dangerous than the opaque varieties, since it results from a high proportion of liquid water flowing over the aircraft surface before freezing and is therefore heavier, tougher, and more tenacious than ice containing a larger amount of air (Meteorological Office, 1957). Glaze ice tends to form when the air temperature rises towards 0 °C, thus allowing an increase in the percentage of supercooled water droplets, which remain liquid immediately after striking the leading edges of the aircraft, and the most severe conditions may be expected in large cumulus or cumulonimbus clouds containing a high concentration of water and large cloud droplets. In comparison, opaque rime is associated with lower temperatures and smaller droplets.

Other forms of icing include freezing rain—which occurs when rain falls through a temperature inversion to enter a colder layer below with temperatures less than 0 °C thus producing supercooled droplets and conditions similar to those for glaze ice—and the accumulation of wet, packed snow on aircraft surfaces. Aviation forecasts of icing conditions normally result in flight paths avoiding thick clouds with temperatures below 0 °C as far as possible, although the fact that isolated cumulus clouds rarely achieve a horizontal diameter greater than 16–24 km means that icing exposure is unlikely to be prolonged, unless the route lies through a belt of cloud lying along a front. However, if glaze ice is encountered, avoiding action is usually taken by changing course or flying below the 0 °C level or above the –40 °C level, where clouds consist almost entirely of ice crystals rather than supercooled water droplets.

*Severe storms* present an amalgam of most aviation weather hazards, including turbulence and icing within well developed cumulonimbus clouds, while strong winds, heavy rain, or hail and lightning constitute additional difficulties. Under normal circumstances, all routine flights seek to avoid the violent airflows within tornadoes or hurricanes, and even mid-latitude depressions include enough changes in wind speed and direction, together with temperature and visibility variations, to merit careful progress as indicated in Fig. 6·5. The convective turbulence found within some of the large cumulonimbus clouds of tropical and subtropical latitudes is perhaps the greatest single atmospheric threat to aircraft safety and Burnham (1968) has shown that almost all losses of large civil jet planes directly attributable to turbulence—averaging about one per year—occur in or near thunderstorms. In the US, updraughts and downdraughts in excess of 30 m/s have been recorded within cumulus clouds, which occasionally reach over 18 000 m in height, and speeds of 90 m/s are possible in the most severe storms. These convective storms tend to form along a NE–SW line across the country and provide a marked obstacle to E–W aviation (Roach, 1967). More recently, Roach and James (1972) have estimated the potential height of giant cumulonimbus clouds over the US, India, Singapore, and parts of the Mediterranean area where future supersonic transport routes are likely to be located.

Structural damage to aircraft by hail has been widely reported for many years, as by Sansom (1954) and Harrower and Evans (1956) over East Africa and France respectively, but, since the risk of damage increases with aircraft speed, there is a continuing need to assess the probabilities of in-flight encounters in relation to design specifications. Crossley (1961) has noted that cases of aircraft damage have been observed with hail as small as 1 mm diameter and has suggested that design strengthening to give protection from impact may be practicable only up to a diameter of 15 mm. However, larger hailstones cannot always be avoided by the use of radar or by restricting flights when there is a high risk of hail and Briggs (1972a) has combined the height and speed pro-

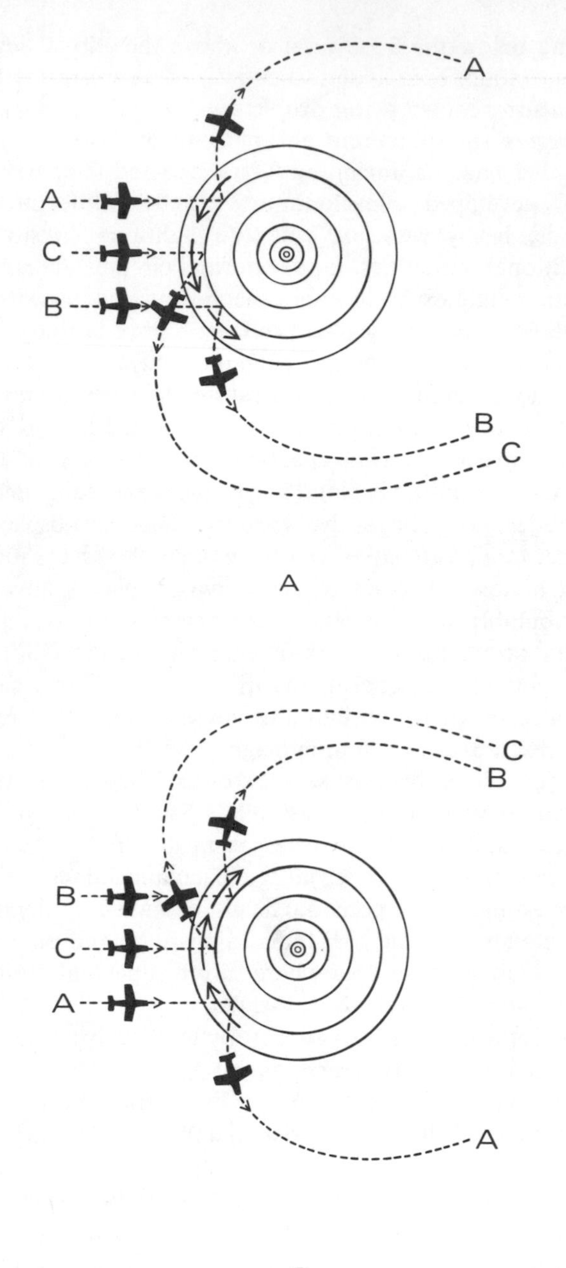


**Fig. 6.5 Variations in course necessary to avoid a storm during flight.**
**(A) Northern hemisphere.**
**(B) Southern hemisphere.**
**The broken lines indicate courses taken to avoid flying near the centre of the storm. After Roberts (1971). Reproduced by permission of the Controller of Her Majesty's Stationery Office.**

files for various types of aircraft with meteorological data based on the incidence of thunderstorms at ground level to calculate the probable hail experience during flight.

As might be expected, there are wide geographical variations in hail risk, with the highest incidence over north-east India and the central plains of the US compared with a low hail risk over North Africa, the Middle East, and most sea routes. Thus, for a Britannia aircraft every 30 000 h of flight over Denver or NE India should give six encounters with 25 mm diameter hail as opposed to only 1·9 encounters with 50 mm hail in 800 000 h over the UK. Similar estimates of flight encounters with heavy rain have been published by Briggs (1972b).

Of all the flying hazards associated with severe storms, lightning strikes to aircraft are relatively unimportant, since they rarely cause more than minor damage and there are very few instances of aircraft being destroyed in flight by lightning (Wallington, 1964). Mason (1964) has stated that, since the introduction of pressurized aircraft and higher cruising levels, the incidence of lightning strikes has decreased, although there is some evidence, as reported by Cobb and Holitza (1968), that an aircraft may actually trigger off a lightning strike especially if it enters a storm at an early stage of dissipation. A reminder that even space travel is dependent on meteorological conditions for at least a small part of the journey occurred in November 1969, when Apollo 12 was struck by lightning twice within one minute of the launch. Although a line of showers and thunderstorms associated with convective activity at a cold front existed in the vicinity, it is possible that the lightning discharges could have been triggered off by the space vehicle itself according to Bosart (1971).

## 6.4 Supersonic transport

The arrival of supersonic transport (SST) introduces a new dimension to aviation-weather relationships although, contrary to some earlier views, this aircraft will in certain ways be more dependent on atmospheric conditions than existing subsonic jets. Thus, despite maintaining a cruising altitude above the tropopause, the most critical phases of flight will be subject to similar meteorological factors affecting other aircraft, while the increased number of passengers to be carried together with the high capital and operating costs involved should make for greater emphasis on safety and efficiency.

As shown by several general surveys, such as those of Jones *et al.* (1967) and Scherhag *et al.* (1967), the weather problems of SST will be different in degree rather than kind from those already discussed in connection with aviation (Fig. 6.6), and much of the changed emphasis

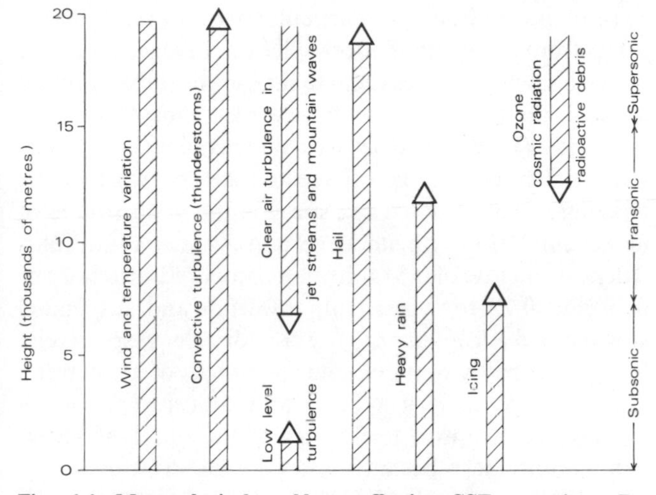


**Fig. 6.6 Meteorological problems affecting SST operations. Reproduced by permission from Jones, R. F. (1966). Meteorology and supersonic flight, *Nature*, Vol. 212, pp. 1181–85.**

will stem from the high cruising speeds which will be some 2·5 times those of current aircraft. For example, the combination of high air speeds and low wind speeds at cruising altitudes in the lower stratosphere means that the influence of wind will be much reduced. Crossley (1969) has claimed that a mean headwind of 4 m/s on the transatlantic crossing will increase the flight time by as little as one minute and it will consequently become less profitable to deviate from great circle routes in order to seek the least-time track. Similarly, it has been suggested by Burnham (1969) that turbulence is likely to affect SST less than subsonic flight and that, on average, passengers in supersonic aircraft will fly 20–100 times as far between encounters with gust-induced accelerations of a given magnitude than at present. This is largely because the SST will fly above the turbulence level for most of the time. On the other hand, severe turbulence has been recorded as high as 15 250 m within mountain waves over the Sierra Nevada and, during the period of transonic acceleration, the discomfort and risk experienced will increase directly with the speed of the aircraft.

The significance of hydrometeors such as hail or heavy rain is comparable to that of turbulence in that, although SST cruising altitudes will lie above the main levels of occurrence, there will be an increased risk due to higher subsonic speeds during the ascent and descent stages through the troposphere in all latitudes. Crossley (1961) has pointed out that flights up to 16 750 m in low latitudes will be exposed to some hail risk and Briggs (1972a) has used height and speed profiles for Concorde to estimate the hailstone diameter which corresponds to one encounter in $10^4$, $10^5$, and $10^6$ h of flight at a given level. Figure 6.7 indicates how the hailstone characteristics vary with height over three areas. In addition, Chaplin (1969) has noted that flight through ice crystals may result in some long-term deterioration of supersonic aircraft.

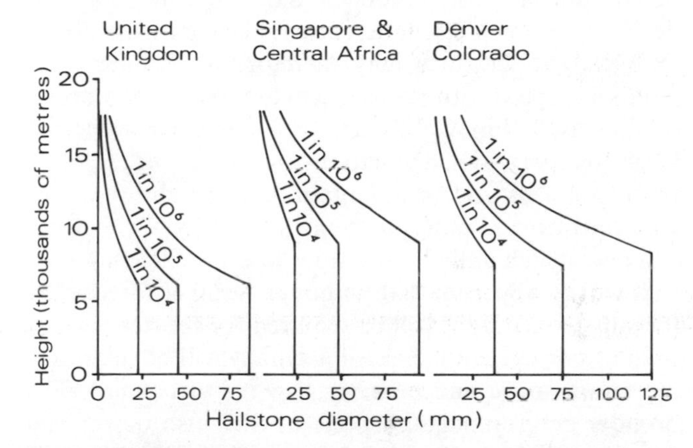


**Fig. 6.7 Estimated hailstone diameter for one encounter in $10^4$, $10^5$, and $10^6$ flight hours. All hours at level of interest. After Briggs (1972a). Reproduced by permission of the Controller of Her Majesty's Stationery Office.**

Supersonic flight will be particularly dependent on atmospheric temperatures for a variety of reasons. In the first place, increasing ambient temperatures reduce the thrust of jet engines, whose performance and fuel consumption are usually based on the mean profile conditions of the International Standard Atmosphere (ISA). Sugden (1969) has quoted evidence that, for a flight of 3200 nautical miles by a Mark 3 aircraft, an ambient temperature of ISA+16 degC would add 5000 kg to the fuel required in ISA conditions. Most of this extra fuel would be needed for the short transonic acceleration phase, which will probably consume at least one-third of the fuel for the entire flight. Similar conclusions have been quoted by Kreitzberg (1967), who claimed that a temperature excess of 11·1 degC in the supersonic climb phase would consume 1960 kg of extra fuel and could displace the equivalent of 20 passengers and their baggage.

Although the transonic acceleration phase is the most temperature-sensitive flight operation, higher than average values will also affect cruising performance and, according to Crossley (1969), the total flight time across the Atlantic will be increased by $0{\cdot}75T$ min if $T$(°C) is the amount by which the mean temperature of the cruise phase exceeds the ISA value. Air temperatures also influence the kinetic heating of the aircraft surface and at a cruise speed of Mach 2·2 the skin temperature of parts of the aircraft will approach limits beyond which a decrease in tensile strength sets in (Freeman, 1970). High ambient temperatures may therefore necessitate a reduction in speed with a consequent increase in the amount of fuel used. Finally, Roach (1970) has reported that large temperature fluctuations, such as occur near thunderstorms and in mountain wave conditions, provide problems relating to the air intake for Concorde's engines, which suffer a resulting loss of efficiency.

It will be apparent that, although weather forecasting for supersonic flight will be required for shorter periods than at present, an increased accuracy will be necessary for temperature and possibly also for turbulence. The broader environmental aspects of SST also depend on suitable atmospheric conditions, as shown by Nicholls (1970) in a study of the effects of weather on the location of the ground area where the sonic bang is heard.

## Water transport climatology

### 6.5 Sea

As in the case of air transport, the hazards of sea travel have been much reduced in modern times but safety is still the main objective and is implicit in both the design and operation of all harbours and vessels. Moorer (1966) has stated that the weather is the greatest peace-time threat to the safety of the 500 ships in the US Atlantic Fleet and instanced the loss of 790 officers and men when three destroyers capsized and sank in a typhoon near the Philippines. Such severe storms have to be avoided by all craft, and many ships will take evasive action from mid-latitude depressions as described by Mayl (1969) for a general cargo ship bound for South Africa from Antwerp.

In higher latitudes, a particular navigational hazard arises from the accumulation of ice on the superstructure of vessels, which causes them to become unstable through a rise in the centre of gravity and then capsize in heavy seas. Hay (1956) has discussed this problem with reference to fishing trawlers off Iceland and shown that the freezing of supercooled sea spray is the main source of ice accumulation with, under certain assumed conditions, a deposition rate of 0·54 tonnes/h compared with rates of 0·19 and 0·22 tonnes/h from snowfall and suspended supercooled droplets of fog or drizzle respectively. Since it probably requires some 50 tonnes of ice to bring about a capsize, it will be clear that trawlers can be endangered within a few hours of the onset of severe icing conditions, especially if they are side-trawlers which have been previously fishing in air temperatures well below zero and thereby starting to ice up by the nature of their work irrespective of the state of the sea. The continuing loss of fishing vessels in northern waters led, in the 1968–69 winter, to the first provision of special meteorological and medical services for British trawlers from a mother-ship based on the fishing grounds, and Smith (1970) reported that, throughout that not-atypical season, there were no trawlers which completed a full voyage of about three weeks without sheltering for at least part of the time.

Although the accumulation of frozen spray represents an important hazard to marine transport in high latitudes, it is the presence of sea-ice which offers the greatest single obstacle to Arctic navigation, and Walker and Penney (1973) have emphasized that ship movements in the Arctic are still effectively limited to a short, late-summer season from July to October. Outside this period, a passage through pack-ice can be achieved only by specially strengthened vessels which either ram the ice directly or slide upwards over the ice so that the ship's weight eventually exerts a breaking load on the ice surface. As a result of the necessarily heavy construction, which involves a good deal of extra steel for plating the shell, ice-breaking vessels are expensive and have such a reduced carrying capacity that they are normally unsuitable for commercial cargo operation. In the North-West Passage, which links the Atlantic and Pacific Oceans through the Canadian Archipelago, winter sea-ice sometimes more than 2 m thick blocks over 4000 km of the route and, since progress is maintained by a series of ramming and reversing manoeuvres, any voyage also results in a very large consumption of fuel (Walker and Scott, 1972). However, an ice-breaking oil tanker, the *S.S. Manhattan*, successfully penetrated through the North-West Passage in 1969 and 1970, and it may be that, despite the formidable difficulties which exist, the demand for Arctic resources such as oil or iron ore may ultimately lead to the establishment of commercial routes throughout the year in this and in similar areas. From a technological viewpoint, given adequate operational support, these voyages are already theoretically possible and Maybourn (1971) has described some of the design and performance features required of large ships operating in the highest latitudes.

After considerations of safety have been fulfilled, the most important application of atmospheric knowledge to ocean transport lies in the weather routeing of ships to achieve the fastest passage compatible with minimum risk or damage. This has been a basic principle of sea navigation at least since the time when sailing ships developed world-wide commercial routes based on the planetary wind system and the avoidance of storm belts, but now technological advances in weather communications combined with the increasing average speed of ships have encouraged practical routeing on a synoptic weather scale. Frankcom (1966) has shown that some synoptic data became available with the start of radio bulletins for shipping around 1920, and the introduction of radio-coded analyses by most meteorological services since 1948 has enabled shipmasters to reconstruct weather maps on board. However, the recent development of the facsimile chart-transmitter and receiver can now provide complete weather maps of oceanic areas without the need for plotting on board.

The progress of a vessel is affected by the wind and Allen (1951) has claimed that a wind from the stern will increase a ship's speed by about 1 per cent compared with a decrease in speed from headwinds by anything from 3 to 13 per cent depending on the size of the ship and its load. But the most significant parameter is the surface wave action and Haltiner and Hamilton (1962) have shown that the speed of any ship can be expressed as an empirical function of wave height and direction. To a large extent, the highest waves—whether sea or swell—are associated with the steepest barometric gradients since wave conditions are a function of wind speed, 'fetch', and the length of time the wind has been blowing. Indeed, the main aim of weather routeing is to avoid high waves which prolong the voyage, increase fuel consumption, and damage cargo. When waves exceed 4 m in height, all ships have to slow down or change course and since 1954 the US naval authorities have attempted to forecast such conditions for their ships.

Other routeing methods have been developed by the Royal Netherlands Meteorological Institute as outlined by Tunnell (1966) and the general procedure for computing the least-time track has been described by Evans (1968). Figure 6.8A represents isopleths of wave heights associated with a depression superimposed on divergent lines approximating to the route direction to be taken from the point of departure at A. The small numbers indicate the predicted speed of the ship at the various points based on the wave chart and information about the speed/wave-height relationships for the ship involved, and Fig. 6·8B is an isopleth chart of these expected speeds. The ship's progress along each divergent line is then plotted every 6 h on the isospeed map for the day in question, and a smooth curve (locus 1 on Fig. 6.8B) is drawn to show the anticipated possible positions of the ship after 24 h. The prognosis wave chart for the second day ahead is similarly converted into isospeed terms and locus 1 is then plotted on this chart (Fig. 6·8C) followed by the computation of a second locus (locus 2) for the period 48 h ahead. Finally, as shown on Fig. 6.8C, the optimum route track for the two days' sailing can be drawn in by moving from the point of origin at A and connecting the concave sections of the loci. There are a number of imperfections in the technique but the chief limitation has been the lack of ocean wave-field forecasts for more extended periods. However, Haltiner *et al.* (1968) have suggested that existing 2-day wave forecasts may be successfully extrapolated up to a week by the use of weather forecasts and this would certainly improve the longer-term efficiency of weather routeing.

According to Verploegh (1967) several countries, such as the German Federal Republic, the Netherlands, Norway, UK, US, and USSR, have already introduced operational weather routeing services but there is still little literature on the practical results which have been achieved. The most experienced user is probably the US Navy, since weather routeing has been regularly available throughout the US Naval Weather Service since September 1958 as reported by Cummins (1967). Most attention is concentrated on voyages by US Military Sea Transport Ships across the North Atlantic and in the fiscal year 1965 the Fleet Weather Facility in Norfolk, Virginia, issued about 800 routeings, on which 63·3 per cent of the ships completed the voyage without slowing for adverse weather, 84·2 per cent of the ships arrived on or ahead of scheduled steaming time, and 98·9 per cent of units completed the voyage without any reported damage to ship or cargo due to weather *en route*.

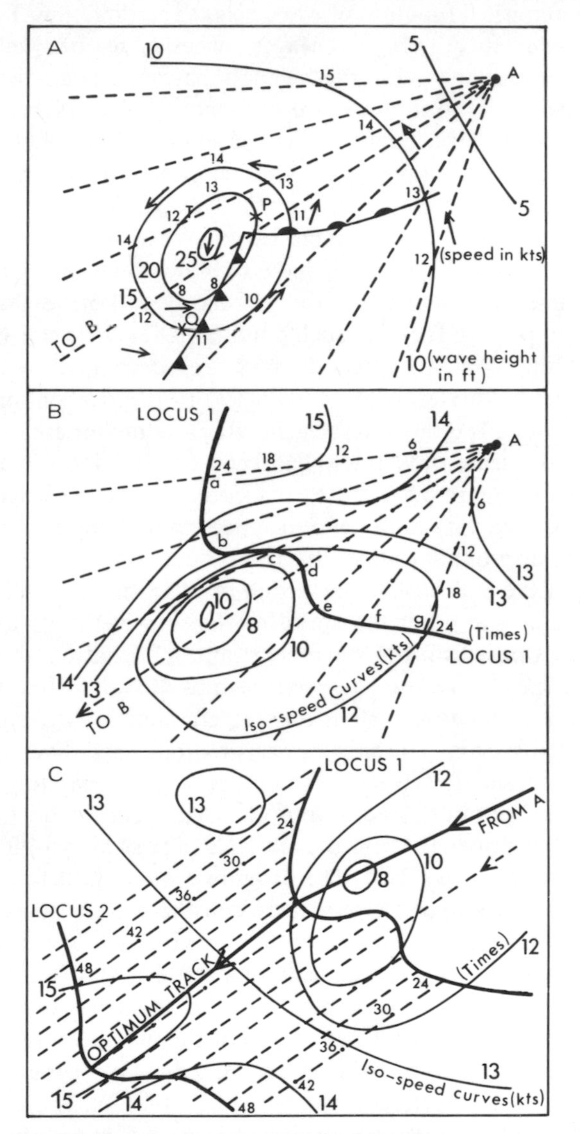


**Fig. 6.8 Weather routeing procedures for ships at sea.**
**(A) Wave-height isopleths and predicted ship speeds.**
**(B) Iso-speed curves and 24-h prediction locus.**
**(C) Iso-speed curves and 48-h prediction locus.**
**After Evans (1968).**

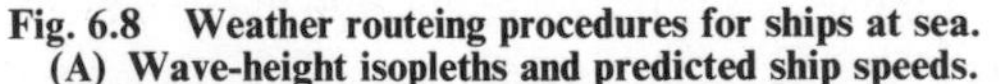

The economic aspects of ship routeing have been considered by Canham (1966) in relation to three different types of ship in service across the North Atlantic. For example, he estimated that for the fast passenger or cargo liner a reduction in fuel consumption of 10 per cent could be achieved in certain circumstances on an annual fuel bill of at least £100 000. A ship of this type would not profit by arriving early compared with the slower cargo vessels, which benefit directly by minimal-time routeing and are more dependent on the accuracy of medium- rather than short-range weather forecasts because they spend more time in passage. Altogether it was concluded that the potential annual saving in fuel costs alone for all British shipping in the North Atlantic would be not less than £2 million if weather routeing was adopted on a widespread scale. Commercial weather routeing has been employed by the Mobil Shipping Company and Spicer (1967) has stated that for an oil tanker of 95 000 tonnes deadweight a saving of only 30 min in vessel time on a trans-Pacific crossing is sufficient to pay for the routeing service, while a time-saving of 12 h represents a financial benefit of £1170.

*En route* delays at sea may lead to the deterioration of perishable cargo and many cargo ships have special design requirements for ventilation or insulation. This is especially important for cargoes such as fresh fruit or frozen meat destined for Europe from New Zealand or South Africa, which have to survive a long voyage through tropical and other climatic zones before reaching port. For some low-latitude countries, there is a constant problem of cargo deterioration, as illustrated by Ramaswamy (1970) in relation to dew-point temperatures and the prevention of sweat damage on Indian shipping routes.

## 6.6 Inland waters

Inland water transport on rivers, lakes, or canals may be adversely influenced by poor visibility or fluctuations in the available water due to floods and droughts, but one of the most regular disruptions to movement occurs with ice formation in northern climates. Losses are both of a direct nature—such as damage to ships, locks, bridges, navigation aids, and port facilities—and indirect as with the short-term and seasonal interruptions to industries, ports, and employment plus the effects on hydropower generation and water supply intakes. Acres (1971) has claimed that the annual net cost of ice formation on the navigable inland waters of Canada is at least $30 million and can be double this amount in severe years. Two of Canada's three main inland waterways (Great Lakes—St Lawrence and Great Slave Lake—Mackenzie systems) are closed to navigation for 4–9 months each year, and navigation on the Lac St Jean—Saguenay system is only possible with specially reinforced ships during winter.

At the present time, it is technically possible to extend the navigational season on the Great Lakes—St. Lawrence system by up to 6 weeks but the cost of $500 million is prohibitive. Nevertheless, there has been a recent increase in interest in the ice climatology of the Lakes area. Snider (1967) and Snider and Linklater (1969) have presented a series of ice synoptic charts; also, better and more rapid reporting of ice conditions has been achieved by the use of automatic picture transmission and black-and-white and infra-red imagery from orbiting satellites (Baker and Tiffany, 1968). Current forecasting procedures now give the dates after which ice-breaker operations will be profitable and the dates beyond which they will be no longer required, and it has even been suggested by Dingman *et al.* (1968) that the St Lawrence Seaway could probably be kept ice-free for the greater part of the winter as a result of the thermal pollution from a number of nuclear reactors.

# Land transport climatology

## 6.7 Roads

Road transport is highly flexible in that it includes a large proportion of private vehicles, each of which tends to operate independently to an individually planned route and timetable schedule compared with the more regular public services. In addition, some of the journeys in the private sector are associated with leisure activities and are non-essential, unlike the commercial movement of passengers or freight, and a combination of these factors leads to complex relationships between weather and traffic flows. Tanner (1952) has demonstrated that the volume of traffic during any period of time is not a simple function of the weather in that period but also depends on the weather when the journey was planned. Generally speaking, adverse weather reduces traffic densities whereas fine weather increases them, especially during summer weekends and other holiday periods. The more exposed forms of transport, such as pedal cycles and motor cycles, are about twice as sensitive to atmospheric conditions as cars; on very foggy days, total traffic may be limited to between half and three-quarters of that recorded on clear days.

To assist in the individual decision-taking process about road journeys, many countries have advisory organizations. In Britain, for example, the Automobile Association maintains a national network reporting on weather and road conditions, handling about 1 million enquiries each winter and up to 10 000 telephone calls per day to the London headquarters alone in exceptionally bad weather (Anon., 1966). Public transport is also weather-dependent, as shown by van Cleef (1917) in a pioneer study of street-car traffic in Duluth, Minnesota, where precipitation accompanied by wind was the most important element in reducing passenger movement and 100 forecasts of average daily traffic during 1914, based on five weather factors, were within 3·2 per cent of the correct figure. The phasing of weather events through the day may be significant and Whiten (1947) has noted that the London taxi-driver does some of his best business when a fine morning is followed by heavy mid-afternoon showers, which force shoppers and others to use cabs. On the other hand, a wet morning deters the public from making trips to town.

Weather conditions influence road safety and the accident frequency by affecting both the volume of traffic and the risks per unit of travel in adverse circumstances such as slippery roads or poor visibility (Road Research Laboratory, 1963). Therefore, poor weather does not necessarily lead to a proportional increase in accidents. Although the total number of casualties during winter in Britain is greater in wet weather than in dry conditions, the total is probably smaller during wet conditions in summer owing to the curtailment of non-essential journeys and the smaller number of road users exposed to risk. In fact, daily totals of winter accidents in Britain increase up to 25 per cent on wet days without icy roads compared to dry days and Robinson (1965) has reported an increase of 30 per cent in road accidents on rainy days in Melbourne, Australia. On the other hand, the incidence of skidding accidents per unit of travel due to wet roads may be as much as 2–3 times higher in summer than in winter.

These accidents are caused by the reduced braking efficiency of motor vehicles, which appears most serious when roads become wet after a spell of warm and comparatively dry weather. As might be expected, a sharp rise in skidding incidents occurs on icy roads, with most accidents taking place on bends on heavily trafficked roads. Many of these accidents involve morning commuter traffic and in one English county 40 per cent of all icy-road accidents took place between 7 am and 11 am (Road Research Laboratory, 1963).

Although rain, snow, ice and fog represent the most important hazards for road transport, Ashmore (1955) has drawn attention to the problem of mirages and Price (1971) has stressed the problem of high-speed motorways in areas suffering from wind exposure. Thus, during 1966–68 in the English county of Yorkshire, 37 vehicles were overturned by wind forces, 20 driven off course to strike safety fences, and 70 two-wheeled vehicles were involved in accidents in which wind was a factor.

The maintenance of safe and efficient road transport during spells of adverse weather is a major concern of the road traffic engineer. Most progress has been achieved against the hazards which affect the road surface itself, but many of these maintenance problems can be minimized at the planning stage of new roads by the selection of the most advantageous route and the provision of adequate design features such as drains or culverts to avoid erosion and flooding (Landsberg, 1947). The extension of motorways into relatively unpopulated or hostile environments may require the extrapolation of existing climatic data combined with design experimentation and Lovell (1966) has described how the location of cuttings and embankments, together with their side gradients, may affect snow accumulation on a major new high-altitude road in Britain.

In the mid-latitudes, the disruption caused to winter road transport by snow and ice is the chief maintenance problem, as outlined by Price (1960), and Rooney (1967) has recognized a five-fold hierarchy of urban snow disruption in the US. Although the frequency of disruption tended to increase with the annual accumulation of snow, Rooney found less disruption in western cities than in urban areas further east, as shown in Fig. 6.9,

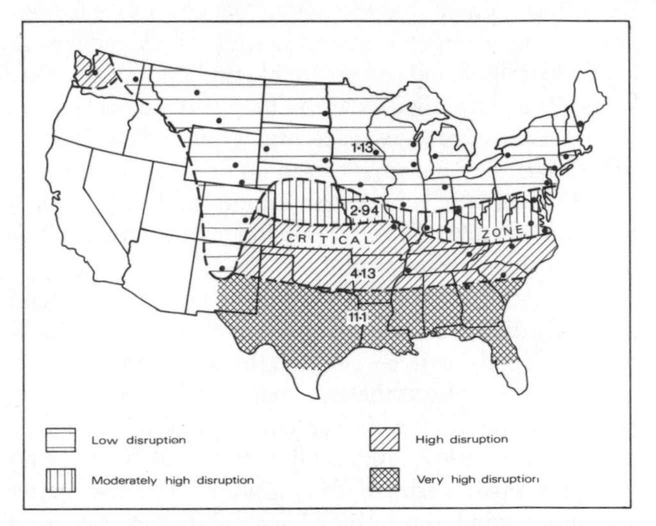


**Fig. 6.9 A generalized pattern of snow-caused disruption in the US. Figures indicate average number of first- and second-order disruptions per 250 mm of snowfall for the sites within each zone. Shading represents the relation of disruption to snowfall. After Rooney (1969).**

and this was attributed to the lower water content of western snowfalls, which made the snow easier to remove, and to the fact that there was a more realistic perception of, and adjustment to, the snow hazard in the west compared to the east and south. Satisfactory winter road maintenance now requires considerable preplanning, as shown by Deetz (1961), and the cost of the snow control programme in New York City alone during 1963–64 totalled $22 million.

The risk of snow and ice disruption can be assessed from estimates of snow-lying frequencies at summit sections of roads, as illustrated by Manley (1938), or, in the short-term, by the use of weather forecasts. Hay (1969) has reported that minimum road surface temperatures may be up to 3 degC lower than minimum air temperatures and there has been a recent emphasis on the forecasting of night minimum temperatures on concrete road surfaces (Parrey *et al.*, 1970 and 1971; Thornes, 1972). In some areas, devices sensitive to moisture and temperature placed in the highway surface may provide warning of freezing conditions. To be effective, these devices, like any other warning system, must allow enough time for preventive action, such as the treatment of roads with salt, to be taken. This raises the question of the most effective alarm temperature for the sensors embedded in the road surface, since a setting which is too low may provide an inadequate warning period, whereas a setting which is too high can lead to an excessive number of false alarms. This problem is illustrated in Table 6.1, taken from Hay and Young (1972), which shows that at alarm settings below 1·5 °C there may well be insufficient time under British conditions in which to complete the salting operation. By assuming that the salting takes place at a uniform rate and that an accident caused solely by the presence of ice can occur anywhere along the stretch of road concerned, it was calculated that the *relative* probability of an accident due to incomplete salting operations increased from 1 at a setting of 1·5 °C to 2·7 at 1·0 °C and rose to 5·2 at a setting of 0·5 °C. Therefore, despite the fact that a setting of 1·5 °C led to false alarms in almost one-third of all cases, it was concluded that this temperature represented the best compromise.

The direct protection of roads from winter hazards may take a variety of forms. An important priority is the use of fences in snow control but, despite comprehensive surveys of the mechanics of snowdrifting in relation to road transport by Pugh (1950) and Schneider (1962), there are few procedures which can be universally applied. The function of a snow fence is to reduce wind velocity enough to cause the snow to fall before it is carried onto the road surface, but each type of fence has its own characteristics. Thus, a solid fence produces drifts on both windward and leeward sides, whereas open fences develop drifts chiefly to leeward, and the leeward drift associated with a solid fence is normally short and deep compared with that produced by an open fence. In general, a good open fence, with a density of about 50 per cent, provides better protection than a solid fence and is also subject to less wind pressure.

Figure 6.10A shows the characteristic airflow pattern and eddy area developed behind a vertical-slat fence 1·2 m high with a density of 50 per cent. As indicated in Fig. 6.10B, the eddy closely influences the deposition of the snow and the shape of the subsequent drift which soon builds up to the top of the eddy area. As more snow is deposited, the particles slide downwind along the top of the drift to fill out the entire area of the eddy. The profile of the completed drift is that of an ichthyoid curve. Nevertheless, disagreement exists about the optimum dimensions, material construction, and precise location of fences in particular areas. The most usual British practice appears to be to use lath and wire fences

| *Alarm temperature settings (°C)* | *Frequency recorded (per cent)* | | | | | | |
|---|---|---|---|---|---|---|---|
| | *Amount of warning (h)* | | | | | | |
| | $<\frac{1}{2}$ | $\frac{1}{2}$–1 | 1–$1\frac{1}{2}$ | $1\frac{1}{2}$–2 | 2–5 | $>5$ | *False alarms* |
| 1·5 | 0 | 10 | 9 | 17 | 27 | 6 | 31 |
| 1·0 | 4 | 23 | 17 | 10 | 18 | 2 | 26 |
| 0·5 | 25 | 41 | 13 | 4 | 6 | 0 | 11 |

**Table 6.1** Percentage frequency of given warning periods and false alarms for different temperature settings of an ice-warning device. (After Hay and Young, 1972.)

perpendicular to the wind direction at a distance about 15 times the fence height from the stretch of road to be protected.

Once snow and ice have reached the road surface, they may be removed by road heating, mechanized methods, or the use of chemicals and abrasives. Road heating is the most expensive technique and is restricted to heavily trafficked locations in urban areas. In Britain, most use has been made of electrical resistance cables, which can be automatically activated, and in the US use has also been made of heated liquids such as mineral oils circulating through pipes buried beneath the highway surface. Mechanical methods of snow clearance by ploughs is naturally undertaken only after there has been a considerable accumulation of snow. For example, the fixed-blade vee plough is usually used for drift-breaking purposes, and rotary-blade ploughs or blowers require a certain minimum snow depth for efficient operation.

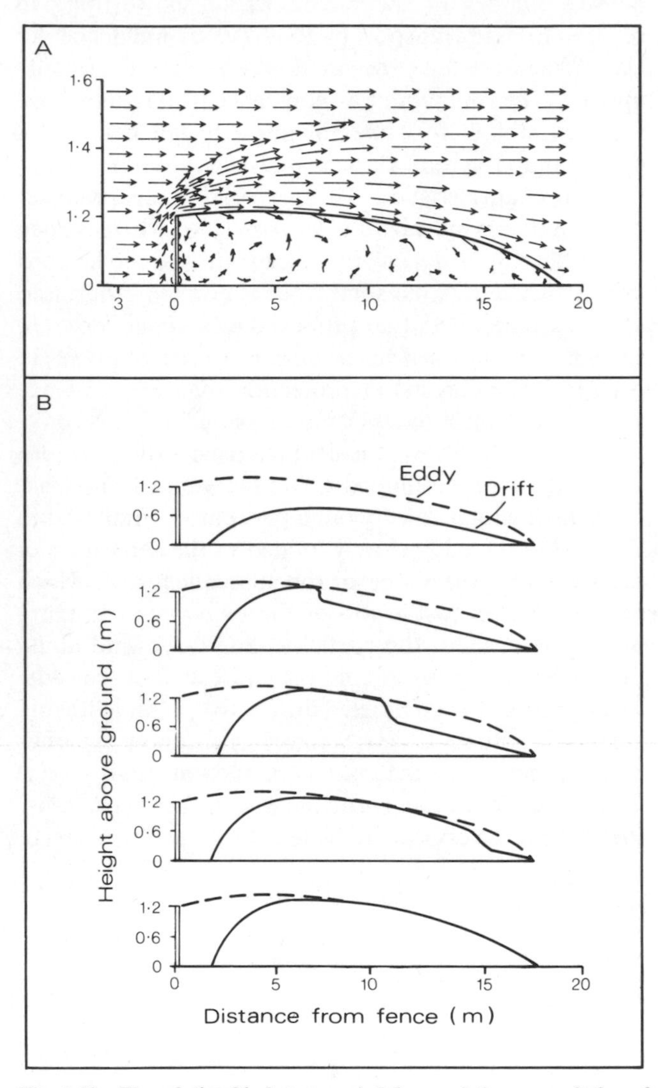


**Fig. 6.10 The relationship between wind flow and the accumulation of a snow drift behind a vertical-slat fence 1·2 m high with a density of 50 per cent.**
**(A) Eddy area behind the fence.**
**(B) Stages in the growth of a drift behind the fence.**
**After Pugh (1950). Reproduced by permission of the Controller of Her Majesty's Stationery Office.**

Despite the continuous improvement of mechanical equipment detailed by Tapley (1963) and Selley (1968), the rapid upsurge in motor vehicle numbers over the last twenty years has led to the demand for higher standards of winter road maintenance based on the widespread application of salt or other de-icing chemicals to prevent the initial accumulation of ice and snow. Abrasives, such as grit, are still used to help vehicular traction on ice or when there is more than 25–50 mm of hard-packed snow, but the recent trend in most countries has been towards an increased use of salt alone, so that in Britain, for example, about 1·5 million tonnes of salt is used annually, mostly in regular applications to the major road networks comprising about one-third of the total road system. The chief reasons for this trend are that rock salt can now be stored in a friable state for long periods of time and may be spread more quickly than when mixed with abrasives without leaving a grit residue to block roadside drainage systems.

Salt is capable of melting ice down to temperatures of −21 °C but, according to the Road Research Laboratory (1968), most snow falls and icing in Britain occur at temperatures above −3 °C and, since thin ice films on road surfaces are rarely more than 0·25 mm thick, the recommended precautionary practice is for salt to be spread at a rate as low as 17 g/m$^2$. If ice has already formed this rate is increased to 68 g/m$^2$ and for snow removal about 17 g/m$^2$ of salt is required per 25 mm of fresh snow per degree Centigrade that the air temperature is below freezing point. If the depth of ice or compacted snow exceeds 50 mm, ploughing is usually started, although very heavy salt applications may be used instead. The increasing use of salt has undoubtedly led to improved winter road transport conditions, but there are attendant problems of vehicle corrosion and local pollution. Although salting is inexpensive compared with some other methods of winter road treatment, Hay and Young (1972) have claimed that in Britain the annual cost of the salt alone has now reached £7 million, and the accelerated corrosion of motor vehicles due to salt applications has been estimated to cost another £50–60 million per year. Hutchinson (1970) has drawn attention to the State of Maine, where average annual applications of almost 25 tonnes of sodium chloride on each 1·5 km of paved highway have produced chloride-ion contamination of wells and farm ponds plus excessive sodium concentrations in soil adjacent to the salted roads.

In certain high-latitude areas, such as the 20 per cent of Canada which is covered in fresh water and muskeg, winter conditions may actually provide easier road trafficability. In the Yukon and North West Territories, much use is made of winter roads built on frozen lakes and rivers, and Acres (1971) has estimated that construction costs average only $100–200 per 1·5 km compared with around $50 000 for each 1·5 km of permanent road in shield and muskeg areas. These seasonal roads bring significant economic benefits to the areas which they serve and the road which is regularly constructed each winter on Great Slave Lake between Yellowknife and Fort Reliance allows a net saving of $6·00 per tonne of goods moved compared with the cost of air freight which is the next cheapest form of transport.

## 6.8 Railways

Although railways often remain operational when other forms of transport are at a standstill, they are by no means immune to weather influences, as shown in the comprehensive review by Hay (1957). Several writers, such as Critchfield (1966), stress the impact of severe storms involving the disruption of services through floods, earthslides, or avalanches but some problems can arise from much less dramatic conditions. For example, Champion (1947) has described the effects of temperature fluctuations on the expansion and contraction of metal rails and linkages, when a kilometre of signal wire may change in length by up to 0·25 m in a 12–h period. Fog is a major hazard to most railway operations and Wintle (1960) claimed that an expenditure of £20 million was necessary to increase safety by the extension of modern colour-light signalling and automatic warning systems to all main lines in Britain.

As in the case of road transport, however, snow and ice are probably the greatest sources of railway disruption over much of the world. In areas regularly experiencing severe winters, such weather disruptions may be reduced to a minimum by adequate capital expenditure on snow-clearing equipment and an informed expectation of the risk. Thus, Rooney (1969) has quoted that the New York Central Railroad normally spends 0·74 per cent of its operating expenses on snow and ice removal and only 12 per cent of the winter train delays on the Chicago, Milwaukee, St Paul, and Pacific Railroad are due to weather conditions, mostly related to snow difficulties in the northern Rockies section.

In comparison, snowfall in Britain is a more variable element and Canovan (1971) has presented a detailed account of the rail transport problems associated with a recent trend towards a longer snow season. Many of the problems are concerned with track maintenance and the icing-up of points is an important cause of delays. This has been counteracted by installing over 8000 point-heaters, many of which use propane gas and are thermostatically controlled. On lines employing top-contact third-and-fourth-rail electric systems, the formation of ice on the conductor rails reduces the voltage reaching the train motors and on 6 January 1970 an accretion of glaze on just one commuter line into London halted trains containing 4000 passengers.

With overhead electric equipment, few problems have arisen, since ice on the wires does not interfere with the pick-up of current, but Parrey (1970) has documented one example of the disruption of train services into London as the combined weight of snow and ice brought down the locomotive pantographs sufficiently to prevent contact with the overhead line equipment. A major source of cancellations is the overnight icing-up of rolling stock in the berthing sidings, the exits of which may be blocked by snow or clogged points and the damp snow characteristic of Britain often freezes after falling, thus welding the moving parts of mechanical equipment and even obliterating signals under wind-driven conditions.

Snow fences are used to reduce the amount of snow reaching the track in the uplands and drifts are also minimized by decreasing the slopes of the sides of railway cuttings as far as possible to about 1 in 10 to reduce the size of the downwind eddy (Fig. 6.11A). In countries where severe drifting occurs the track is constructed on low embankments, as indicated in Fig. 6.11C. Minor drifts build up on either side but soon reach saturation and the track is then swept clear by the wind. The slopes of the embankment must not be too steep, and the line must not be raised too high above the surrounding ground, as otherwise a marked downwind eddy similar to that in Fig. 6.11B will develop, leading to the formation of a cornice and the accumulation of snow on the downwind line.

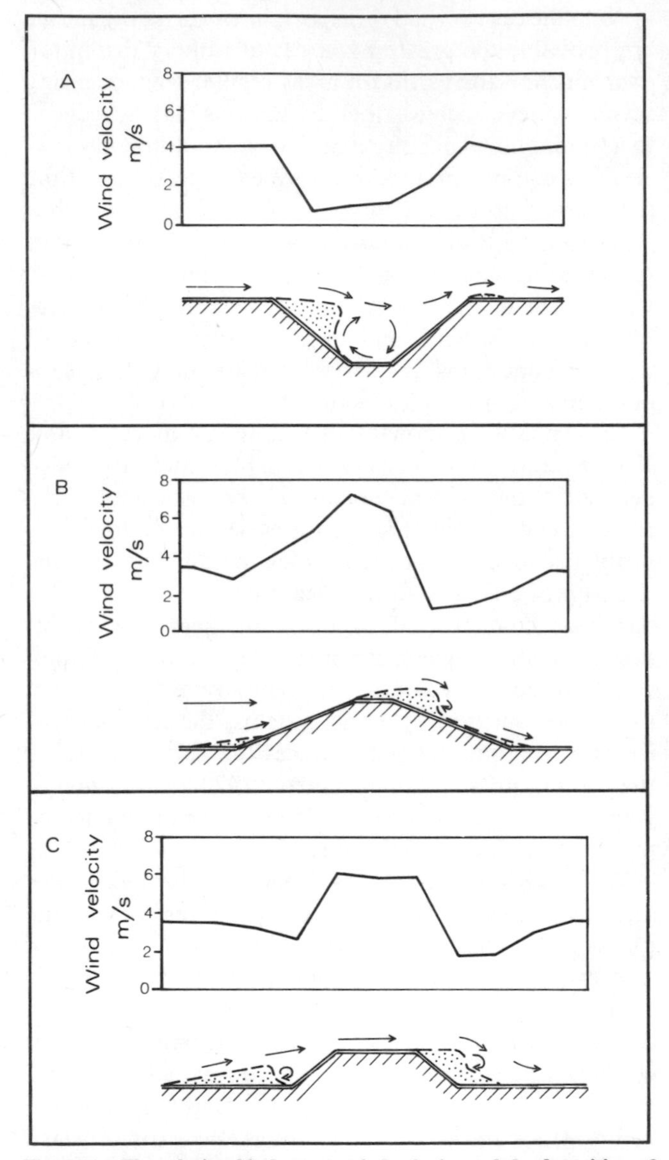


**Fig. 6.11 The relationship between wind velocity and the deposition of snow around various obstacles.**
**(A) A cutting.**
**(B) A high bank.**
**(C) A low bank.**
**After Pugh (1950). Reproduced by permission of the Controller of Her Majesty's Stationery Office.**

Delays to rail services may cause the deterioration of perishable freight as well as passenger inconvenience and, even under normal running conditions, the movement of certain goods requires special protection from the weather. Since most chemical and bacteriological reactions approximately double for each 10 degC temperature increase, insulated vehicles are normally used for meat and fresh fish, and in summer fruit and vegetables must be provided with ample ventilation and space must be left for liquids in transit to expand with a rise in temperature (Henley, 1951). In winter, condensation damages sheet steel and sugar confectionery, and frost can affect goods such as seed potatoes. Technology is helping to overcome many of the climatic difficulties affecting railway operations but there are always new problems to be solved. For example, the advent of the advanced passenger train (APT) capable of speeds above 240 km/h raises the possibility of trains being overturned by wind forces (Price, 1971). Important effects are also created by the movement of such trains past nearby objects and train windows can be sucked out by the pressure changes at high speeds.

## References

ACRES, H. G. LTD (1971). Review of current technology and evaluation of research priorities. *Report Series No. 17, Inland Waters Branch*, Department of the Environment (Ottawa), 299 pp.

ALLEN, L. (1951). Navigating with the weather. *Bull. Amer. Met. Soc.*, **32**:245–50.

ANONYMOUS (1966). How the AA reports on weather. *Weather*, **21**:429–32.

ASHMORE, S. E. (1955). North Wales road mirage. *Weather*, **10**:336–42.

BAKER, J. L. and TIFFANY, O. L. (1968). Using satellites to look at the Great Lakes ice. *Limnos*, **1**:12–16.

BARRY, W. S. (1965). *Airline Management*. Allen and Unwin Ltd, London, 352 pp.

BECKWITH, W. B. (1966). Impacts of weather on the airline industry: the value of fog dispersal programmes. In Sewell, W. R. D. (ed.) *Human dimensions of weather modification*. University of Chicago, Research Papers in Geography No. 105, 195–207.

BOSART, L. F. (1971). Weather at the launch of Apollo 12. *Weather*, **26**:19–23.

BRIGGS, J. (1972a). Probability of aircraft encounters with hail. *Met. Mag.*, **101**:33–8.

BRIGGS, J. (1972b). Probabilities of aircraft encounters with heavy rain. *Met. Mag.*, **101**:8–13.

BURNHAM, J. (1968). Atmospheric gusts: A review of the results of some recent R.A.E. research. *Royal Aircraft Establishment, Tech. Rep. 68244*, 25 pp. (Reported by Roach, W. T., 1970.)

BURNHAM, J. (1969). Atmospheric turbulence at supersonic transport cruise altitudes. *Q. Jl. R. Met. Soc.*, **95**:782–3.

CANHAM, H. J. S. (1966). Economic aspects of weather routeing. *Marine Observer*, **36**:195–9.

CANOVAN, R. A. (1971). Wintry prospects for British Rail. *Weather*, **26**:472–91.

CHAMPION, D. L. (1947). Weather and railway operation in Britain. *Weather*, **2**:373–80.

CHAPLIN, J. C. (1969). Atmospheric effects on the supersonic transport. *Q. Jl. R. Met. Soc.*, **95**:779–81.

COBB, W. E. and HOLITZA, F. J. (1968). A note on lightning strikes to aircraft. *Mon. Weath. Rev.*, **96**:807–8.

CORBY, G. A. (1957). Air flow over mountains. Notes for forecasters and pilots. *Meteorological Reports No. 18*, Met Office, HMSO, London.

CRITCHFIELD, H. J. (1966). *General climatology* (2nd Edn). Prentice-Hall Inc., New Jersey, 420 pp.

CROSSLEY, A. F. (1961). Hail in relation to the risk of encounters in flight. *Met. Mag.*, **90**:101–10.

CROSSLEY, A. F. (1969). The atmospheric environment of supersonic transport aircraft. *Q. Jl. R. Met. Soc.*, **95**:784–8.

CUMMINS, W. E. (1967). Practical results of weather routeing. *Marine Observer*, **37**:23–6.

DEETZ, R. S. (1961). Winter maintenance policies on the Ohio turnpike. *Amer. Road Builders' Ass. Tech. Bull.*, **229**:5–14.

DINGMAN, S. L., WEEKS, W. F., and YEN, Y. C. (1968). The effects of thermal pollution on river-ice conditions. *Water Resour. Res.*, **4**:349–62.

DUBOIS, E. (1965). Fog dispersal on runway approaches. *Weather*, **20**:313–14.

EVANS, S. H. (1968). Weather routeing of ships. *Weather*, **23**:2–8.

FRANKCOM, C. E. N. (1966)' The general problem of weather routeing by the shipmaster himself or as advised by the meteorologist ashore. *Marine Observer*, **36**:186–91.

FREEMAN, M. H. (1970). Weather forecasting for supersonic transport. *Met. Mag.*, **99**:138–43.

GLOYNE, R. W. (1966). Some features of the climate of the immediate area served by the Forth and Tay road bridges. *Scott. Geogrl. Mag.*, **82**:111–18.

HALTINER, G. J., BLEICK, W. E., and FAULKNER, F. D. (1968). A proposed method for ship routeing using long range weather forecasts. *Mon. Weath. Rev.*, **96**:319–22.

HALTINER, G. J. and HAMILTON, H. D. (1962). Minimal-time ship routeing. *J. Appl. Met.*, **1**:1–7.

HARROWER, T. N. S. and EVANS, D. C. (1956). Damage to aircraft by heavy hail at high altitude. *Met. Mag.*, **85**:330–8.

HAY, J. S. (1969). Some observations of night minimum road temperatures. *Met. Mag.*, **98**:55–9.

HAY, J. S. and YOUNG, C. P. (1972). Weather forecasting for the prevention of icy roads in the United Kingdom. In Taylor, J. A. (ed.) *Weather forecasting for agriculture and industry*, David and Charles, 155–65.

HAY, R. F. M. (1956). Ice accumulation upon trawlers in northern waters. *Met. Mag.*, **85**:225–29.

HAY, W. W. (1957). Effects of weather on railroad operation, maintenance and construction. *Meteor. Monog.*, **2**:10–36.

HENLEY, E. D. (1951). Weather and the carriage of freight by rail in Britain. *Weather*, **6**:233–6.

HICKS, J. R. (1967). Improving visibility near airports during periods of fog. *J. Appl. Met.*, **6**:39–42.

HUTCHINSON, F. E. (1970). Environmental pollution from highway de-icing compounds. *J. Soil Water Cons.*, **25**:144–6.

JONES, R. F., MCINTURFF, R. M., and TEWELES, S. (1967). Meteorological problems in the design and operation of supersonic aircraft. *WMO Tech. Note* No. 89, 71 pp.

KEDDIE, B. (1971). Some aspects of wind information required in the landing of aircraft. *Met. Mag.*, **100**:134–43.

KREITZBERG, C. W. (1967). Mesoscale temperature changes in the supersonic transport climb region. *J. Appl. Met.*, **6**:905–10.

LANDSBERG, H. (1947). Include climate in highway plans. *Better Roads*, **17**:25–8.

LOVELL, S. M. (1966). Design of roads to minimize snow, fog and ice problems. *Municipal Engng.*, **143**, 26:1369, 1371, and 1373.

LUDLAM, F. H. (1967). Characteristics of billow clouds and their relation to clear air turbulence. *Q. Jl. R. Met. Soc.*, **93**:419–35.

MCGINNIGLE, J. B. (1970). An occurrence of highly localised clear-air turbulence over the southern Mediterranean. *Met. Mag.*, **99**:208–11.

MANCUSO, R. L. and ENDLICH, R. M. (1969). Analysing and forecasting clear-air turbulence probabilities over the United States. *Mon. Weath. Rev.*, **97**:527–33.

MANLEY, G. (1938). Snowfall and its relation to transport problems, with special reference to northern England. *Geogrl. J.*, **92**:522–26.

MASON, B. J. (1966). The role of meteorology in the national economy. *Weather*, **21**:382–93.

MASON, D. (1954). Hill standing waves and safety heights. *Weather*, **9**:45–55.

MASON, D. (1964). Lightning strikes on aircraft—II. *Weather*, **19**:253–5.

MAYBOURN, R. (1971). Problems of operating large ships in the Arctic. *J. Inst. Nav.*, **24**:135–46.

MAYL, S. D. (1969). Instant weather dodging. *Weather*, **24**: 196–9.

METEOROLOGICAL OFFICE (1957). *Elementary meteorology for aircrew*. HMSO, London, 239 pp.

MOORER, T. H. (1966). Importance of weather to the modern seafarer. *Bull. Amer. Met. Soc.*, **47**:976–9.

NICHOLLS, J. M. (1970). Meteorological effects on the sonic bang. *Weather*, **25**:265–71.

NICHOLLS, J. M. (1973). Aircraft measurements of disturbed airflow over mountains. *Weather*, **28**:141–52.

OSMUN, W. G. (1969). Airline warm fog dispersal programme. *Weatherwise*, **22**:48–53.

PARREY, G. E. (1970). A railway problem during the heavy snowfall of 4 March 1970. *Met. Mag.*, **99**:299–304.

PARREY, G. E., RITCHIE, W. G., and VIRGO, S. E. (1970). Comparison of methods of forecasting night minimum temperatures on concrete road surfaces. *Met. Mag.*, **99**: 349–55.

PARREY, G. E., RITCHIE, W. G., and VIRGO, S. E. (1971). Minimum temperatures at the surfaces of concrete roads and concrete slabs. *Met. Mag.*, **100**:27–31.

PLANK, V. G. (1969). Clearing ground fog with helicopters. *Weatherwise*, **22**:91–8.

PRICE, B. T. (1971). Airflow problems related to surface transport systems. *Phil. Trans. R. Soc., Series A*, **269**:327–33.

PRICE, W. I. J. (1960). Modern trends in winter road maintenance. *J. Instn. Highway Engrs.*, **7**:319–26.

PUGH, H. L. D. (1950). *Snow Fences*. Road Research Laboratory, Tech. Paper No. 19, HMSO, London, 52 pp.

RAMASWAMY, C. (1970). Dew point temperatures over the world and their importance to Indian shipping. *Current Science*, **39**:291–6.

ROACH, W. T. (1967). On the nature of the summit areas of severe storms in Oklahoma. *Q. Jl. R. Met. Soc.*, **93**:318–36.

ROACH, W. T. (1969). Some aircraft reports of high-level turbulence. *Met. Mag.*, **98**:65–78.

ROACH, W. T. (1970). Weather and Concorde. *Weather*, **25**:254–64.

ROACH, W. T. and JAMES, B. F. (1972). A climatology of the potential vertical extent of giant cumulonimbus in some selected areas. *Met. Mag.*, **101**:161–81.

ROAD RESEARCH LABORATORY (1963). *Research on road safety*. HMSO, London, 602 pp.

ROAD RESEARCH LABORATORY (1968). Salt treatment of snow and ice on roads. *Road Note No. 18* (2nd Edn.), HMSO, London.

ROBERTS, E. D. (1971). *Handbook of aviation meteorology*. HMSO, London, 404 pp.

ROBINSON, A. H. O. (1965). Road weather alerts. In: *What is weather worth?* Australian Bureau of Meteorology, Melbourne, 41–3.

ROONEY, J. F. (1967). The urban snow hazard in the United States: an appraisal of disruption. *Geogrl. Rev.*, **57**:538–59.

ROONEY, J. F. (1969). The economic and social implications of snow and ice. In Chorley, R. J. (ed.) *Water, earth and man*. Methuen, London, 389–401.

SANSOM, H. W. (1954). Damage to aircraft by hailstones. *Weather*, **9**:125.

SCHAEFER, V. J. (1946). The production of ice crystals in a cloud of supercooled water droplets. *Science*, **93**:239–40.

SCHERHAG, R., WARNECKE, G., and WEHRY, W. (1967). Meteorological parameters affecting supersonic transport operations. *Inst. Navigation J.*, **20**:53–63.

SCHNEIDER, T. R. (1962). *Snowdrifts and winter ice on roads*. Tech. Translation No. 1038. National Research Council of Canada, Ottawa, 200 pp.

SELLEY, J. S. (1968). Progress and improvements in winter-maintenance equipment. *Municipal Engng.*, **145**, 23:115 and 117.

SMITH, D. P. (1970). Trawler safety off Iceland. An account of the weather service provided from the 'Orsino' during the winter of 1968–9. *Marine Observer*, **40**:24–31.

SNIDER, C. R. (1967). Great Lakes ice season of 1967. *Mon. Weath. Rev.*, **95**:685–96.

SNIDER, C. R. and LINKLATER, G. D. (1969). Great Lakes ice season of 1968. *Mon. Weath. Rev.*, **97**:315–32.

SPICER, H. (1967). Mobil's experience with commercial weather routeing. *Marine Observer*, **37**:30–3.

SPILLANE, K. T. (1967). Clear air turbulence and supersonic transport. *Nature*, **214**:237–9.

SUGDEN, L. (1969). Provision of meteorological service for supersonic transport operations. *Q. Jl. R. Met. Soc.*, **95**: 789–94.

TANNER, J. C. (1952). Weather and road traffic flow. *Weather*, **7**:270–5.

TAPLEY, W. P. (1963). Development of winter maintenance equipment for motorways. *The Surveyor and Municipal Engineer*, **122**, 3722:1237–40.

THORNES, J. E. (1972). An objective aid for estimating the night minimum temperature of a concrete road surface. *Met. Mag.*, **101**:13–24.

TUNNELL, G. A. (1966). An account of the ship routeing methods developed by the KNMI (The Royal Netherlands Meteorological Institute). *Marine Observer*, **36**:192–5.

van Cleef, E. (1917). The influence of weather on street-car traffic in Duluth, Minnesota. *Geogrl. Rev.*, **3**:126–34.

Verploegh, G. (1967). Weather routeing of ships. *World Met. Org., Bull.*, **16**:139.

Vonnegut, B. (1947). The nucleation of ice formation by silver iodide. *J. Appl. Phys.*, **18**:593–5.

Wallington, C. E. (1964). Lightning strikes on aircraft—I. *Weather*, **19**:206–8.

Walker, J. M. and Penney, P. W. (1973). Arctic sea-ice and maritime transport technology. *Weather*, **28**:358–71.

Walker, J. M. and Scott, K. C. (1972). The North-West Passage: A marine highway? *Weather*, **27**:326–32.

Ward, A. (1954). Sea breezes at North Front, Gibraltar. *Meteorological Office, Met. Reports*, **2**, 14, HMSO, 8 pp.

Whiten, A. J. (1947). Weather and the cabby. *Weather*, **2**: 308–9.

Wintle, B. J. (1960). Railways versus the weather. *Weather*, **15**:137–9.

Woods, J. A. (1972). Satellite radiation measurements and clear air turbulence probability. *Science*, **177**:1100–2.

Wycoff, P. H. (1966). Evaluation of the state of the art. In Sewell, W. R. D. (ed.) *Human dimensions of weather modification*, University of Chicago, Research Papers in Geography No. 105, 27–39.

# 7. Climate and the community

### 7.1 Introduction

The main purpose of this chapter is to identify and discuss three further areas of social and economic life into which climate intrudes, namely: biometeorology, building climatology, and employment and recreation climatology. It should be stressed, however, that the selection of these largely independent themes is somewhat arbitrary and that, even when considered with the contents of the previous chapters, these topics are not necessarily intended to cover the complete range of climatic influences on the community. Indeed, it will be apparent that there are other aspects of corporate life that involve an appreciation of atmospheric events, although the role of climate cannot perhaps be seen as clearly, or quantified as easily, as in the case of some weather-dependent economic activities. This is not to say that certain of these aspects are unimportant and, for example, several workers have interested themselves in the artistic and aesthetic implications of weather and climate.

Bonacina (1972) noted that a recurrent theme within this field has been the study of the perception and portrayal of skyscapes in literature and art, and, in many instances, it can be shown that poets and artists have provided a surprisingly accurate record of landscape meteorology. In a detailed appraisal of Shelley's poem *Ode to the West Wind*, Ludlam (1972) concluded that the description of an approaching thunderstorm written near Florence towards the end of October 1819 was not only unrivalled in English literature but, until very recently, compared quite favourably with existing scientific accounts of this complex phenomenon. Furthermore, in addition to the actual representation of the cloud forms, it was also clear that Shelley understood how the storm related to the sudden onset of the Mediterranean winter.

In the sphere of landscape painting, the work of John Constable has long been praised for having a realistic approach to skyscapes (Bonacina, 1937) and even criticisms levelled by Badt (1950) at the structure of cumulus clouds in a specific painting have been refuted by Tyldesley (1968) on the basis that the picture represents a particular airmass situation with exceptional visibility. Some meteorologists have used landscape paintings as a means of a better understanding of past regional climates and Lamb (1966 and 1967) was able to show that the percentage of cloudiness depicted in certain European paintings provided corroborative evidence of the variations in summer climate over north-western Europe during several centuries. In a much more comprehensive survey, Neuberger (1970) studied more than 12 000 paintings in the US and eight European countries covering the period from 1400 to 1967. More than half of all the paintings contained some meteorological information and significant differences were observed in the blueness of the sky and the degree of visibility or cloudiness, with the clearest atmospheres represented by the Mediterranean schools and the cloudiest by the British school. The paintings also provided convincing evidence of climatic changes, such as the period of the so-called Little Ice Age between 1550 and 1850 when there was a marked increase in the cloudiness of the pictures.

## Biometeorology

### 7.2 Physiological comfort

The average human body is most efficient at a core temperature of 37 °C and the principal effect of the weather on physiological comfort occurs through modifications to the heat balance, as recognized in early review papers by Gold (1935) and Brunt (1943). Following Landsberg (1969), bodily heat balance can be expressed as:

$$M \pm R \pm C - E = 0$$

This formula indicates that, to maintain thermal equilibrium, the metabolic heat ($M$) created chemically within the body together with the atmospheric heat gained or lost by radiation ($R$) and convection ($C$) and lost by evaporation ($E$) must add up to zero. To achieve this state, the body has to dissipate approximately $2 \cdot 5 \times 10^6$ calories daily derived from food conversion, which otherwise would lead to an increase in body temperature of almost 2 degC/h, while the heat generated in internal organs and muscles is partly advected in the bloodstream and partly conducted to the surface of the skin (Green, 1967).

Compared with the natural fluctuations of air temperature, the human body is comfortable within only a narrow thermal range, and beyond an internal temperature of 26–40 °C irreversible deterioration often results, although recovery has been recorded from falls as low as 18 °C and rises as high as 43·5 °C. In practice, bodily comfort is experienced on the more sensitive areas of the skin surface, which is naturally 2–5 degC cooler than the core temperature, and usually lies between 31 and 35 °C. Outside these limits, the body adjusts the rate of heat production or loss in various ways. Thus, at a skin temperature of around 30 °C, shivering occurs in an attempt to generate more heat, whereas sweating is employed as the main cooling mechanism at high levels of heat and humidity. The relative importance of the various heat transfer processes differs markedly according to the prevailing climatic conditions. For example, in a temperate climate a resting individual clothed normally loses about 60 per cent of bodily heat production by radiation (Millington, 1964). When the ambient temperature reaches 32 °C, the radiative heat loss may drop to zero, whereas in arctic temperatures radiation will probably account for more than 60 per cent of the loss.

Physiological reactions to the atmospheric environment are further complicated by differences in tolerance exhibited by individuals or certain ethnic groups and by adaptation achieved through prolonged acclimatization. Tolerance to cold conditions can be measured by means of the *critical temperature*, which was defined by Hardy and Dubois (1940) as the lowest temperature needed to maintain the resting metabolism without any increment in heat production. Experiments with naked humans have shown that the critical temperature for a white man living in a temperate climate is about 27 °C as opposed to 25–27 °C for Norwegian Lapps and 24 °C for Japanese in winter (Yoshimura and Yoshimura, 1969). These lower values are attributed to a lower metabolic rate during cold exposure and may also reflect differences in nutritional background. Similarly, Wyndham *et al.* (1964) demonstrated that native Bantu had higher rates of heat conductance than Caucasians living in South Africa at temperatures less than 10 °C and differences in metabolic rates and in skin and rectal temperatures were also observed.

According to Edholm (1966), physiological studies of acclimatization were begun in the eighteenth century and have been mainly concerned with heat or cold stress and the effects of high altitude. For example, bodily adaptation to tropical conditions occurs with a progressive increase in the sweat rate, although Portig (1968) has claimed that the purely climatic stress associated with the humid tropics is less than is often believed. High-altitude acclimatization involves a complex physiological response to reduced barometric pressure and oxygen levels often combined with high radiative and thermal stress. The tropical and sub-tropical Andes represent a unique bioclimate formed by high altitude and low latitude (Prohaska, 1970). The role of elevation becomes dominant when the threshold of *anoxia* or oxygen deficiency is approached and, although individuals may show a fairly wide altitudinal tolerance to the associated mountain sickness, acclimatized Indians living in the highest permanent settlements have adjusted to oxygen values which are only half those found at sea level. In the desert areas and during the dry season, additional stresses are imposed by the large diurnal range of temperatures and the considerable differences in net radiation received by insolated and shadowed parts of the body, and the drying power of the air is also marked. It is probable that the largest solar radiation intensities on earth occur in high, arid sub-tropical mountains during summer. Alexander (1965) has stated that a sailor permanently on the oceans will receive only 20 per cent of the natural radiation income of the inhabitants of the high Tibetan plateaux and Terjung (1970) has made observations in mid-July above 3000 m in the White Mountains of California in order to investigate physiological reactions under conditions of high radiation input.

The degree of bodily discomfort arising from excessive heat is usually measured by one or more of the many biometeorological temperature indices which have been specially derived for the purpose. One of the most common indices is that of *effective temperature* (ET), which was introduced by Houghten and Yaglou (1923) and may be defined as the temperature of still air saturated with water vapour in which subjects experience a subjectively equivalent sensation of comfort. Usually the effective temperature is determined directly, either for persons stripped to the waist or naturally clad, from nomograms expressing various combinations of dry-bulb and wet-bulb temperatures. In a series of controlled experiments, over 90 per cent of the test subjects found an ET of 25·6 °C too warm for comfort whereas only about 10 per cent considered an ET of 22·2 °C to be uncomfortable.

Rather later, Thom (1959) developed another measure of discomfort called the *temperature–humidity index* (THI), which was designed to assess summer conditions in the US, and is frequently written:

$$\text{THI} = 0{\cdot}4(Td + Tw) + 15$$

where $Td$ and $Tw$ are respectively dry-bulb and wet-bulb temperatures in degF. When the THI reaches 70, about 10 per cent of exposed subjects become uncomfortable. This percentage rises to 50 at a THI of 75, and to 100 at THI 79. Other authors, such as Brooks (1950), have argued that wet-bulb temperatures above 25 °C constitute a simple method of defining the limit of human comfort. Jauregui and Soto (1967) have applied this criterion to Mexico during the summer season, when excessive wet-bulb temperatures exist at the hottest part of the day over most of the coastal plains and lowlands compared with wet-bulb values less than 20 °C in the high plateau region.

The concept of effective temperature can be used to illustrate some of the geographical variations of human comfort within the tropics, bearing in mind that the 'comfort zone' for acclimatized persons living in low latitudes is normally taken as 18·9–22·4 °C ET. Stephenson (1963) has demonstrated that, although forced ventilation by fans or other means is incapable of providing 'comfortable' conditions during the April–June period 130 km north of the equator at Singapore, the ET only rarely exceeds 29·4 °C and outdoor sports can be safely indulged in throughout the year. This contrasts markedly with conditions at Bahrein, latitude 26 °N, where the ET is some 3 degC higher than at Singapore during the least comfortable part of the year. Forced ventilation is inadequate to deal with conditions during the greater part of the summer and excessive exercise could prove dangerous at times between July and September but especially in August, which had a mean monthly ET of 28 °C over the 1962–66 period (Watt, 1967).

Most studies emphasize the significance of local factors. For example, in Malaysia the ET declines both with altitude, which keeps the hill stations comfortable throughout the year, and with proximity to the coast where exposure to sea breezes is important in the lowlands (Wycherley, 1967). Coastal circulations may be at least partially responsible for maintaining the relatively comfortable conditions at sea level on Gan, near the centre of the Indian Ocean, described by McLeod (1965). On the other hand, Hounam (1967) has computed average values of ET for the hottest hour of the day at Alice Springs in central Australia and concluded that, although Alice Springs is uncomfortable for much of the warm season, the large diurnal range of temperature associated with such a continental location ensures much more comfortable nights than exist near the tropical coast of Australia.

In complete contrast, a temperate station, like London, experiences cold rather than warm discomfort and, despite the fact that ET values may occasionally rise to become similar to those in the tropics, forced ventilation can easily restore comfortable conditions (Foord, 1968). However, there is some evidence that the 'heat island' effect may contribute to discomfort in some urbanized areas within the temperate zone. During seven days in August 1969, Clarke and Bach (1971) measured air temperatures between 1330 and 2200 h above grass and paved surfaces in downtown Cincinnati and simultaneously at a suburban site 27 km east of the city. As shown in Fig. 7.1, the mean temperature was higher over the paved rather than the grass surfaces and also tended to be higher at the urban rather than the suburban location. The major difference in effective temperature was noted during the evening when the average urban values remained above 25·6 °C (90 per cent of persons uncomfortable), whereas the suburban site had mean values less than 22·2 °C (10 per cent of persons uncomfortable). This discrepancy suggests that, in such areas, planners might be advised to give some thought to the spacing of parks or other green areas within towns, together with the selection of building materials with lower heat storage properties and higher albedos, in order to reduce summer discomfort.

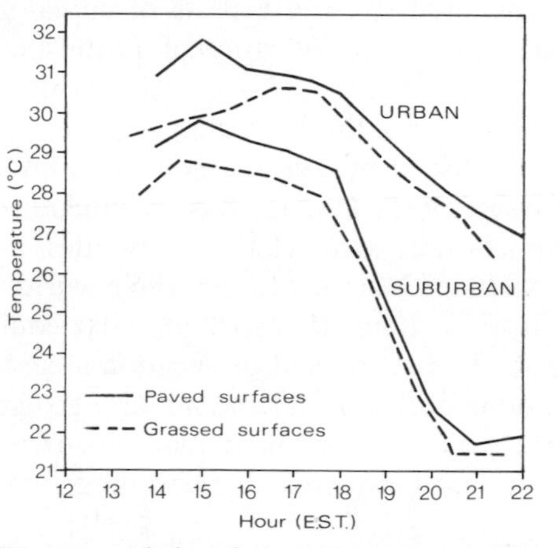


**Fig. 7.1 The mean variation of air temperature recorded over paved and grassed surfaces in downtown Cincinnati and at a nearby suburban site during August 1969. After Clarke and Bach (1971).**

Stone (1943) has reviewed a large number of instruments and formulae suitable for the assessment of cold discomfort, but the most useful estimates are often achieved through the *wind-chill index* of Siple and Passel (1945), which is defined as a measure of the quantity of heat which the atmosphere is capable of absorbing within an hour from an exposed surface 1 m square. The formula was devised following experiments conducted in Antarctica on the freezing rate of water sealed in small plastic cylinders and may be expressed as:

$$K_o = (100v + 10{\cdot}45 - v)(33 - t_a)$$

where $K_0$ is wind chill in Kcal/m$^2$ h, $v$ is wind speed in m/s, and $t_a$ is air temperature in °C. The index measures the cooling power of wind and temperature in complete shade without regard to evaporation, and the cooling rate is based on a neutral skin temperature of 33 °C. A number of different combinations of wind speed and temperature result in the same amount of cooling power. For example, a wind of 22 m/s and a temperature of −6·7 °C or a wind of 4·6 m/s and a temperature of −18·9 °C would both produce a value of 1400 Kcal/m$^2$ h (Falconer, 1968).

Because the wind-chill index represents only the dry convective cooling power of the atmosphere, it has been criticized by Court (1948) and others. The main objection is that, at low temperatures, convective heat exchange together with radiative and conductive heat transfer from the skin are much reduced by protective clothing. In these circumstances, heat loss occasioned by the evaporation of moisture from the lungs increases in relative importance and may amount to 20 per cent of the total bodily heat loss. Similarly, a really accurate calculation of heat losses from a fully clothed man should also take some account of his own heat production, the insulation offered by the clothing, and the evaporative heat losses from perspiration as well as expired air.

On the other hand, Wilson (1967) has found a very satisfactory correlation between the empirical wind-chill index and recorded cases of frostbite in Antarctica. Figure 7.2 illustrates that, with the exception of one borderline case, all the actual frostbites investigated occurred at wind-chill values between 1400 and 2100, thus confirming the assumption that exposed skin starts to freeze at an index of 1400. The isolated frostbite case at an index of 1380 occurred during sledging into wind-driven snow and it has been reported by Massey (1959),

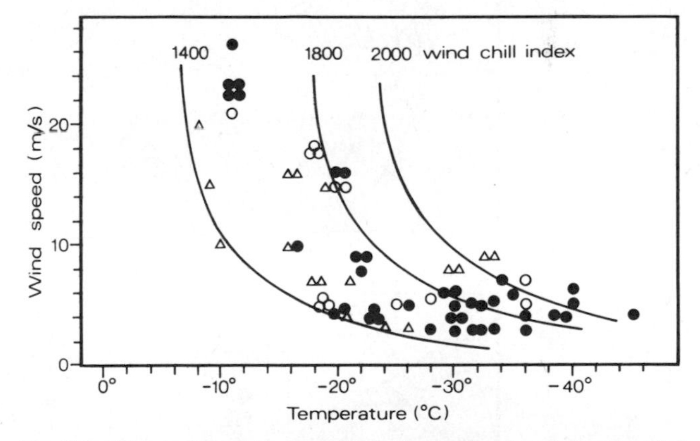


**Fig. 7.2 Cases of frostbite recorded in Antarctica in relation to wind chill based on temperature and wind speed. Dots indicate frostbites with well established meteorological data; circles mark cases that occurred during shifting meteorological conditions; triangles represent frostbites incurred during sledging with wind speed estimated. After Wilson (1967).**

among others, that the resultant wetting of the skin by melting snow often accelerates frostbite.

In the context of Antarctica, where the strength of the mean wind is greater than anywhere else on earth, the wind speed is implicated at least as much as the low temperatures in the very high rates of cooling which are experienced. Under extreme conditions, with a temperature of −25 °C and a wind speed of 49 m/s, the cooling power can attain 16–20 cal/cm$^2$ min and, during a stay outdoors, as much heat energy can be lost from the uncovered part of the face as is required for the metabolism of a person at rest (Loewe, 1972). High wind-chill values may occasionally be recorded even in temperate, maritime areas. Thus, Howe (1962) has documented the cold Christmas period of 1961 when strong easterly winds produced wind-chill indices exceeding 1100 in parts of southern England. This value is comparable with the mean wind-chill index for the coldest month of the year at such continental locations as Leningrad or Calgary.

There have been many attempts to classify and map the range of physiological comfort. Gregorczuk and Cena (1967) computed the effective temperature on a global scale for the months of January and July, and Gregorczuk (1968), following the work on a *sensation scale* in Argentina by Brazol (1954), approached the problem through the concept of air enthalpy. This is a quantity which indicates the heat content of the mixture of air and water vapour and is advantageous in that it is expressed in the same physical units as the metabolic rate for warm-blooded animals.

A rather different approach was made by Maunder (1962), who used thirteen aspects of climate, suitably weighted, to give some indication of the most favourable and least desirable areas of New Zealand from the viewpoint of man. One of the most comprehensive studies has been accomplished by Terjung (1966a), who devised a *comfort index* for the US, based on an integration of dry-bulb temperatures with relative humidities, and subsequently extended his work to the whole world (Terjung, 1968a). It was claimed that such an index could help in locating the best retirement or vacation areas and be of application in both medical and military geography. On the world scale, it was found that the Afro-Eurasian region in the northern hemisphere has the greatest physio-

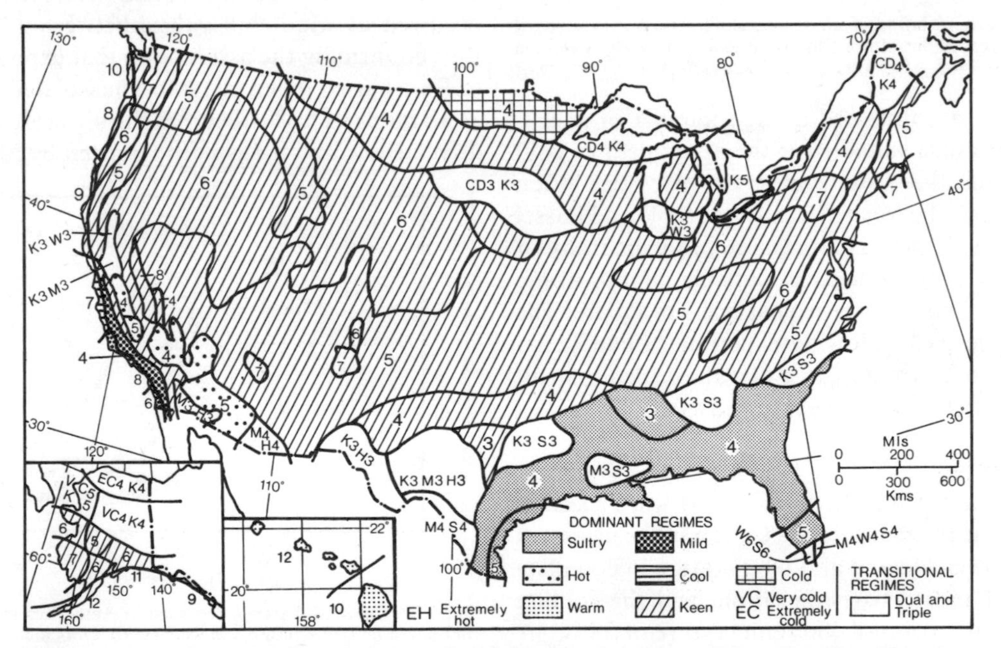


**Fig. 7.3 Annual physioclimatic regimes over the U.S. based on day-time conditions of comfort. The values in the Dominant regime areas indicate the number of months during an average year when that regime prevails. After Terjung (1967b). Copyrighted by the American Geographical Society of New York.**

climatic extremes, whereas areas with cool-current littorals and the low-latitude highlands have optimum conditions for man. The importance of physiological stress in the African continent was confirmed in a series of further papers (Terjung, 1966b, 1967a and 1968b). In general, it was found that some areas considered 'tropical' had less climatic stress than either very hot–dry regions or transition areas, such as the Sudan, between very hot–dry and very hot–humid conditions, and the major stress area in summer was identified around the southern Red Sea and in the interior of southern Somalia.

In an attempt to synthesize regional comfort patterns in the US, Terjung (1967b) developed the concept of the *annual physioclimatic regime* (APR) from previously computed values of the monthly daytime comfort index. The detailed classification is fairly complex but the simplified map in Fig. 7.3 shows that two basic regimes occur: a Dominant regime where one physioclimatic type occurs more often than others and a Transitional regime where several types exist. Each Dominant regime is divided according to the number of 'type-months', and $K_5$, for instance, denotes an area with a *keen* regime for five months of the year. Broadly speaking, it can be seen that the eastern two-thirds of the US is latitudinally aligned from south to north with respect to increasing stress, whereas the more mountainous west has a more complicated pattern. Large parts of the south and southeast have relatively low values of climatic stress because the mild winters go some way to balance the oppressive summers. No such seasonal compensation exists further north and areas in the midwest and elsewhere have high values.

## 7.3 Human disease and mortality

The field of medical climatology is both large and complex, and consequently little more than a brief appraisal can be attempted in this book. Many of the associations between climate and disease, as in the case of rheumatic diseases, have been thought to exist since Classical times (Lawrence, 1967), and for a more definitive treatment the reader is referred to the exhaustive survey by Tromp (1963), which lists no fewer than 4400 separate references, and the rather more compact review by Tromp and Sargent (1964).

Some of the most convincing correlations between disease and the atmosphere have been established for respiratory diseases, which are the commonest form of illness and account for more world-wide morbidity than any other form of disease (Hope-Simpson, 1967). In particular, considerable attention has been devoted to the study of asthma, and Menger (1967) has shown how the frequency of attacks varies with different air masses in Germany. Similar long-term variations, on a seasonal and annual scale, have been demonstrated for Australia by Derrick (1965 and 1966), but it is also apparent that important short-term fluctuations also exist. For example, from an investigation of emergency clinic visits for asthma at three major New York City hospitals during the autumn period in 1957, 1961, and 1962, Greenburg *et al.* (1967) found a statistically significant increase in visits resulting from the onset of either the first or second cold spells of the winter half-year.

Low temperatures in autumn have also been implicated in asthma outbreaks at Brisbane, Australia, by Derrick (1969), who concluded that high-asthma weeks tended to show a decrease in temperature, humidity, and rainfall plus an increase in sunshine duration. The increase in illness generally occurred within 48 h of the change to colder, drier weather as shown in Fig. 7.4, which portrays the meteorological conditions surrounding the asthma wave of 4–7 May 1961. This outbreak resulted from the arrival of a cold front, which replaced warm, moist air originating off the Tasman Sea by a cold, dry Antarctic

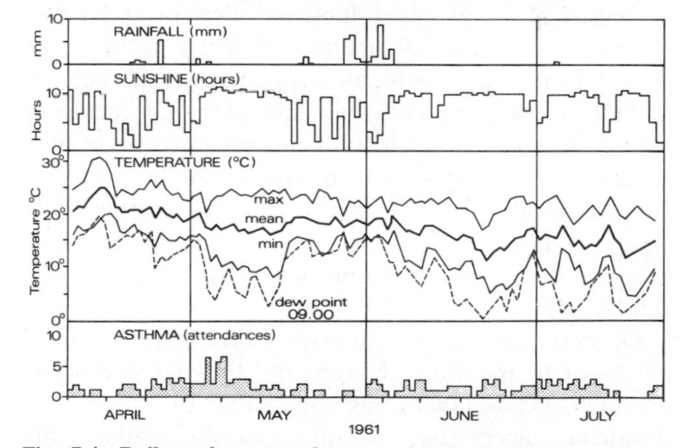


**Fig. 7.4 Daily asthma attendances and related weather data for Brisbane, Australia, during the period April to July 1961. After Derrick (1969).**

air mass, with the effect that on 4 May the dew point fell from 15 °C at 0300 h to −3·9 °C at 1200 h. Conversely, the low-asthma period between 18–31 May 1961 followed a rise in minimum temperature and dew point associated with the return of cloudy days.

It should be noted that asthma and certain other respiratory diseases may also be dependent to some extent on air pollution. Thus, Paulus and Smith (1967) showed that attacks of bronchial asthma among students periodically exposed to air-borne pollutants from grain-processing and -storage installations were twice as frequent in students who were grain-sensitive or allergic to the dust. Hay-fever pollen can cause distress amongst people who react to it, and Dingle (1957) has stressed the role of precipitation in flushing out pollen from the atmosphere.

In addition to the existence of weather-related diseases, specific weather parameters or synoptic situations can affect human health adversely. A notable example occurs when the passage of a warm front, a fall in barometric pressure, or the arrival of föhn conditions brings about a variety of subjective complaints such as a feeling of malaise, irritability, sleeplessness, or nausea (Dordick, 1958). More objective clinical responses, such as increased pulse rate and a fall in blood pressure, have also been found to accompany the other symptoms. Work in Denmark by Hansen (1970) and Hansen and Pedersen (1972) has indicated that changes in atmospheric pressure alone are sufficient to induce cases of peripheral arterial embolism and perforated duodenal ulcer provided that the rate of pressure change is sufficiently rapid. Some infectious diseases are highly dependent on the weather and Olson (1969 and 1970) has claimed that the incidence of bubonic plague in Vietnam is inversely proportional to rainfall. This appears to be due to the fact that the average number of fleas per rat decreases exponentially when the monthly rainfall exceeds 100 mm and increases exponentially when the monthly rainfall is less than this total.

Other studies have sought to isolate the various medical effects of temperature changes and thermal stress. For example, Davis (1958) has found that the number of duodenal ulcer cases admitted to the Philadelphia General Hospital tended to increase following exceptional short-term fluctuations of air temperature. Since haemorrhage from ulcers is intimately linked to the circulatory system, which is known to react to temperature changes, it was felt that the rise in ulcer cases was a direct reflection of thermal stress. In the Netherlands, Tromp (1967) has reported important temperature-related changes in the blood sedimentation rates of patients, and in Israel climatic heat stress due to the hot, dry *sharav* wind has been implicated in adverse changes to the human endocrine system and to the blood circulation of cardiac patients by Sulman *et al.* (1970). There is some evidence that mental stress may also be high under the influence of warm, dry, high-velocity winds such as the föhn of mountain areas or winds like the sirocco or sharav which occur in Mediterranean lands (Winstanley, 1972). These additional mental stresses may have sociological repercussions and Miller (1968) has correlated the daily homicide rate in southern California with the incidence of the hot, dry Santa Ana wind. Defining a 'Santa Ana' day on the basis of a relative humidity less than 15 per cent at noon, it was found that during 1964 and 1965 the Santa Ana wind prevailed on 53 days. Of these 53 days, 34 had a higher-than-average incidence of homicidal crimes, 3 had the normal rate and 16 days recorded a number of homicides below the normal rate.

As might be expected, many studies have been conducted into the relationships between human mortality and the atmosphere on a variety of time-scales. On the seasonal scale, Goldsmith and Perkins (1967) showed that, although excessive mortality occurs during the winter months in places as far apart as Britain, Sweden, and Australia, the seasonal fluctuations are less marked in areas such as Hawaii or California, where there is only a small difference between winter and summer temperatures. However, Momiyama (1968) has pointed out that seasonal mortality peaks have not remained constant through time, especially in the industrialized countries which enjoy highly developed public health and medical services. Evidence from the US and elsewhere suggests that many countries originally had a summer mortality peak, especially for infants, which has been eliminated for several decades by improvements in public health and preventive medicine. In the US, the remaining winter peak in infant mortality has been declining since the 1940s as a result of the wider adoption of central heating

| Weather element(s) or accord | Cardio-vascular system (1) | Autonomic nervous system (2) | Respiratory system (3) | Mortality (4) |
|---|---|---|---|---|
| Föhn | — | — | 2–2–0 | 4–1–2 |
| Air mass | — | — | 3–1–1 | 2–2–0 |
| Weather type | 16–16–5 | 4–3–1 | 11–9–4 | 6–5–2 |
| Fronts | 7–4–2 | 3–3–1 | 9–8–2 | 16–15–9 |
| Temperature | 3–3–1 | — | 1–1–0 | 3–3–1 |
| Temperature + humidity | — | — | 4–4–0 | 1–1–0 |
| Pressure | 7–6–1 | — | — | 1–1–0 |
| Atmospheric electricity, magnetic phenomena, and radiation | 3–2–1 | 2–1–1 | 3–3–1 | 3–1–0 |

First number: number of investigations included.
Second number: cases in which a numerical relationship was established.
Third number: cases in which the statistical significance of the results was determined.
(1) Includes thromboses, blood pressure variations, heart failure (non-specific), cardiac infarct, apoplexy, angina pectoris.
(2) Includes eclampsia.
(3) Includes asthma, embolism, haemorrhage, influenza, pneumonia, diphtheria.
(4) From all causes, including suicide.

**Table 7.1** Weather influences on morbidity and mortality 1935–68. (After Driscoll, 1971.)

and the improvement of general living standards. There is now little seasonal variation in infant mortality in the US, and the same trend towards the deseasonalization of disease can be found in other advanced countries such as Britain and Japan, and will presumably appear eventually elsewhere.

Tromp (1963) was at pains to demonstrate the importance of short-term weather variations on mortality and, more recently, Driscoll (1971) has confirmed the suitability of this emphasis. Table 7.1 summarizes the results of more than 100 research papers, published mainly in Europe during the 1935–68 period, which attempted to associate some aspect of morbidity or mortality with atmospheric parameters. Altogether, 83 per cent of the studies showed some positive relationship and most were concerned with the effects of different

| | Minneapolis | | Chicago | | Pittsburgh | | New York | |
|---|---|---|---|---|---|---|---|---|
| | *n* (days) | Change (%) | *n* (days) | Change (%) | *n* (days) | Change (%) | *n* (days) | Change (%) |
| *JANUARY* | | | | | | | | |
| *T* | 43 | +5·66 | 44 | +0·07 | 41 | +5·54 | 36 | +0·96 |
| *F* | 39 | −4·49 | 37 | −0·47 | 26 | −2·48 | 30 | −0·43 |
| *C* | 20 | −1·12 | 10 | −3·48 | 26 | −3·11 | 31 | −0·64 |
| *B* | 11 | −3·77 | 23 | +2·24 | 24 | −1·29 | 19 | −2·14 |
| *APRIL* | | | | | | | | |
| *T* | 39 | +1·00 | 37 | +2·38 | 35 | +2·03 | 33 | +1·91 |
| *F* | 18 | −6·47 | 24 | −2·52 | 25 | −1·51 | 32 | −1·69 |
| *C* | 27 | −1·06 | 23 | −1·41 | 25 | −2·15 | 20 | +0·57 |
| *B* | 31 | +4·55 | 30 | −1·04 | 30 | +2·74 | 28 | −0·94 |

*T* = trough day—area in frontal zone or closed low.
*F* = area on frontside of anticyclone, winds W to N.
*C* = at or near centre of high pressure.
*B* = area at backside of anticyclone, winds S to SW.

**Table 7.2** Mortality changes by weather types in four cities in the US, 1962–65. (After Driscoll, 1971.)

weather types or frontal passages. In a comprehensive study of daily weather-related deaths in 10 large cities of the US in the months January, April, July, and October for 1962–65, it was found that excessive mortality occurred mainly in north-central and north-eastern US and, although total mortality correlated best with weather features, it was closely followed by deaths over age 70 years (Driscoll, 1971).

As shown in Table 7.2, which lists mortality changes according to weather types in four areas, increases in mortality were coincident with pre-frontal weather and decreases with post-frontal weather. Although mortality is normally assumed to increase during periods of great atmospheric turbulence caused by the passage of active fronts, this is not always so; a pronounced cold front which crossed the US in the autumn of 1963 apparently failed to increase mortality in any discernible way (Driscoll and Landsberg, 1967).

By comparison, severe heat stress will almost certainly provoke fatalities, especially when temperatures rise suddenly in spring or early summer before the body has had an opportunity for thermal adjustment, or during any period when conditions enter the heat-stroke danger zone. One of the best documented examples in the latter category occurred in Illinois during July 1966 when, as a result of the highest temperatures recorded for over 10 years, at least 70 heat-stroke deaths were reported (Bridger and Helfand, 1968). The situation for the worst affected part of Illinois State is shown in Fig. 7.5 where it can be seen that, in the major heat-wave from 10 to 14 July, temperatures were well above the heat-stroke danger line, whether expressed on a daily average basis or in terms of the temperature–humidity index. Over the same five-day period, there was a 36 per cent increase in deaths, mainly through cardiovascular disease in persons 65 or older, and it would appear that there is a real need to forecast heat-stress conditions so that adequate air-conditioning in nursing homes and elsewhere can be provided to alleviate the suffering of the aged and chronically infirm.

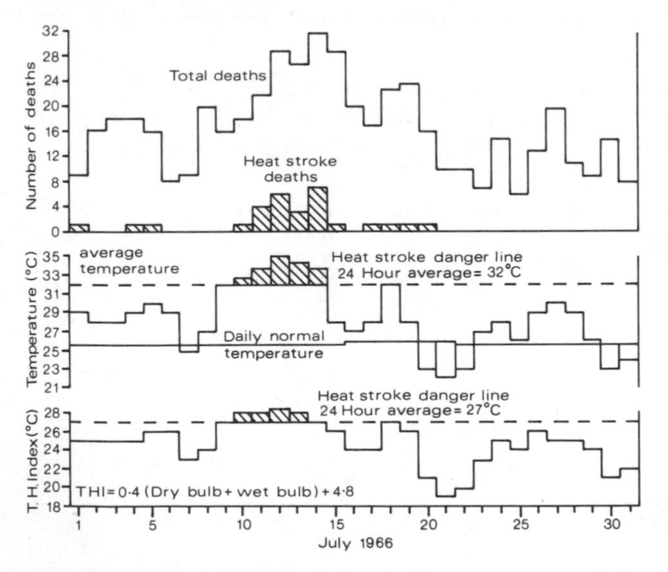


**Fig. 7.5 Daily heat-stroke deaths in relation to temperature during July 1966 in Madison and St Clair Counties, Illinois. After Bridger and Helfand (1968).**

It has already been noted that some additional heat stress may be produced in certain city environments but there is also evidence that urbanization is responsible for more widespread mortality. Thus, Padmanabhamurty (1972) has examined the growth of urbanization and industrialization in the metropolitan areas of Toronto and Montreal from 1947 to 1970 and has concluded that the resulting urban climate can be implicated in the trend to increasing mortality in both areas. Most of the additional deaths were due to chronic bronchitis, plus neoplasm of the trachea, lung and bronchus, and the analysis suggests that the increase in fog and smoke-haze days provides the best guide to the deterioration of the biometeorological environment.

## Building climatology

Most buildings are required to perform two basic functions. The first is to provide some protection against direct atmospheric conditions and create an artificial climate suitable for living accommodation, storage, or some other specified purpose. Second, buildings must be structurally safe and able to withstand the stresses of the prevailing climatic environment during the anticipated lifetime of the structure. These requirements, and the ways in which they are fulfilled, will be recurrent themes throughout this section but the broad relationships between climate and building can perhaps best be illustrated by reference to the characteristic three-stage process within the industry, involving planning and design, construction, and the subsequent maintenance of a satisfactory indoor climate through thermal and ventilation control.

## 7.4 Planning and design

Lawrence (1954) has claimed that architects and engineers need a greater awareness of climatology in order to achieve some of the symbiotic relationships found in the natural habitats of the animal world. However, one of the initial problems is to define which climatic parameters or conditions are relevant, and Lacy (1972) has listed 25 meterological factors, mostly required in the form of hourly observations, which have varying degrees of significance for the different operations within the building industry. As shown in Table 7.3, apart from the specific measurements, the scale of relevant climatic reference tends to diminish with the spatial extent of building interest. At the planning stage it is necessary to organize a city layout in order to optimize regional and local climatic conditions. For example, since the city of Chandigarh was located in the Indian Punjab on a low-lying site with excessive temperatures and humidities, the streets were laid out to attract as much ventilation as possible (Holford, 1971). In higher latitudes, natural illumination may be deficient and a corridor-type pattern of streets may be inefficient from a day-lighting viewpoint. It is well known that primitive dwellings are usually well adapted to minimize such climatic difficulties, and Zohar *et al.* (1971) have shown how the entrances to Bedouin shelters in the north-east Sinai desert are invariably oriented away from the prevailing west wind.

Successful planning also requires a knowledge of the climatic conditions which will result from the actual construction of the city itself (Page, 1972). An awareness of pollution distribution and heat-island effects is important here but probably the clearest illustration of man-made problems arises from the unfavourable airflows created by high-rise towers or slab buildings. Wise (1971) has presented case studies of a number of town centres in Britain and Fig. 7.6 shows typical conditions around a slab and an adjacent low building. Although the main airstream is deflected over and around the large block, air also flows down the face of the slab between the buildings and is drawn into the areas of separated flow at the sides. Tall buildings create high winds in the shaded zones in Fig. 7.6, so that mean velocities may be

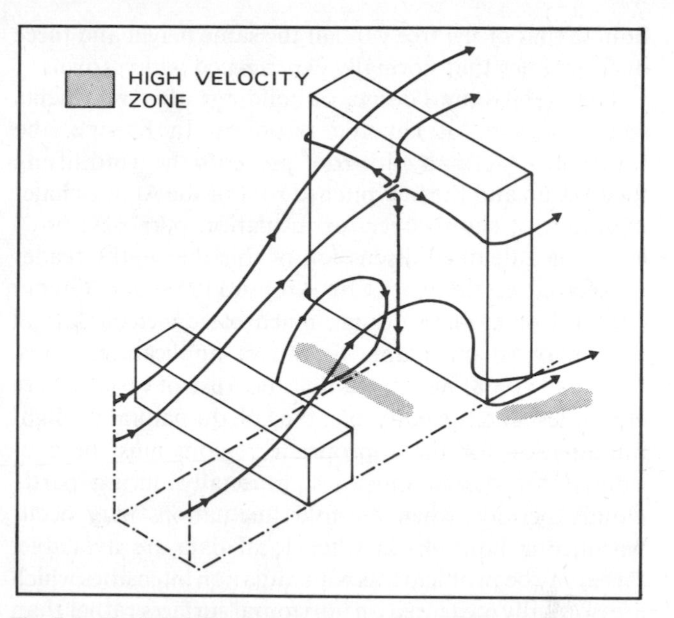


**Fig. 7.6 Typical airflow patterns around a slab and an adjacent low building based on wind-tunnel experiments. After Wise (1971). Reproduced by permission of the Royal Society of London.**

| | *Macroscale (inc. regional)* | *Local scale (town and site)* | *Microscale (around, on, and within building)* |
|---|---|---|---|
| Regional planning | Dominant | Important† | Less important |
| Town planning | Important | Dominant | Important |
| Site selection and building design | Important* | Important | Dominant |

*Macroscale may be dominant for standardization and building regulations.
†Local scale may be dominant in areas of rugged topography.

**Table 7.3** Relative importance of different meteorological scales on stages of decision-making within architecture and construction. (After Lacy, 1972.)

double that of the free wind at the same height and three or four times that normally experienced within towns.

The architectural design of buildings often represents some measure of climatic response. In Europe, the traditional German curb-roof prevents the wind lifting the roof up and the low-pitched roof of the Alpine chalet maintains a snow cover for insulation purposes; however, for a detailed discussion of this theme, the reader is referred to the works by Aronin (1953) and Givoni (1969). For modern design, much basic meteorological data may be unsuitable for direct application. Thus, Godshall (1968) has stated that, because of variations in cloudiness and turbidity over the US, the natural daylight illumination for the appropriate regions must be considered for design purposes, especially during partly cloudy periods when ten-fold fluctuations may occur within one hour. Even when local data are available, there may be problems, as with radiation intensities which are normally measured on horizontal surfaces rather than the vertical or inclined faces necessary for calculations of solar heat gain to buildings. In such cases, data-conversion procedures have to be employed (Loudon, 1967).

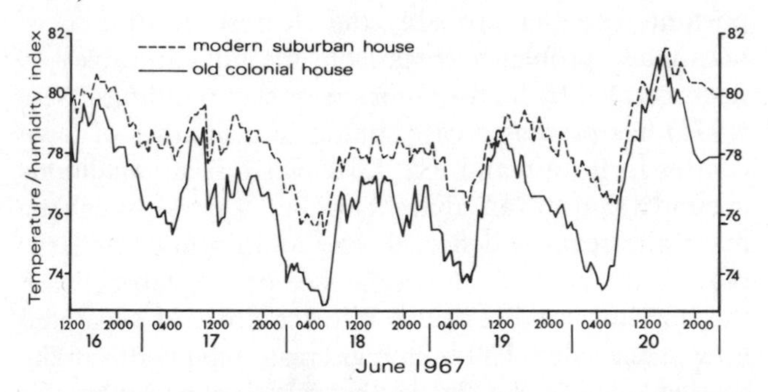


**Fig. 7.7 Variations in the Temperature–Humidity Index within an old colonial-style house and a modern suburban house in Singapore during part of June 1967. After Greenwood and Hill (1968).**

For many buildings, the choice of construction materials may be critical. The thermal conductivity of masonry may exert an influence on indoor comfort, and work by Ball (1968) and Arnold (1969) indicated how thermal conductivity can be measured and related to the moisture content of the materials. In the equatorial climate of Singapore, modern concrete living units have been found to be less comfortable than more traditional structures made of wood and palm thatch since the concrete heats up rapidly under direct insolation and then slowly releases the stored heat. Only by ensuring good cross-ventilation can the temperature in the modern houses be reduced to more tolerable levels and Fig. 7.7 shows the differences in temperature–humidity index for an old colonial-style house compared with a modern suburban house.

Building design is often a delicate compromise between opposing advantages and disadvantages. Thus, within recent decades the trend towards higher buildings with larger glazed areas has led to a problem of summer over-heating in temperate latitudes. Loudon (1970) has quoted evidence that in pre-war offices in London only 9 per cent of office workers complained of overheating compared with up to 40 per cent in modern blocks. The maximum internal temperatures may accumulate over a few days of anticyclonic weather and Fig. 7.8 illustrates the tem-

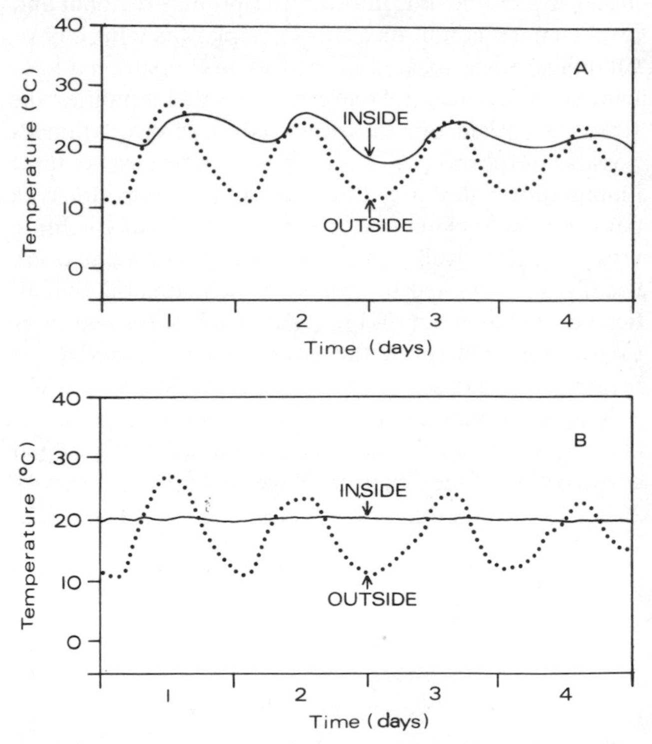


**Fig. 7.8 Air temperature variations recorded within different office rooms during sunny August weather in London.**
**(A) South-facing room, 50 per cent glazed.**
**(B) East-facing room, 20 per cent glazed.**
**After Loudon (1970).**

perature characteristics for two different rooms during a spell of settled summer weather. An east-facing office of massive construction, 20 per cent glazed and shaded by trees remained at 20·5 °C throughout the period when the outside temperature varied between 10 and 27 °C. In contrast, the temperature in a neighbouring room with large south-facing windows reached 26 °C during the same period. Such overheating cannot be easily overcome either by the installation of cooling plant or double glazing, and Hardy and Mitchell (1970) have suggested that, for the occupation of a building during office hours, there are thermal advantages in orienting buildings of oblong plan with the long facade facing east–west and buildings of square plan so that the diagonals run north–south and east–west. Alternatively, peak temperatures can be calculated as a function of window size and the area of glazing can be reduced accordingly (Loudon, 1968).

The nature of structural stresses experienced by buildings differs according to the climatic environment but lightning conductors are a widespread means of counteracting the obvious hazard of electrical storms (Chalmers, 1965). Rather less obvious is the damage to buildings, decorations, and even contents which arises from rain carried along at an angle to the vertical for, if this rain penetrates the outer walls, it increases the thermal conductivity of the building material, causes higher heat losses and raises the risk of internal condensation. Lacy (1951) and Lacy and Shellard (1962) have computed a driving-rain index for Britain, which is proportional to the total amount of rain which would be driven onto a vertical surface always facing the wind. Figure 7.9 has been compiled by using a combination of average annual rainfall with mean wind speed and it can be seen that values range from three in south-east England to over twenty in exposed western districts where greater weatherproofing is needed.

Nevertheless, most structural damage occurs as a result of wind stress. The Tay Bridge disaster, which claimed 75 lives in Scotland in 1879, stimulated the first studies of wind loading on structures but it was not until the collapse of the Tacoma suspension bridge in Washington in 1940 under marked oscillatory stress that wind loading was seen to take an unsteady as well as a steady form. The static wind loading approach is still probably adequate for most buildings but Scruton and Rogers (1971) have emphasized that the trend towards lighter, more slender structures increases the importance of the dynamic response.

For design purposes, the engineer needs to know the estimated return periods for various wind speeds, whether the velocities are taken as the maximum hourly mean speed or—as in the case of the small, lightweight structure—the maximum speed in a gust of about 3 s duration, which will probably be between 50 and 100 per cent greater than the equivalent mean hourly loading (Scruton and Newberry, 1963; Shellard, 1967a). The height of a building is important because in a gale the mean speed at only 52 m over built-up areas may be half

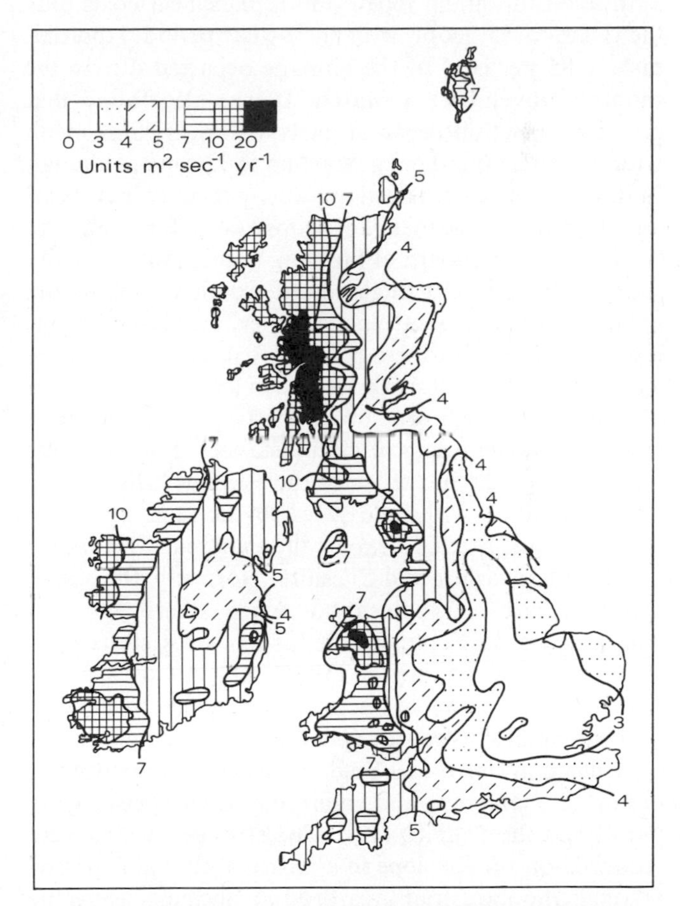


**Fig. 7.9 The distribution of an annual driving-rain index computed for the British Isles. After Lacy and Shellard (1962). Reproduced by permission of the Controller of Her Majesty's Stationery Office.**

as much again as the speed at the standard instrumental height of 10 m and Davis and Newstein (1968) have examined variations in gust behaviour up to a height of 280 m above Philadelphia.

On the other hand, more accurate data cannot compensate for poor design. Shellard (1967b) has described how three cooling towers in Yorkshire collapsed during a westerly gale with a return period of at least once in 5 years, since the arrangement of the towers caused wind funnelling which created large stresses on the leeward structures. In a comprehensive survey of wind damage to buildings in the UK between 1962 and 1969, Menzies (1971) reported that, on average, more than 100 000 buildings were affected each year, involving at least £7 million in annual repair and replacement costs plus the death of 35 people over the 8-year period. Approximately 85 per cent of the damage occurred during the months November to March. It was also found that minor destruction began at gusts as low as 20 m/s with widespread, major damage starting at 32·5 m/s, although buildings in southern England were generally less resistant to wind damage than those constructed further north.

The higher standard of building regulations normally found over northern England and Scotland reflects the greater frequency of adverse weather, including damaging winds, in the northern parts of Britain, but even here local variations in design criteria can lead to a relaxation of standards which can subsequently prove inadequate in certain meteorological circumstances. For example, buildings erected in the lee of the Pennine hills are not usually built to withstand gales to the same extent as buildings constructed near to the exposed, west-facing coasts of Lancashire and Cheshire. Normally, the design standards are satisfactory but when severe gales do occur, as on 16 February 1962, such areas suffer considerable losses (Aanensen, 1965). On this occasion, a strong westerly gale was actually intensified to the east of the Pennines by lee-wave phenomena which produced, over much of north-east England, a repeating pattern of bands of maximum and minimum wind speeds lying parallel to the Pennine crest. The strongest winds were recorded on the lee slope in a narrow belt which passed through the industrial area around Sheffield. Here the gale attained an hourly mean speed of 20 m/s, with gusting up to 43 m/s, and had a tentatively estimated return period of 150 years. Little building loss was experienced to the west of the Pennines but, in Sheffield, wind stresses arising from the combination of pressure and suction effects caused damage to 101 500 dwellings out of a total housing stock of 161 000 in the area. An account of building damage caused by a tornado in south-east England has been presented by Eaton (1971).

### 7.5 Construction

The construction phase of the building industry is highly dependent on the weather. According to Russo (1966), a total of $88 billion was spent on construction in the US during 1964 of which $39·7 billion or 45 per cent was spent on weather-sensitive operations. This weather-sensitive sector produces losses varying from $3 to 10 billion per year depending on the severity of conditions and it was found that 43 major construction operations are influenced by the weather to some extent. In this situation, accurate, relevant forecasts are essential and Beebe (1967) has stated that potentially, even with the existing forecast accuracy, the US industry could save between $0·5 and 1·0 billion annually using both Weather Bureau and private consulting meteorological services. This represents an increase in profits of between 50 and 100 per cent and the maximum savings, assuming a 100 per cent accuracy for forecast periods up to 24 h, have been estimated at $300 million more than are obtainable with the present forecast accuracy.

The economic justification for a more specific construction-oriented forecast service has been made by Broome (1966) and Mason (1966) for Britain, where up to 1½ million people produce a total output valued at £2900 million in 1966. The value of the time lost through adverse weather amounts to £100 million or 3·5 per cent of the production, equivalent to the loss of 10 working days per year. No service yet provides the detailed on-site forecasting service required, either at the tender stage or for day-to-day operations, but it is estimated that large sites could save at least one working day through a better application of forecast information.

In many mid-latitude countries, such as Britain, Canada, and Russia, there is a tradition of completing the basic constructional work during the summer and autumn so that winter activity may take place under cover but there is now growing pressure to maintain

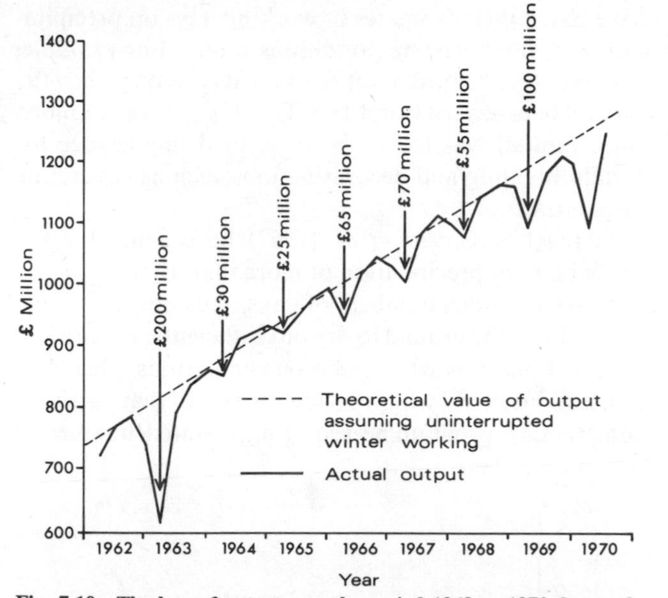


**Fig. 7.10 The loss of output over the period 1962 to 1970 due to the interruption of construction operations during winter in Britain. After Winter Building Advisory Committee (1971). Reproduced by permission of the Controller of Her Majesty's Stationery Office.**

| *Phenomenon* | *In conjunction with* | *Effect* |
|---|---|---|
| Rain | | 1. Affects site access and movement.<br>2. Spoils newly finished surfaces.<br>3. Delays drying out of buildings.<br>4. Damages excavations.<br>5. Delays concreting, bricklaying, and all external trades.<br>6. Damages unprotected materials.<br>7. Causes discomfort to personnel.<br>8. Increases site hazards. |
| | High wind | 1. Increases rain penetration.<br>2. Reduces protection offered by horizontal covers.<br>3. Increases site hazards. |
| High wind | | 1. Makes steel erection, roofing, wall sheeting, scaffolding, and similar operations hazardous.<br>2. Limits or prevents operation of tall cranes and cradles, etc.<br>3. Damages untied walls, partially fixed cladding, and incomplete structures.<br>4. Scatters loose materials and components.<br>5. Endangers temporary enclosures. |
| Low and sub-zero temperatures | | 1. Damages mortar, concrete, brickwork, etc.<br>2. Slows or stops development of concrete strength.<br>3. Freezes ground and prevents subsequent work in contact with it, e.g., concreting.<br>4. Slows down excavation.<br>5. Delays painting, plastering, etc.<br>6. Causes delay or failure in starting of mechanical plant.<br>7. Freezes unlagged water pipes and may affect other services.<br>8. Freezes material stockpiles.<br>9. Disrupts supplies of materials.<br>10. Increases transportation hazards.<br>11. Creates discomfort and danger for site personnel.<br>12. Deposits frost film on formwork, steel reinforcement, and partially completed structures. |
| | High wind | Increases probability of freezing and aggravates effects of 1–12 above. |
| Snow | | 1. Impedes movement of labour, plant, and material.<br>2. Blankets externally stored materials.<br>3. Increases hazards and discomfort for personnel.<br>4. Impedes all external operations.<br>5. Creates additional weight on horizontal surfaces. |
| | High wind | Causes drifting which may disrupt external communications. |

**Table 7.4** Effects of weather on construction operations. (After Winter Building Advisory Committee, 1971.)

output during the winter months in order to avoid many of the economic and social problems associated with a seasonal run-down. Figure 7.10 shows that in Britain the annual loss of production each winter is around £70 to 100 million which is about 2½ per cent of the total production or equivalent to a fall of 10 per cent in the winter quarter (Winter Building Advisory Committee, 1971). Every year, between 30 000 and 50 000 men are laid off, and in the exceptional winter of 1962–63 nearly 160 000 men were laid off entailing pay losses of some

£30 million, attendant social hardship and a reduction of purchasing power in the economy. Some idea of the wide-ranging effects of the weather on constructional activities can be gained from Table 7.4.

Rainfall is one of the most disrupting elements. The critical limit for most outdoor activities appears to be a rate of about 0·5 mm/h which affects operations extending from initial site surveys to final exterior painting. Major problems are also created by low or freezing temperatures and, although green concrete will increase in strength down to −11 °C, temperatures below 10 °C reduce the

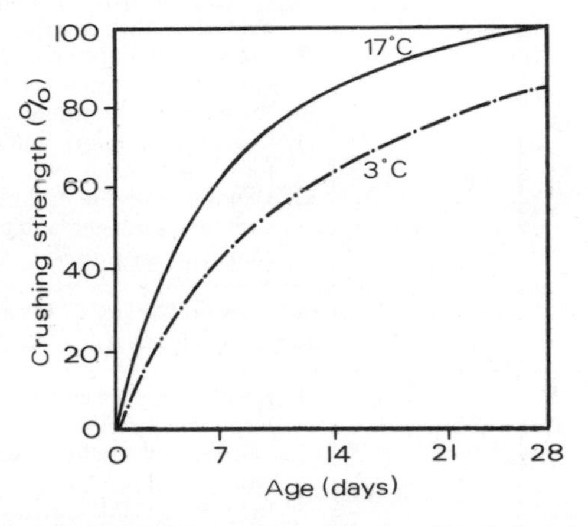


**Fig. 7.11 The effect of temperature on the rate of hardening of a typical concrete made with ordinary Portland cement. Crushing strength shown as a percentage of 28-day strength of 1:2:4 concrete cured at 17 °C. After Winter Building Advisory Committee (1971). Reproduced by permission of the Controller of Her Majesty's Stationery Office.**

rate of hardening, as shown in Fig. 7.11, and there is a danger of frost damage below about −2 °C. High winds compound the difficulties associated with both precipitation and low temperatures and may restrict all operations on tall structures.

The construction industry requires weather information in a form which permits estimates of the number of workable days per month or construction season, since these data will determine when a job can be started and whether it can be completed by a pre-specified date (Musgrave, 1968). Frisby (1957) gave a general definition of a working day as having less than 6 mm of rain with less than 50 mm during the previous two days, an average temperature above −8·9 °C, a wind of less than 13 m/s and less than 150 mm of snow on the ground. Many authors have based their estimates of working days on precipitation and soil moisture conditions alone. For example, Foord (1972) defined a wet working day as one when the rainfall between 0900 and 1800 GMT was 1 mm or more, since rainfall was felt to be of critical importance for certain building and decorating jobs such as concreting or painting.

In the US, Covert *et al.* (1967) have simulated the probability of precipitation of more than 10 mm per day in runs of various numbers of days, plus one additional day to allow the ground to dry out sufficiently for work to restart. Emphasis was placed on the periods when rain did not interfere with outside work so that, given a construction programme requiring a stated number of

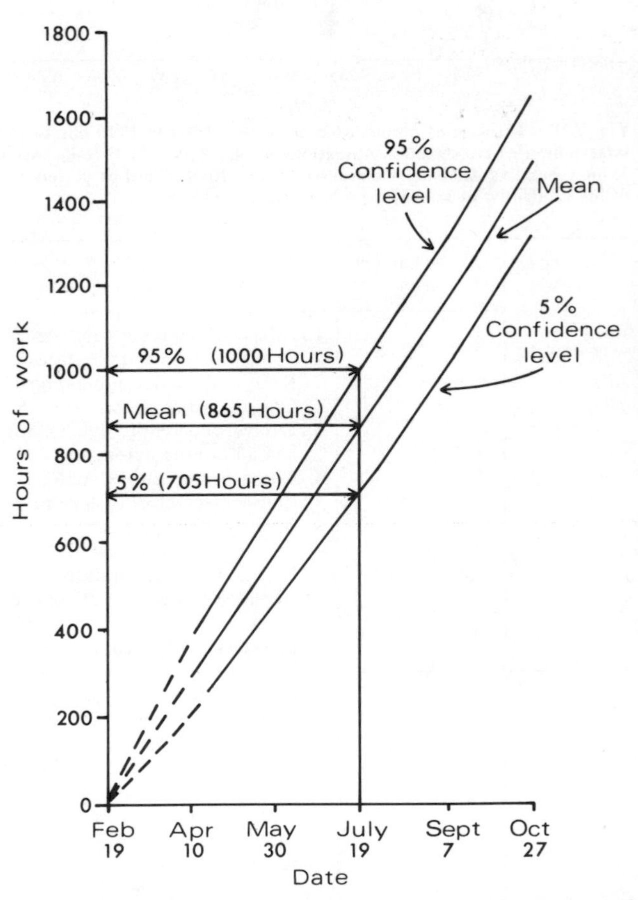


**Fig. 7.12 Cumulative hours of worktime, including weekends, for road construction at Jefferson City, Missouri, 1918–65. After Maunder *et al.* (1971b).**

work-days, the likelihood of meeting overtime costs or financial penalties imposed as a result of late completion could be evaluated.

More recently, Maunder *et al.* (1971a and b) have combined daily engineering records from two construction projects with soil moisture and precipitation data to develop a simulation model capable of describing workability conditions for road building near Jefferson City, Missouri. Using long-term rainfall figures and assuming an 8-h work-day during the April–October construction season, they found that, on average, 70–80 per cent of the total possible time could have been worked between 1918 and 1967 with individual monthly means ranging from 69 per cent in April and May to 82 per cent in July. As shown in Fig. 7.12, the results suggest, for example, that a road construction operation needing 1000 h of work is most likely to be completed with the least delay if work is started in late June or early July.

### 7.6 Indoor climates

Environmental control within buildings can be maintained within precise limits by artificial heating, cooling, and ventilation to create optimum bodily comfort and efficiency. Where prevailing temperatures are well below blood heat, additional heating may be required but with ambient dry bulb temperatures near blood heat, as in parts of the humid tropics, evaporative cooling should be enhanced either by increasing air movement or by the extraction of water vapour through air conditioning (Fry and Drew, 1956). In low latitudes, it is also important to exclude radiant heat as far as possible using either internal controls such as curtains or blinds or external controls like awnings, canopies, and deep balconies (Petherbridge and Loudon, 1966). In addition, indoor comfort has an important influence on efficiency; Angus (1968) quoted the results of an investigation into a light sedentary task in industry involving manual dexterity, which took 12 per cent longer to complete in a room at 10 °C compared with a room temperature of 16·7 °C. Different activities have different environmental requirements and the level of thermal comfort within operating theatres, offices and schools has been discussed respectively by Wyon *et al.* (1969), Humphreys and Nicol (1970), and Langdon and Loudon (1970).

The maintenance of a suitable indoor climate depends on the successful interpretation of atmospheric behaviour both inside and outside the building. For example, Daws (1970) has shown how air entering rooms mixes with, and transfers momentum to, air already present and has outlined the action of micro-convection streams above heat sources, including the human body, which loses about one-third of its heat by convection. On the other hand, the total interior heating requirement, including the length of the heating season, is normally calculated on the basis of heating degree–days or the accumulated temperatures below a selected desirable base value. The base temperature, taken from standard meteorological observations, is often either 15·6 °C or 18·3 °C, which represent conditions 2·7 degC below the desirable comfort level, the difference being accounted for by solar radiation receipts through windows and incidental heat gains from lighting or cooking. However, it is important to ensure that degree–day calculations are made from directly relevant local data. Thus, Parry (1957) suggested that, even in a moderate-size town in an area of low relief, the combination of lower windspeeds and higher mean temperatures within the town may produce local variations of 20–30 per cent in annual heat requirements. Similarly, Chandler (1964) has estimated that the urban influence of London accounts for a reduction of some 400 degree–days or about 10 per cent of the heating values obtained at comparable heights outside the city.

## Employment and leisure

### 7.7 Commerce and industry

Although the operation and well-being of all commercial and industrial enterprises which are based primarily indoors is governed essentially by a complex of political, economic, and social factors, certain spheres of business activity also depend to some extent on the nature of the atmospheric environment. In some circumstances, the broad regional climate may define much of the demand for goods such as air-conditioning equipment, and in areas with a large proportion of the population engaged in seasonal occupations, such as agriculture or construction, the cyclical fluctuation of personal incomes may affect purchasing power throughout the economy. Other areas may have a climate which encourages industry.

According to Wilson (1966), Tucson in southern Arizona is one of the fastest-growing cities in the US and it appears that a dry climate with abundant sunshine has played a significant role in attracting light industry and other economic activities to the area.

However, because much of the shorter-term dependency may easily be obscured by other factors, the weather problems in industry frequently go unrecognized and unexplored by either the industrialist or the professional meteorologist. Part of the difficulty may stem from the fact that the general weather information available from public sources, such as the Meteorological Office or the US Weather Bureau, has little direct application in the commercial world, and both Hallanger (1963) and Boyer (1966) have stressed the need for specialist advice obtained from private meteorological consulting services with the meteorologist working in close co-operation with a team of management executives. Consequently, the published material in commercial and industrial climatology is heavily concentrated in the spheres of retailing and insurance.

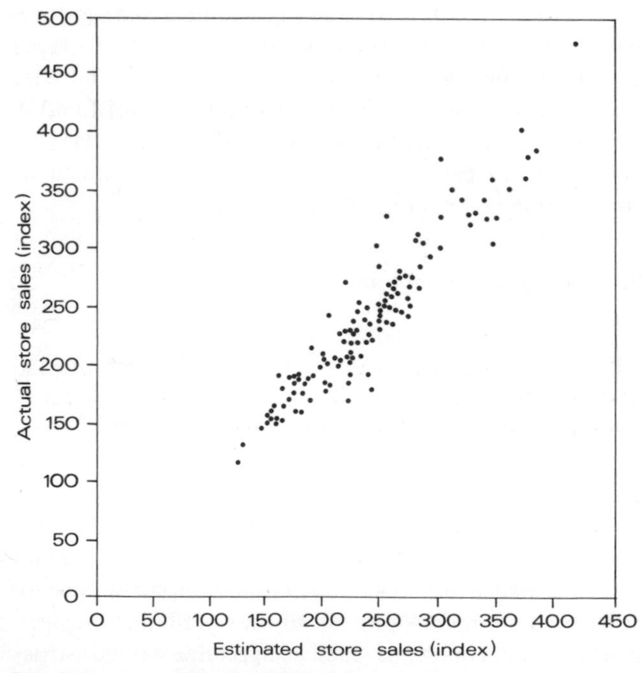


**Fig. 7.13 The relationship between estimated store sales and actual store sales in Des Moines, Iowa, during the Easter period in 1945, 1947, and 1948. After Steele (1951). Reprinted from *Journal of Marketing* published by the American Marketing Association.**

One of the most obvious influences of weather on retail sales occurs when meteorological conditions are sufficiently adverse to deter or even physically prevent people from visiting shops. Thus, in a study of department store sales in Des Moines, Iowa, during the seven weeks before Easter over the 1940–48 period, Steele (1951) found that the amount of snow cover was the outstanding factor leading to depressed sales, even if the snow fell in the previous week. Total sales variation was approached by means of a multiple regression equation developed between daily sales and selected weather parameters. Figure 7.13 illustrates the relationship between estimated and actual store sales for three years in the period and it can be seen that a high degree of correlation exists with 88 per cent of the variance explained by weather factors.

In a similar study of a large department store in New York City, Linden (1959a) concluded that precipitation reduced daily sales by an average of 8 per cent. The reduction could be as much as 15 per cent when the rain lasted longer than about 6 h and retail activity was especially low as a result of a wet morning. On the other hand, as shown in Fig. 7.14, rainfall led to a considerable

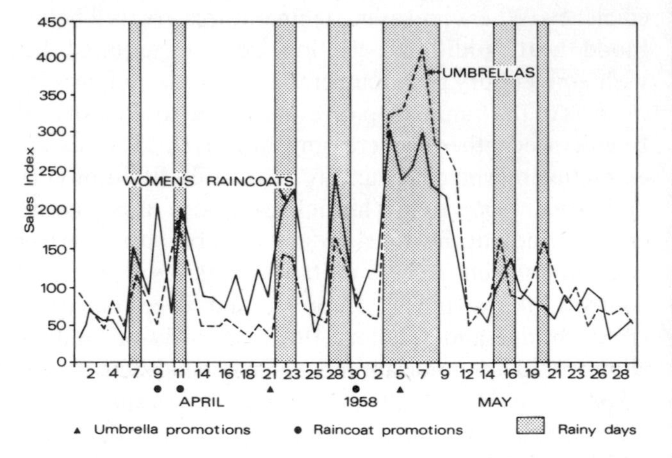


**Fig. 7.14 Daily sales of women's raincoats and umbrellas in a large New York department store during April and May 1958. After Linden (1959a).**

stimulation of sales in women's raincoats and umbrellas. Of the 51 trading days depicted in April and May 1958, 14 were subjected to rainfall at some time during shopping hours and these rainfall days accounted for 40–50 per cent of all raincoat and umbrella sales. On especially wet days, the sales of women's raincoats and umbrellas were respectively double and treble the level on dry days and it can be seen from Fig. 7.14 that rainfall incidence was a more dominant factor than the advertising campaigns mounted in the store for either product.

Linden (1959b) has demonstrated that extreme temperatures in excess of 29·4 °C can lower sales in a New York department store by 1·5 per cent, but the real significance of air temperature lies with the many weather-sensitive retail goods, such as refrigerators, air conditioners, fans, and outer clothing, which generate a marked seasonal rhythm of demand. The critical seasons are spring and autumn and, according to Petty (1963), the ideal retailing weather is warmer than usual in spring and cooler than usual in autumn. The reverse pattern depresses sales, which are never fully recovered later because the season for the use of the particular commodity is then shortened and people tend to make do with what they already possess. For example, unseasonably cold weather during May 1960 in the US led to reduced sales of summer clothing, air conditioners, fans, and outdoor furniture (Anon., 1960).

Clothing is one of the most weather-sensitive retail items and, in the warm spring of 1957, Linden (1959b) has reported that sales of women's suits, boys' wear, and girls' wear during March and April were respectively 10, 10, and 20 per cent higher than on a year-ago basis for a New York store. In September 1956, which was the coldest September in New York for over 20 years, sales of women's coats were up by 22 per cent compared with the almost normal September of the preceding year.

Sales of cosmetics depend partially on the weather, with suntan lotion moving slowly in cold summers or deodorants being used more quickly in hot, humid weather, and Kestin (1967) has argued the case for a weather-sales index in the industry which, combined with long-range weather forecasting, would help the manufacturing, distribution, and advertising processes.

The demand for beer is known to be weather-sensitive and in a survey of beer sales in the state of Rhode Island a strong link with temperature was established by Zeisel (1950). A more recent investigation of beer consumption in southern England by Gander (1972) has indicated the possibility of more complex relationships, although no details were published because of commercial security. However, it was estimated that the weather accounts for 5 per cent of the total sales volume. This 5 per cent is worth £34 million per year and, regarded as wholesale stock which is unnecessarily locking-up capital at an average return of 10 per cent, represents an annual waste of around £3·4 million. For a recent analysis of retail trade data in the US, the reader should consult Maunder (1973).

| *Storm type* | *Range of losses (millions of dollars)* |
|---|---|
| Hurricanes | 250– 500 |
| Tornadoes | 100– 200 |
| Hail/wind thunderstorms | 125– 250 |
| Extra-tropical wind storms | 25– 50 |
| Total | 500–1000 |

**Table 7.5** Estimated range of annual insured property losses by storm type in the United States. (After Hendrick and Friedman, 1966.)

The insurance profession is partially concerned with the financial consequences of severe weather conditions and several insurance companies now employ a meteorologist. In the US alone, the liability on property insurance from storm damage was estimated at more than $500 billion by Hendrick and Friedman (1966) with paid claims amounting to between $0·5 to 1 billion per year.

Table 7.5 indicates that hurricanes account for half of the total property insurance losses in the US and, in the twenty years following 1949, out of 105 catastrophes attributable to a single event and involving insurance claims of $1 million or more, no less than 97 were due to either wind or hail. The most expensive storm was the East Coast cyclone of November 1950 which created a total insurance loss of $175 million (Friedman and Hendrick, 1960). Insurance claims in Britain are rather smaller, although, after the exceptional winter of 1962–63, insurers paid out over £20 million for weather damage claims compared with only £5 million following a normal winter. Most of the claims were for burst pipes and water damage (Anon., 1963a and b). A specialized form of weather insurance is the so-called 'Pluvius' scheme which deals principally with the effect of rain on outside activities and has been described by Chase (1970). The scheme was started in 1920 and there is now a constant demand for this cover in areas throughout the world, where no reliable dry season exists in which events can be held. In the case of winter sporting events, which are not usually affected by rainfall, the policy can be extended to include cancellation or abandonment resulting from any kind of adverse weather.

### 7.8 Recreation climatology

It can be shown that some indoor hobbies may have an atmospheric basis, as described by Carr (1970) and Beaver (1972) for meteorological philately, which is a form of thematic stamp collecting. It was estimated that weather-related topics appear on over 700 postage stamps issued by some 130 countries and that with the inclusion of peripheral themes, such as famous personages linked to the development of meteorology, the number of stamps could well exceed 1000. Most meteorological stamps are issued by African countries, mainly on the annual 'World Meteorological Day', and it was concluded that these stamps are produced as much for the philatelic market as for genuine postal use. However, it is in connection with the growing demand for outdoor recreation, including sports and other pastimes, that the weather most frequently assumes a dominant role.

Throughout the developed world there is a general trend towards more leisure time resulting from shorter working hours. Together with the associated affluence and greater mobility due to car ownership and package holiday flights, this change in social behaviour has placed increasing importance on recreation climatology and in 1970 holiday weather enquiries were the most numerous category received by the British Meteorological Office (Anon., 1971). The tourist industry is experiencing a similar upsurge in the rate of growth and many countries rely on tourism for a significant proportion of their foreign exchange. In view of the wide range of outdoor recreational pursuits now available, it will be convenient to review the broad characteristics of holiday weather before moving on to consider the particular requirements of some of the more specialized sporting activities.

Some recreational opportunities, notably the planned annual vacation, may be some distance from the tourist's home area and in these cases there is a need for adequate advance information about the probable weather conditions to be expected. For example, restricted baggage allowances imposed by all airline companies mean that tourists have to make a careful choice of clothes, and the final selection will depend largely on the expected weather. The prospective visitor may not always have access to reliable and impartial information, as shown by Perry (1972b), who commented on the presentation of climatological data to the layman and criticized some Mediterranean holiday brochures which emphasize relatively high mid-winter temperatures but make no mention of rainfall at this time of year. Those people who take a holiday for health or convalescence purposes are often guided by the desire for either a bracing or relaxing climate and there have been several attempts to analyse the mainly subjective motivation of holiday-makers towards an area providing bodily comfort, ranging from the work of Tyler (1935) in Britain to more recent work on the coastal climate of Belgium by de Jaeger (1970).

In some parts of the world bodily discomfort may seriously limit the tourist potential. Thus, Terjung (1968c) has divided Alaska into various physio-climatic zones based on the 'Comfort Index' which shows that, not only does the littoral area remain either cool or keen even in July, but also indicates that the major tourist attractions such as the forest and national parks generally lie outside the most comfortable parts of the State. When the holiday has begun, it may be necessary to make regular use of weather forecasts in order to optimize the local conditions. This is especially true of areas with variable weather patterns such as Britain, and Holford (1967) has demonstrated how flexible planning helps the selection of the most suitable days for hill-walking, beach activities, and picnics. Where resort areas are near to large population centres, there may be an important element of impulse travel on public holidays. This type of recreation is highly sensitive to the weather, which can lead to marked seasonal variations in the tourist industry, as illustrated for Australia by Tucker (1965).

According to Clawson (1966), the ideal climate for outdoor recreation would be one where '... it never rained, was always pleasantly warm but not hot, was always mildly sunny, was never too humid, had only gentle breezes, etc.'. Although this ideal exists nowhere in the world, there have been numerous attempts to quantify summer weather by the application of indices which permit some comparison from place to place and season to season. Some progress was achieved in France by Peguy (1961) but, following the work of Poulter (1962) in deriving an index of summer weather for Kew, most attention appears to have been devoted to Britain with the proposal of further indices by Fergusson (1964), Rackliff (1965), and Hughes (1967).

Fergusson was specifically concerned with the problem of where to holiday in England. Believing that most tourists were mainly seeking sunshine, he increased the weighting for sunshine duration in Poulter's Index and concluded that, compared with the south coast of England, few of the northern resorts could be considered climatically attractive with values of the Fergusson Index ranging from 89° for Ryde (Isle of Wight) down to only 76° at Tynemouth on the north-east coast.

In a review of existing summer weather indices, Davis (1968) has shown that, apart from differences in the choice of either mean or mean maximum daily temperatures, all the formulae relate to the three months June, July, and August and take a common form:

$$I = W_1 \text{ temp} + W_2 \text{ sun} - W_3 \text{ rain}$$

whereby summer temperatures and sunshine are weighted and added together and summer rainfall is weighted and subtracted. Davis proposed an Optimum Index ($I_o$) so that:

$$I_o = 10T_x + 20S^1 - 7R^1$$

where $T_x$ is the mean daily maximum temperature (°F) for the three months June, July, and August, $S^1$ is the daily mean sunshine in hours for those three months and $R^1$ is the total rainfall in inches. In this formula, the weights were chosen to make the weighted standard deviations of the three elements in the Index almost equal and it was argued that such weights are applicable to most areas in Britain. Figure 7.15 shows the distribution of the mean Optimum Index over Britain and it can be seen that values range from over 800 around the Thames Estuary to less than 500 in the Pennines and Scottish Highlands with the lowest sea-level values in west Scotland. In detail, the temperature factor appears to be the most important, and the Index falls with increasing altitude at around 100 points per 305 m, largely because of temperature.

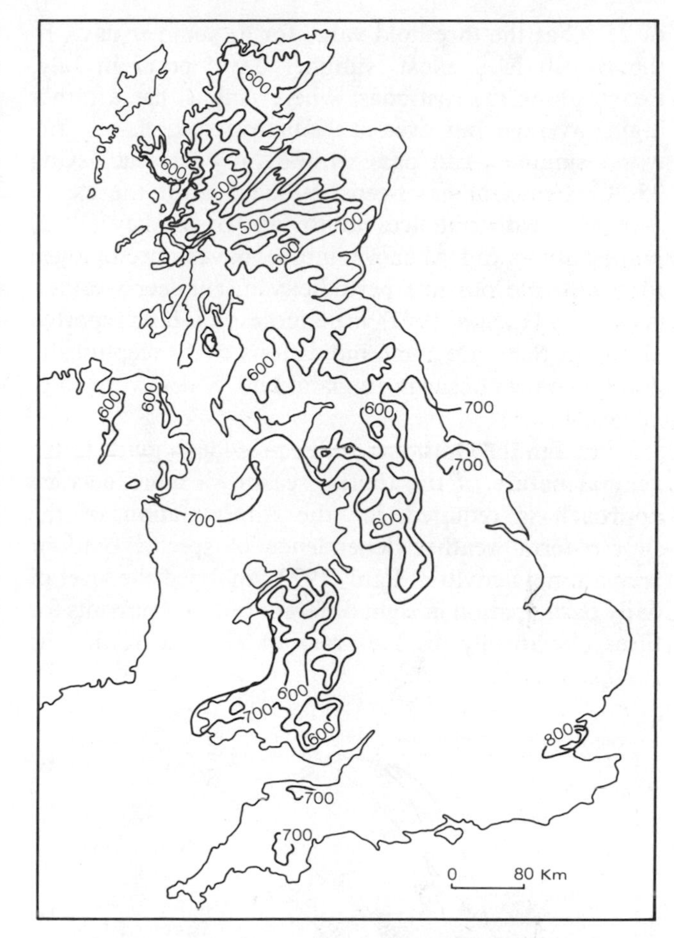


**Fig. 7.15 Isopleths of the Optimum Index for summer weather calculated for Britain over the 1916–50 period. After Davis (1968).**

In terms of seaside holidays, it should be noted that, although coastal resorts have more sunshine and marginally less rainfall than places at similar elevations a short distance inland, they have lower values on the Index. This is because the sea breeze lowers the maximum coastal temperature in light winds when high values are recorded inland and this outweighs the effects of increased sunshine.

Apart from the Optimum Index, Murray (1972) has suggested a so-called 'simple' summer index which is generally applicable to any period from May to September whereas Perry (1968) has drawn attention to the international recognition of the maximum temperature of 25 °C as the threshold value for a 'summer day'. In the British Isles, most 'summer days' occur in July, except along the east coast where August has a rather higher average, but, even in the favoured south-east, the worst summers can pass without any days achieving 25 °C. Criticism has been levelled against the use of summer weather indices in Britain by Laskey (1969) among others and, whenever indices have been employed to investigate possible periodicity in summer weather, such as by Hughes (1972), little success has been reported although there are some indications that exceptionally good summers occur in England and Wales every 10 to 12 years.

Although the Optimum Index provides a guide to the general nature of the tourist season, a more detailed approach is required for the interpretation of the shorter-term weather dependence of specific outdoor recreational activities. Paul (1972) analysed the level of daily participation in eight different outdoor pursuits for three climatically diverse areas of Canada during the

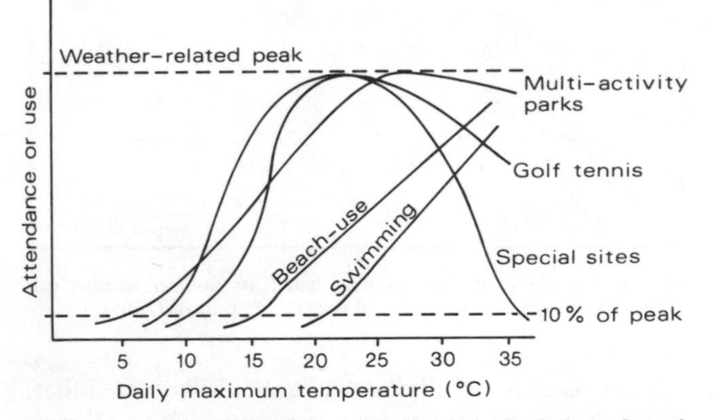


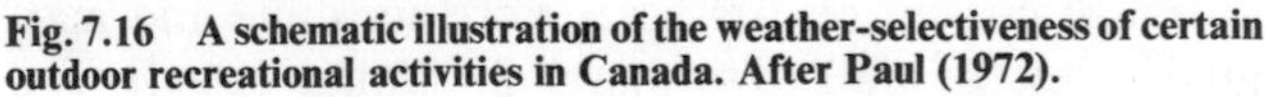
**Fig. 7.16 A schematic illustration of the weather-selectiveness of certain outdoor recreational activities in Canada. After Paul (1972).**

summer of 1969 and found that the effect of the weather varied markedly according to the type of recreation available. For example, swimming and beach-use activities were mainly dependent on the daily maximum temperature and the total sunshine hours, whereas attendance at multi-activity parks, picnicking, and pleasure driving showed only a slight relationship with any of the selected weather parameters. Figure 7.16 is a schematic illustration of the relationships between five forms of outdoor recreation and daily maximum temperature based on the conclusions reached from these Canadian data. It can be seen that, with maximum temperatures below 16 °C, only multi-activity parks, golf courses, tennis courts, and special sites, such as zoos or sites of outstanding geological interest, received any appreciable attendance. Moving up the temperature scale, attendance at the multi-activity parks peaked-out at around 27 °C, compared with only 24 °C for golf, tennis, and special sites, while swimming and beach-use activities became increasingly popular with outdoor recreationists beyond about 21 °C.

It would appear that models similar to this could be developed to the stage where daily attendance at outdoor recreation facilities, and the associated traffic flows, could be predicted especially when, as in the case of swimming and beach-use, a linear relationship can be established with easily obtainable weather variables. On the other hand, the Canadian study revealed certain regional discrepancies in the response of swimmers to daily maximum temperatures and swimmers in all three areas accepted temperatures 3 to 4 degC lower in May, June, and September than during the school holiday period when temperatures are normally higher.

Maunder (1970) has claimed that the effect of weather on particular sports can be considered either in terms of the effect on the sport itself or the impact on the attendance figures at the sports ground or arena. In the former category, Shaw (1963) has detailed some of the effects of weather on baseball, and sports stadium managers attempt to provide as much insulation as possible from the elements for both playing areas and spectators to safeguard attendance figures during adverse weather. Perhaps the best example of this is the Astrodome at Houston, Texas, where there is an entirely enclosed stadium complete with air conditioning and capable of holding 60 000 people. On the other hand, some outdoor sports such as gliding, yachting, and skiing have quite specialized weather requirements.

Just as aviation is the most weather-dependent form of transport, so *gliding* is arguably the most weather-sensitive of all sports and Wallington (1961) has devoted an entire book to the meteorological principles underlying successful gliding. In Britain, gliding began as a sport in 1933 and, by 1966, there were 70 clubs in existence with

300 high-performance gliders and at least five times as many skilled pilots (Wickham, 1966). The atmospheric uplift necessary to support gliders is principally associated with thermal convection currents, sea breezes, or lee-wave effects and, under the most favourable conditions, flights may last for hours and cover hundreds of miles. It was during 1930 in the US that a glider pilot first used thermals for soaring and it is now known that the best breeding ground for thermals is the super-adiabatic layer in which incipient thermals grow rapidly before accelerating upwards. Many thermals have an internal motion comparable to a vortex ring and this provides more uplift than if the thermal were simply an internally inert bubble of warm air.

Over Britain, thermals powerful enough for gliding occur only from mid-April to the end of September and, although inland thermals may achieve heights of 1500 m, conditions deteriorate within 15–30 km of the coast. This deterioration is due to the inland penetration of the sea breeze which, in a small country like Britain, can be a serious handicap to pilots attempting long-distance flights to qualify for proficiency badges. However, the sea-breeze front itself may be most helpful to a glider and Wallington (1965) has described an occasion when good gliding conditions were found, although the turbulent mixing zone between the land and sea air was only about 90 m wide and the general prospects for thermal convection appeared poor. Gliders have attained heights of over 7300 m in lee waves over Britain but the possibility of lee-wave soaring varies markedly with season, time of day, and synoptic type. Most lee waves occur during winter but the poor visibility created by the associated low cloud means that spring and autumn tend to produce the best soaring weather. During summer the amplitude of lee waves will be affected by the diurnal heating and cooling cycle. Generally speaking, the best soaring prospects are found between 1 and 3 h after sunrise or in the evening, when the increasing amplitude and the shortening of the lee wavelength combine to produce strong vertical currents compared to the suppressed waves which occur as a result of afternoon convection. The most favourable synoptic situation for lee waves is often the warm sector of depressions, when fresh west to south-west winds show increased speed with height but little change in direction.

When national and international gliding championships are staged, the competitors are set different tasks each day by a small panel of experts, who are guided by the very detailed weather forecasts which are specially provided (Wallington, 1968). These forecasts are concerned with the fine structure of the atmosphere, such as the probable depth of the convective layer hour by hour, the possibility of cloud development reducing the strength of thermals, or the chance of uplift in lee waves. In addition, warnings of hail, downdraughts from large convective storms, or the danger from severe turbulence associated with rotor flow in lee waves must also be given.

Other airborne sports exhibit a weather-dependence comparable to that of gliding. For example, *parachuting* is a growing sport in many areas and in Britain there are now more than 20 clubs with about 2000 members who make around 30 000 jumps each year (d'Allenger, 1970). Although it is possible for an experienced parachutist wearing a sport parachute to make a safe landing with surface winds in excess of 10 m/s, the incidence of injuries on landing has been found to increase appreciably when winds are stronger than 7 m/s and this latter figure is normally regarded as the permissible maximum for safe parachuting by day. In order to assess the downwind drift towards the target landing area, a parachutist needs information about wind conditions in both the upper, free-fall part of the jump, which often takes place at a height of 2000 m, and in the atmospheric layer below parachute-opening height. The ideal weather for parachuting is an anticyclonic combination of light winds, good visibility, and the absence of cloud for, although precipitation affects the sport very little, at least one person has been frozen to death as a result of opening a parachute within a cumulo-nimbus cloud and subsequently becoming covered in ice. *Pigeon racing* is another sport which is very much under the influence of changes in windspeed and direction, and Rice (1969) has reported how the journey time and recovery rate of birds racing from Sicily or Italy to Malta depends on the weather conditions. The weather is also important for *hot-air ballooning*. In particular, Samuel (1972) has stated that ambient temperatures above 25 °C can prove hazardous since the excess temperature required inside the balloon to achieve lift-off under such conditions may well be more than the balloon fabric can withstand.

Watts (1965) has detailed the extent to which *yachting* is sensitive to the atmosphere and, here again, the emphasis lies on the characteristics of the wind. More information is required on the detailed structure of the wind over coastal waters, where most yachting takes place, and Watts (1967) has distinguished between the artificial steadiness of the airflow achieved during wind-tunnel tests of yachts and the fluctuations experienced in natural airstreams. The recurrent gustiness found in unstable polar airstreams can be related to the gust-cell model, first outlined by Giblett (1932) and his collaborators, in which the wind veers as it gusts and backs as it lulls. Since the periodicity of these gusts is often of the right order to be used tactically by dinghies, and even by deep-keel yachts under highly unstable conditions, it follows that the veering and backing rhythm of the gust-cell can be applied directly by yachtsmen when beating to windward. Indeed, Watts (1968) has claimed that much of the success in competitive yachting hinges on the interpretation of the micro-structure of the wind and, after designating Poole Bay as the area for the British Trials, the Royal Yachting Association authorized meteorological investigations in the area in preparation for the Mexico Olympics off Acapulco in October 1968.

A professional meteorologist accompanied the British Olympic Yachting Team to Acapulco and Houghton (1969) has described the contribution he was able to make to the team's achievements over five weeks of racing. In particular, a more variable gust behaviour was found at Acapulco, owing to different wind-deflection forces arising from the relatively low latitude, while the local sea breezes tended to be weaker than on a summer's day in Britain. This was attributed to the fact that, despite the existence of land temperatures which climbed to over 32 °C, the sea-surface temperature in autumn was as high as 31 °C and consequently a really strong thermal gradient was never established between land and sea.

A predictive knowledge of the sea breeze is important for competitive yachting, because the Olympic rules require a starting line which will allow a beat to windward for the first 1·5 nautical miles of the race. Many races start between 1000 and 1200 h local time, which is just about when the wind direction will shift as a result of the developing sea breeze, and the organizers must be prepared for this as the yachts come up to the starting line. Where races take place over long distances, such as in the single-handed trans-Atlantic event, the forecasting problems are similar to those involved in commercial ship-routeing. Thus, Zobel (1969) has reported how the winner of the 1968 trans-Atlantic race from Plymouth to Rhode Island relied on computer forecasts of the best course to steer so that, on one occasion, the yachtsman was routed north of a severe depression and benefited from following easterly winds, whereas his competitors 80–95 km further south received a Force 10 gale in their faces.

*Skiing* is another outdoor sporting activity, which is clearly weather-sensitive insofar as it necessitates the presence of a suitable snow cover, and the degree of sensitivity becomes especially high in areas where climatic conditions are marginal. This is the case in the Scottish Highlands where, despite uncertain snow prospects in a variable season that usually runs from mid-December to mid-April, Perry (1971 and 1972a) has noted how the investment in skiing facilities such as chairlifts, ski roads, and snow fences has increased since the early 1960s to the extent that an estimated 25 000 people now participate during a normal winter. In effect, reliable ski slopes are found only on the longer-lasting snow-beds that accumulate in incipient north-facing corries above 760 m and, apart from the frequent lack of snow, poor visibility and high winds often interrupt the sport.

Safety regulations close the chairlifts when the cross-wind velocity exceeds 15 m/s and the ski tows are taken out of operation when the wind reaches 27 m/s. This means that the upper part of the Cairngorm ski lift is inoperable for about 35 per cent of the total time, compared with the adjacent ski tow, which has a winter operating efficiency of 85 per cent, and in some seasons almost half the days are unsuitable for skiing. Avalanches represent a further atmospherically dependent hazard to which skiers are exposed and there is a need for improved avalanche-forecasting services. According to Langmuir (1970), snow conditions in Scotland are comparable with those on the Pacific Coast of the US and the dominant avalanche is the so-called direct-action type, which occurs either during or immediately following snowfall in the form of a soft or medium-hard slab. On the other hand, most avalanches in Switzerland take place in spring when the temperature first begins to rise towards 0 °C (Plasschaert, 1969). This temperature increase weakens the bonds between the ice granules and ultimately surrounds each granule with a film of water which tends to lubricate any movement within the snow mass.

## References


AANENSEN, C. J. M. (ed.) (1965). Gales in Yorkshire in February 1962. *Meteorological Office, Geophys. Mem. No. 108,* **14**, HMSO, London, 44 pp.

ALEXANDER, P. (1965). *Atomic radiation and life* (2nd Edn.). Pelican, Harmondsworth.

ANGUS, T. C. (1968). *The control of indoor climate*. Pergamon Press, Oxford, 110 pp.

ANONYMOUS (1960). Chilly May for shops. *The Economist,* **196**:36, 39.

ANONYMOUS (1963a). Cold weather costs. *The Economist,* **206**:824.

ANONYMOUS (1963b). The big freeze-up. *The Economist,* **208**:193.

ANONYMOUS (1971). Weather services for the community. *Met. Office. Annual Report, 1970,* 1–10.

ARNOLD, P. J. (1969). Thermal conductivity of masonry materials. *J. Instn. Heat. Vent. Engrs.,* **37**:101–8 and 117.

ARONIN, J. E. (1953). *Climate and architecture*. Rheinhold Pub. Corp., New York, 304 pp.

BADT, K. (1950). *John Constable's clouds*. Routledge and Kegan Paul, London.

BALL, E. F. (1968). Measurements of thermal conductivity of building materials. *J. Instn. Heat. Vent. Engrs.,* **36**:51–6.

BEAVER, S. H. (1972). Meteorology and philately. *Weather,* **27**:290–4.

BEEBE, R. G. (1967). The construction industry as related to weather. *Bull. Amer. Met. Soc.,* **48**:409.

BONACINA, L. C. W. (1937). John Constable's centenary: his position as a painter of weather. *Q. Jl. R. Met. Soc.,* **63**: 483–90.

BONACINA, L. C. W. (1972). Meteorology and art. *Weather,* **27**:300–1.

BOYER, A. (1966). Expanding industrial meteorology. *Bull. Amer. Met. Soc.,* **47**:528.

BRAZOL, D. (1954). Bosquejo bioclimatico de la Republica Argentina. *Meteoros (Buenos Aires),* **4**:381–94.

BRIDGER, C. A. and HELFAND, L. A. (1968). Mortality from heat during July 1966 in Illinois. *Int. J. Biometeor.,* **12**:51–70.

BROOKS, C. E. P. (1950). *Climate in everyday life*. E. Benn, London.

BROOME, M. R. (1966). Weather forecasting and the contractor. *Weather,* **21**:406–10.

BRUNT, D. (1943). Some reactions of the human body to its physical environment. *Q. Jl. R. Met. Soc.,* **69**:77–114.

CARR, T. R. (1970). Meteorological philately. *Weather,* **25**: 560–1.

CHALMERS, J. A. (1965). The action of a lightning conductor. *Weather,* **20**:183–5.

CHANDLER, T. J. (1964). An accumulated temperature map of the London area. *Met. Mag.,* **93**:242–5.

CHASE, P. H. (1970). Fifty years of weather insurance. *Weather,* **25**:294–8.

CLARKE, J. F. and BACH, W. (1971). Comparison of the comfort conditions in different urban and suburban microenvironments. *Int. J. Biometeor.,* **15**:41–54.

CLAWSON, M. (1966). The influence of weather on outdoor recreation. In Sewell, W. R. D. (ed.) *Human dimensions of weather modification*. Dept. of Geography, University of Chicago Research Paper No. 105, 183–93.

COURT, A. (1948). Wind chill. *Bull. Amer. Met. Soc.,* **29**:487–93.

COVERT, R. P., GOLDHAMER, M. M. and LEWIS, G. F. (1967). An estimation of the effects of precipitation on the scheduling of extended outdoor activities. *J. Appl. Met.,* **6**:683–7.

D'ALLENGER, P. K. (1970). Parachuting and the weather. *Weather,* **25**:188–92.

DAVIS, F. K. (1958). Ulcers and temperature changes. *Bull. Amer. Met. Soc.,* **39**:652–4.

DAVIS, F. K. and NEWSTEIN, H. (1968). The variation of gust factors with mean windspeed and with height. *J. Appl. Met.,* **7**:372–8.

DAVIS, N. E. (1968). An optimum summer weather index. *Weather,* **23**:305–17.

DAWS, L. F. (1970). Movement of airstreams indoors. *J. Instn. Heat. Vent. Engrs.,* **37**:241–53.

DERRICK, E. H. (1965). The seasonal variation of asthma in Brisbane: its relation to temperature and humidity. *Int. J. Biometeor.,* **9**:239–51.

DERRICK, E. H. (1966). The annual variation of asthma in Brisbane: its relation to the weather. *Int. J. Biometeor.,* **10**:91–9.

DERRICK, E. H. (1969). The short-term variation of asthma in Brisbane: its relation to the weather and other factors. *Int. J. Biometeor.,* **13**:295–308.

DE JAEGER, E. (1970). Het Belgisch Kustklimaat. Kenmerken en fysiologische invloed. *Aardvijkskunde/Geographie,* **22**: 237–53.

DINGLE, A. N. (1957). Hayfever pollen counts and some weather effects. *Bull. Amer. Met. Soc.,* **38**:465–9.

DORDICK, I. (1958). The influence of variations in atmospheric pressure upon human beings. *Weather,* **13**:359–64.

DRISCOLL, D. M. (1971). The relationship between weather and mortality in ten major metropolitan areas in the United States, 1962–65. *Int. J. Biometeor.,* **15**:23–39.

DRISCOLL, D. M. and LANDSBERG, H. E. (1967). Synoptic aspects of mortality. A case study. *Int. J. Biometeor.,* **11**: 323–8.

EATON, K. J. (1971). Tornadoes in south-east England, 25 January 1971. *Weather*, **26**:344–7.

EDHOLM, O. G. (1966). Problems of acclimatization in Man. *Weather*, **21**:340–50.

FALCONER, R. (1968). Wind-chill, a useful wintertime weather variable. *Weatherwise*, **21**:227–9.

FERGUSSON, P. (1964). Summer weather at the English seaside. *Weather*, **19**:144–6.

FOORD, H. V. (1968). An index of comfort for London. *Met. Mag.*, **97**:282–6.

FOORD, H. V. (1972). An investigation into consecutive wet working days at London Weather Centre. *Met. Mag.*, **101**: 362–6.

FRIEDMAN, D. G. and HENDRICK, R. L. (1960). The role of a meteorologist in an insurance company. *Weatherwise*, **13**:141–5.

FRISBY, E. M. (1957). Weather engineering in the service of industry. *Adv. Sci.*, **13**:341–4.

FRY, M. and DREW, J. (1956). *Tropical architecture in the humid zone*. B. T. Batsford, London.

GANDER, R. S. (1972). The demand for beer as a function of the weather in the British Isles. In Taylor, J. A. (ed.) *Weather forecasting for agriculture and industry*. David and Charles, Newton Abbot, 184–94.

GIBLETT, M. A. *et al.* (1932). The structure of wind over level country. *Met. Office Geophys. Memoir 54*, HMSO, London.

GIVONI, B. (1969). *Man, climate and architecture*. Applied Science Publishers, Barking, 364 pp.

GODSHALL, F. A. (1968). Natural daylight illumination in north-eastern United States. *J. Appl. Met.*, **7**:499–503.

GOLD, E. (1935). The effect of wind, temperature, humidity and sunshine on the loss of heat of a body at temperature 98 °F. *Q. Jl. R. Met. Soc.*, **61**:316–31.

GOLDSMITH, J. R. and PERKINS, N. M. (1967). Seasonal variations in mortality. In Tromp, S. W. and Weihe, W. H. (eds) *Biometeor. Vol. 2, Pt. 1.* Pergamon Press, Oxford, 97–114.

GREEN, J. S. A. (1967). Holiday meteorology: reflections on weather and outdoor comfort. *Weather*, **22**:128–31.

GREENBURG, L., FIELD, F., REED, J. I. and ERHARDT, C. L. (1967). Asthma and temperature change. In Tromp, S. W. and Weihe, W. H. (eds) *Biometeor. Vol. 2, Pt. 1.* Pergamon Press, Oxford, 3–6.

GREENWOOD, P. G. and HILL, R. D. (1968). Buildings and climate in Singapore. *J. Trop. Geog.*, **26**:37–47.

GREGORCZUK, M. (1968). Bioclimates of the world related to air enthalpy. *Int. J. Biometeor.*, **12**:35–9.

GREGORCZUK, M. and CENA, K. (1967). Distribution of effective temperature over the surface of the earth. *Int. J. Biometeor.*, **11**:145–9.

HALLANGER, N. L. (1963). The business of weather: its potentials and uses. *Bull. Amer. Met. Soc.*, **44**:63–7.

HANSEN, J. B. (1970). The relation between barometric pressure and the incidence of peripheral arterial embolism. *Int. J. Biometeor.*, **14**:391–7.

HANSEN, J. B. and PEDERSEN, S. A. (1972). The relation between barometric pressure and the incidence of perforated duodenal ulcer. *Int. J. Biometeor.*, **16**:85–91.

HARDY, A. C. and MITCHELL, H. G. (1970). Building a climate—the Wallsend project. *J. Instn. Heat. Vent. Engrs.*, **38**:71–81.

HARDY, J. D. and DUBOIS, E. F. (1940). Differences between men and women in their response to heat and cold. *Proc. Nat. Acad. Sci.* (Wash.), **26**:389–98.

HENDRICK, R. L. and FRIEDMAN, D. G. (1966). Potential impacts of storm modification on the insurance industry. In Sewell, W. R. D. (ed.) *Human dimensions of weather modification.* University of Chicago, Dept. of Geography Research Paper No. 105, 249–60.

HOLFORD, I. (1967). Planning your days according to the weather. *Weather*, **22**:132–3.

HOLFORD, LORD (1971). Problems for the architect and town planner caused by air in motion. *Phil. Trans. R. Soc. Ser. A.*, **269**:335–41.

HOPE-SIMPSON, R. E. (1967). Seasonal effects on respiratory disease in relation to specific virus infections. In Tromp, S. W. and Weihe, W. H. (eds) *Biometeor. Vol. 2, Pt. 1.* Pergamon Press, Oxford, 77–82.

HOUGHTEN, F. C. and YAGLOU, C. P. (1923). Determining lines of equal comfort. *Trans. Amer. Soc. Heat. Vent. Eng.*, **29**:163–76.

HOUGHTON, D. (1969). Acapulco '68. *Weather*, **24**:2–18.

HOUNAM, C. E. (1967). Meteorological factors affecting physical comfort (with special reference to Alice Springs, Australia). *Int. J. Biometeor.*, **11**:151–62.

HOWE, G. M. (1962). Windchill, absolute humidity and the cold spell of Christmas 1961. *Weather*, **17**:349–58.

HUGHES, G. H. (1967). Summers in Manchester. *Weather*, **22**:199–200.

HUGHES, G. H. (1972). Periodicity of 34 years in even summers. *Weather*, **27**:241–6.

HUMPHREYS, M. A. and NICOL, J. F. (1970). An investigation into thermal comfort of office workers. *J. Instn. Heat. Vent. Engrs.*, **38**:181–9.

JAUREGUI, E. and SOTO, C. (1967). Wet bulb temperature and discomfort index: areal distribution in Mexico. *Int. J. Biometeor.*, **11**:21–8.

KESTIN, M. (1967). The effect of weather on cosmetic sales. *Amer. Perfume and Cosmetics*, **82**:28.

LACY, R. E. (1951). Distribution of rainfall round a house. *Met. Mag.*, **80**:184–9.

LACY, R. E. (1972). Survey of meteorological information for architecture and building. *Current Paper CP 5/72. Building Research Station*, Department of the Environment (London), 22 pp.

Lacy, R. E. and Shellard, H. C. (1962). An index of driving rain. *Met. Mag.*, **91**:177–84.

Lamb, H. H. (1966). Climate in the 1960's. *Geogrl. J.*, **132**:183–212.

Lamb, H. H. (1967). Britain's changing climate. *Geogrl. J.*, **133**:445–66.

Landsberg, H. E. (1969). *Weather and health*. Doubleday and Co., New York, 148 pp.

Langdon, F. J. and Loudon, A. G. (1970). Discomfort in schools from overheating in summer. *J. Instn. Heat. Vent. Engrs.*, **37**:265–74.

Langmuir, E. (1970). Snow profiles in Scotland. *Weather*, **25**:205–9.

Laskey, L. S. (1969). Is the 'simple index' misleading? *Weather*, **24**:473–5.

Lawrence, E. N. (1954). Microclimatology and town planning. *Weather*, **8**:227–32.

Lawrence, J. S. (1967). Climate and rheumatic diseases. In Tromp, S. W. and Weihe, W. H. (eds) *Biometeor. Vol. 2, Pt. 1*. Pergamon Press, Oxford, 130–9.

Linden, F. (1959a). Weather in business. *The Conference Board Business Record*, **16**:90–4.

Linden, F. (1959b). Merchandising with the weather. *The Conference Board Business Record*, **16**:144–9.

Loewe, F. (1972). The land of storms. *Weather*, **27**:110–21.

Loudon, A. G. (1967). The interpretation of solar radiation measurements for building problems. *Research Paper 73, Building Research Station*, Ministry of Public Building and Works (London).

Loudon, A. G. (1968). Window design criteria to avoid overheating by excessive solar heat gains. *Current Paper 4/68. Building Research Station*, Ministry of Public Building and Works (London).

Loudon, A. G. (1970). Summertime temperatures in buildings without air-conditioning. *J. Instn. Heat. Vent. Engrs.*, **37**:280–92.

Ludlam, F. H. (1972). The meteorology of the 'Ode to the West Wind'. *Weather*, **27**:503–14.

McCleod, C. N. (1965). An index of comfort for Gan. *Met. Mag.*, **94**:166–71.

Mason, B. J. (1966). The role of meteorology in the national economy. *Weather*, **21**:382–93.

Massey, P. M. O. (1959). Finger numbness and temperature in Antarctica. *J. Appl. Physiol.*, **14**:616–20.

Maunder, W. J. (1962). A human classification of climate. *Weather*, **17**:3–12.

Maunder, W. J. (1970). *The value of the weather*. Methuen, London, 388 pp.

Maunder, W. J. (1973). Weekly weather and economic activities on a national scale: an example using United States retail trade data. *Weather*, **28**:2–18.

Maunder, W. J., Johnson, S. R. and McQuigg, J. D. (1971a). Study of the effect of weather on road construction: a simulation model. *Mon. Weath. Rev.*, **99**:939–45.

Maunder, W. J., Johnson, S. R. and McQuigg, J. D. (1971b). The effect of weather on road construction: application of a simulation model. *Mon. Weath. Rev.*, **99**:946–53.

Menger, W. (1967). Klimatische Gesichtspunkte in der Behandlung des Asthma Bronchiale. In Tromp, S. W. and Weihe, W. H. (eds) *Biometeor. Vol. 2, Pt. 1*. Pergamon Press, Oxford, 7–14.

Menzies, J. B. (1971). Wind damage to buildings. *Building*, **221**:67–76.

Miller, W. H. (1968). Santa Ana winds and crime. *Prof. Geogr.*, **20**:23–7.

Millington, R. A. (1964). Physiological responses to cold. *Weather*, **19**:334–8.

Momiyama, M. (1968). Biometeorological study of the seasonal variation of mortality in Japan and other countries on the seasonal disease calendar. *Int. J. Biometeor.*, **12**:377–93.

Murray, R. (1972). A simple summer index with an illustration for summer 1971. *Weather*, **27**:161–9.

Musgrave, J. C. (1968). Measuring the influence of weather on housing starts. *Construction Rev.*, **14**:4–7.

Neuberger, H. (1970). Climate in art. *Weather*, **25**:46–56.

Olson, W. P. (1969). Rat-flea indices, rainfall and plague outbreaks in Vietnam, with emphasis on the Pleiku area. *Amer. J. Trop. Med. Hyg.*, **18**:621–8.

Olson, W. P. (1970). Rainfall and plague in Vietnam. *Int. J. Biometeor.*, **14**:357–60.

Padmanabhamurty, B. (1972). A study of biotropism of climate in two Canadian cities. *Int. J. Biometeor.*, **16**:107–17.

Page, J. K. (1972). The problem of forecasting the properties of the built environment from the climatological properties of the green-field site. In Taylor, J. A. (ed.) *Weather forecasting for agriculture and industry*, David and Charles, Newton Abbot, 195–208.

Parry, M. (1957). Local climate and house heating. *Adv. Sci.*, **13**:326–31.

Paul, A. H. (1972). Weather and the daily use of outdoor recreation areas in Canada. In Taylor, J. A. (ed.) *Weather forecasting for agriculture and industry*. David and Charles, Newton Abbot, 132–46.

Paulus, H. J. and Smith, T. J. (1967). Association of allergic bronchial asthma with certain air pollutants and weather parameters. *Int. J. Biometeor.*, **11**:119–27.

Peguy, C. (1961). *Précis de climatologie*. Masson, Paris.

Perry, A. (1968). 'Summer Days' in the British Isles. *Weather*, **23**:212–14.

Perry, A. H. (1971). Climatic influences on the development of the Scottish skiing industry. *Scott. Geogrl. Mag.*, **87**:197–201.

Perry, A. H. (1972a). The weather forecaster and the tourist—the example of the Scottish skiing industry. In Taylor, J. A. (ed.) *Weather forecasting for agriculture and industry*. David and Charles, Newton Abbot, 126–31.

PERRY, A. H. (1972b). Weather, climate and tourism. *Weather*, **27**: 199–203.

PETHERBRIDGE, P. and LOUDON, A. G. (1966). Principles of sun control. *The Architects' Journal*, **143**: 143–9.

PETTY, M. T. (1963). Weather and consumer sales. *Bull. Amer. Met. Soc.*, **44**: 68–71.

PLASSCHAERT, J. H. M. (1969). Weather and avalanches. *Weather*, **24**: 99–102.

PORTIG, W. H. (1968). The humid warm tropical climate and man. *Weather*, **23**: 177–8.

POULTER, R. M. (1962). The next few summers in London. *Weather*, **17**: 253–7.

PROHASKA, F. (1970). Distinctive bioclimatic parameters of the subtropical-tropical Andes. *Int. J. Biometeor.*, **14**: 1–12.

RACKLIFF, P. G. (1965). Summer and winter indices at Armagh. *Weather*, **20**: 38–44.

RICE, R. W. (1969). Weather and pigeon racing in Malta. *Weather*, **24**: 281–2.

RUSSO, J. A. (1966). The economic impact of weather on the construction industry of the United States. *Bull. Amer. Met. Soc.*, **47**: 967–72.

SAMUEL, G. A. (1972). Some meteorological and other aspects of hot-air ballooning. *Met. Mag.*, **101**: 25–9.

SCRUTON, C. and NEWBERRY, C. W. (1963). On the estimation of wind loads for building and structural design. *Proc. Inst. Civ. Eng.*, **25**: 97–126.

SCRUTON, C. and ROGERS, E. W. E. (1971). Steady and unsteady wind loading of buildings and structures. *Phil. Trans. R. Soc. Lond. Ser. A*, **269**: 353–83.

SHAW, E. B. (1963). Geography and baseball. *J. Geogr.*, **62**: 74–76.

SHELLARD, H. C. (1967a). Wind records and their application to structural design. *Met. Mag.*, **96**: 235–43.

SHELLARD, H. C. (1967b). Collapse of cooling towers in a gale, Ferrybridge, 1 November 1965. *Weather*, **22**: 232–40.

SIPLE, P. A. and PASSEL, C. F. (1945). Measurements of dry atmospheric cooling in subfreezing temperatures. *Proc. Amer. Phil. Soc.*, **89**: 177–99.

STEELE, A. T. (1951). Weather's effect on the sales of a department store. *J. Marketing*, **15**: 436–43.

STEPHENSON, P. M. (1963). An index of comfort for Singapore. *Met. Mag.*, **92**: 338–45.

STONE, R. G. (1943). On the practical evaluation and interpretation of the cooling power in bioclimatology. *Bull. Amer. Met. Soc.*, **34**: 295–305.

SULMAN, F. G., DANON, A., PFEIFER, Y., TAL, E. and WELLER, C. P. (1970). Urinalysis of patients suffering from climatic heat stress (Sharav). *Int. J. Biometeor.*, **14**: 45–53.

TERJUNG, W. H. (1966a). Physiologic climates of the coterminous United States: a bioclimatic classification based on man. *Ann. Ass. Amer. Geogrs.*, **56**: 141–79.

TERJUNG, W. H. (1966b). The seasonal march of physiological climates and cumulative stress in the Sudan. *J. Trop. Geog.*, **22**: 49–62.

TERJUNG, W. H. (1967a). The geographical application of some selected physio-climatic indices to Africa. *Int. J. Biometeor.*, **11**: 5–19.

TERJUNG, W. H. (1967b). Annual physioclimatic stresses and regimes in the United States. *Geogrl. Rev.*, **57**: 225–40.

TERJUNG, W. H. (1968a). World patterns of the distribution of the monthly comfort index. *Int. J. Biometeor.*, **12**: 119–51.

TERJUNG, W. H. (1968b). Bi-monthly physiological climates and annual stresses and regimes of Africa. *Geograf. Ann.*, **50A**: 173–92.

TERJUNG, W. H. (1968c). Some thoughts on recreation geography in Alaska from a physio-climatic viewpoint. *The Californian Geographer*, **9**: 27–39.

TERJUNG, W. H. (1970). The energy budget of man at high altitudes. *Int. J. Biometeor.*, **14**: 13–43.

THOM, E. C. (1959). The discomfort index. *Weatherwise*, **12**: 57–60.

TROMP, S. W. (1963). *Medical Biometeorology*. Elsevier Publishing Co., New York.

TROMP, S. W. (1967). Blood sedimentation rate patterns in the Netherlands during the period 1955–1965. *Int. J. Biometeor.*, **11**: 105–17.

TROMP, S. W. and SARGENT, F. (1964). *A survey of human biometeorology*. W.M.O. Tech. Note 65, Geneva, 113 pp.

TUCKER, G. W. L. (1965). The weather and the holiday-maker. In *What is weather worth?* Bureau of Meteorology, Melbourne, 49–51.

TYLDESLEY, J. B. (1968). Weather in Constable's painting. *Weather*, **23**: 344–8.

TYLER, W. F. (1935). Bracing and relaxing climates. *Q. Jl. R. Met. Soc.*, **61**: 309–15.

WALLINGTON, C. E. (1961). *Meteorology for glider pilots*. J. Murray, London, 284 pp.

WALLINGTON, C. E. (1965). Gliding through a sea-breeze front. *Weather*, **20**: 140–4.

WALLINGTON, C. E. (1968). Forecasting for gliding. *Weather*, **23**: 236–45.

WATT, G. A. (1967). An index of comfort for Bahrein. *Met. Mag.*, **96**: 321–7.

WATTS, A. (1965). *Wind and sailing boats*. Adlard Coles Ltd, London.

WATTS, A. (1967). The real wind and the yacht. *Weather*, **22**: 23–9.

WATTS, A. (1968). Winds for the Olympics. *Weather*, **23**: 9–22.

WICKHAM, P. G. (1966). Weather for gliding over Britain. *Weather*, **21**: 154–61.

WILSON, A. (1966). The impact of climate on industrial growth: Tucson, Arizona: A case study. In Sewell, W. R. D. (ed.) *Human dimensions of weather modification*. University of Chicago, Dept. of Geography Research Paper No. 105: 249–60.

WILSON, O. (1967). Objective evaluation of wind chill index by records of frostbite in the Antarctic. *Int. J. Biometeor.*, **11**:29–32.

WINSTANLEY, D. (1972). Sharav. *Weather*, **27**:146–60.

WINTER BUILDING ADVISORY COMMITTEE (1971). *Winter building: A review of winter building techniques*. Department of the Environment, HMSO, London. 35 pp.

WISE, A. F. E. (1971). Effects due to groups of buildings. *Phil. Trans. R. Soc. Lond. Ser. A*, **269**:469–85.

WYCHERLEY, P. R. (1967). Index of comfort throughout Malaysia. *Met. Mag.*, **96**:73–7.

WYNDHAM, C. H. *et al.* (1964). Physiological reactions of Caucasian and Bantu males in acute exposure to cold. *J. Appl. Physiol.*, **19**:583–92.

WYON, D. P., LIDWELL, O. M. and WILLIAMS, R. E. O. (1969). Thermal comfort during surgical operations. *J. Instn. Heat. Vent. Engrs.*, **37**:150–8, 168.

YOSHIMURA, M. and YOSHIMURA, H. (1969). Cold tolerance and critical temperature of the Japanese. *Int. J. Biometeor.*, **13**:163–72.

ZEISEL, H. (1950). How temperature affects sales. *Printers Ink*, **233**:40–2.

ZOBEL, R. F. (1969). Forecasting for 'The Observer' single-handed trans-Atlantic yacht race. *Weather*, **24**:152–5.

ZOHAR, Y., SCHILLER, G. and KARSCHON, R. (1971). Determination of the direction of the prevailing wind from the orientation of crescent-shaped Beduin shelters in north-eastern Sinai. *Agric. Met.*, **8**:319–23.

# 8. Weather forecasting and weather modification

## 8.1 Introduction

Throughout this book there has been an attempt to show that man is no longer relegated to an entirely passive relationship with the atmosphere, and two of the most positive human responses to the atmospheric environment are those of weather forecasting and weather modification. In certain circumstances, weather forecasts have had historic significance and Stagg (1971) has described the background to the forecasting conducted for the critical D-day invasion of Normandy by the Allies during the Second World War. Initially, most emphasis was placed on the development of the scientific aspects of weather forecasting but, as stated by Moorman (1968), forecasting is now recognized as being a service as well as a science. Accordingly, in recent years more attention has been paid to the requirements of the weather forecast consumer and a comparative analysis of two questionnaire surveys made twenty years apart led Maunder (1969) to conclude that the attitudes of the general public towards weather forecasting terms, as well as the actual needs for forecast information, may change from time to time and place to place. Similarly, there is a growing realization that, although weather modification is based on physical science, the implications involved raise many social issues (Droessler, 1966; Malone, 1967). In view of these trends, the following sections deal with both the physical-science and the social-science aspects of weather forecasting and weather modification.

## 8.2 Early developments in weather forecasting

The origins of practical weather forecasting date back to the mid-nineteenth century and the invention of the electric telegraph which, for the first time, enabled meteorological data to be assembled and transmitted in advance of the movements of the weather itself. In Britain, daily forecasts deriving from an organized network of observing stations were published as early as 1848, but the first *Daily Weather Report* to be issued by the newly formed Meteorological Office did not appear until September 1860 (Anon., 1956). At this time, the information available was limited to surface observations of pressure, temperature, wind, and prevailing weather recorded by a few land stations. Although there followed a steady growth in the exchange of meteorological data between Britain and the adjacent European countries, it was not until 1907 that the development of wireless telegraphy led to the receipt in London of the first weather reports from Atlantic shipping and thus allowed the compilation of reasonably comprehensive weather maps.

During the nineteenth century, weather forecasting made increasing use of the observed relationships between pressure patterns, wind behaviour, and subsequent weather events. However, despite the fact that the surface isobaric chart became established as the basic forecasting tool almost from the outset, the real significance of atmospheric pressure was recognized only in the 1920s and 1930s when the concepts of air masses and fronts, developed earlier by the influential Norwegian school of meteorologists, were incorporated into operational forecasting (Sawyer, 1967a). From this time onwards, short-term forecasting in the mid-latitudes was to rely heavily on the classic model of the warm-sector depression with its characteristic life cycle and associated weather.

This is not to deny, of course, that further substantial improvements in techniques and understanding took place later. For example, the introduction of the radiosonde in 1939 led directly to significant advances in upper-air forecasting. In addition, a more complete knowledge of the three-dimensional structure of the synoptic-scale weather systems such as depressions and anticyclones was obtained and Ludlam (1966) has outlined the major developments surrounding the scientific understanding of the mid-latitude depression. Nevertheless, the fundamental principles of weather forecasting were to remain unchanged for over a century.

As shown in the general reviews by Sutcliffe (1952), Miles (1961), and Houghton (1965), the traditional method of weather forecasting was based on a largely subjective, two-stage operation. First, the forecaster prepared a prognostic or *prebaratic* chart indicating the location of isobars and fronts at the end of the forecast period and then, second, plotted the expected weather details on the map in the light of accumulated experience relating to the development of the relevant travelling weather systems. Wickham (1970) has emphasized that the success of this method depended on the skill of the individual forecasters in identifying models of the most common weather systems and then modifying the idealized model to meet the requirements of the real-life atmosphere. For forecast periods of up to 24 h ahead, this type of flexible approach has proved most reliable and the subjective extrapolation of existing weather trends will probably always retain some importance in forecasting, not least for the short-term, specialized forecast which relies heavily on an intimate knowledge of local topography and weather phenomena.

On the other hand, it has been known for many years that there are serious limitations inherent in the subjective approach. Thus, the basic synoptic models involved were nothing more than relatively simple descriptive accounts of typical atmospheric systems with an individual life-expectancy rarely exceeding a few days and, consequently, even the best forecasts could only provide qualitative predictions, which not only became very much less accurate beyond about 24–48 h ahead but also formed no reliable foundation for more extended-range forecasting. Another major weakness was the restricted data-handling capability of the traditional methods. Any forecast can only be as good as the input data employed and every forecaster has to interpret his weather map in the context of raw meteorological information coming in from a wide geographical area.

The human forecaster was placed, therefore, in an insoluble dilemma, since he required more and more observations in order to improve his forecasts but was unable to assemble, analyse, and plot the information in sufficient time to issue a prediction far enough in advance of the evolving weather. It eventually became clear that substantial future progress could be achieved only by more objective scientific methods arising from a deeper understanding of atmospheric processes and geared to a highly efficient technology of data analysis and information distribution.

From a purely scientific viewpoint, it has always seemed more satisfying to replace empirical forecasting methods by a more numerical approach based on physical laws and, after solving some of the initial mathematical problems created by the non-linear differential equations involved, Richardson (1922) outlined a fully objective forecasting scheme. Richardson's plan revolved around the solution of the time-dependent dynamical equations of motion, conservation of mass, and energy computed for five atmospheric layers at horizontal intervals of about 200 km but, in order to keep ahead of the weather, it was estimated that 64 000 people operating desk calculators would be required. Such computational problems, both practical and theoretical, ensured that numerical forecasting remained little more than a dream for two more decades. However, the advent of the electronic computer in the early 1950s provided a new stimulus to the theoreticians, who were attempting to simulate the behaviour of the atmosphere with the aid of mathematical models (Knighting, 1958). Over the last twenty years or so, important advances have taken place simultaneously in the three related fields of meteorological observation, data processing, and numerical modelling. These technical improvements have been applied to atmospheric prediction on progressively enlarging scales of space and time, and it will be convenient to review the developments in the context of both short-term and long-range forecasting.

### 8.3 Short-term forecasting

Computers have been used for forecasting in the British Meteorological Office since November 1954 and the short-term forecast has become increasingly computer-based. Essentially, the computer assembles all the incoming data from land stations, weather ships, and satellites and derives the best-fit values for a regular grid network covering most of the northern hemisphere. After charts detailing the existing state of the atmosphere have been drawn, the computer works in terms of contour heights and thickness values to predict future patterns at fixed intervals of time and, even with the early computers, numerical forecasts for 48 h ahead were available within $1\frac{1}{4}$ h of the receipt of the data (Sutcliffe, 1964; Mason, 1966a). The computer soon proved itself superior to the manual forecaster in the preparation of prognostic charts of upper air conditions but surface charts still require a certain amount of human modification. Some of the progress achieved in the first six years of numerical forecasting in the US has been described by Fawcett (1962).

Satellite meteorology began in April 1960 with the launching of TIROS 1 from what was then the Cape Canaveral site in Florida, and an assessment of the early contribution of satellites to forecasting and other aspects of meteorology has been provided by Barrett (1967). As with the previous development of radar, satellites have proved valuable in locating and tracking the highly destructive meso-scale systems such as hurricanes, squall lines, and hailstorms. Satellite television pictures have been received daily at Bracknell since March 1966 from the APT facility (Pothecary and Ratcliffe, 1966) and the forecaster can now see weather systems evolving over large areas of the globe within only an hour or so of the photograph being taken. Cloud photographs taken over the Atlantic have been most useful for short-term forecasting in Britain, as demonstrated by Ratcliffe (1966), since the structure and movement of new depressions can be deduced several hours before surface and upper air data become available for the area. In other cases, satellite surveillance allows the monitoring of small secondary depressions, which might have been undetected by the sparse network of ocean weather ships.

The original Meteorological Office numerical simulation model assumed the existence of a dry atmosphere and was, therefore, incapable of forecasting either humidity or precipitation. More recently, however, further equations have been added to produce a high-resolution, 10–level model which can simulate the physical processes in a moist, baroclinic atmosphere on the synoptic scale (Bushby and Timpson, 1967; Benwell and Timpson, 1968). The computations are made over an area of $6500 \times 5000$ km centred on the British Isles with over 12 000 grid points spaced 50 km apart, with dependent variables calculated at alternate grid points and for ten vertical levels between 1000 and 100 mb. These computations include the heights of the various pressure surfaces, the mean temperatures of the nine vertical layers, the horizontal and vertical components of the wind plus the humidity and condensed water content at seven levels below 300 mb. The model allows for various extraneous factors—such as topographic effects, frictional drag, and horizontal eddy diffusion—and has been employed to study the dynamics of fronts as well as to make quantitative forecasts of frontal rainfall for periods of up to 36 h ahead.

According to Mason (1971), the preservation of computational stability requires the integration of the hydrodynamic equation in time steps of only 90 s, which means that a forecast for a 24-h period necessitates about $10^{10}$ numerical operations. On the most powerful computer then existing, and installed by the Meteorological Office in 1971, this exercise could be accomplished within 30 min but it was suggested that the growing sophistication of mathematical models may eventually produce a need for machines 100 times as powerful with a capability of $10^9$ operations per second.

Even if quantitative forecasts of rainfall eventually become commonplace in the forecasting routine, there will still be some scope for synoptic interpretation and Lowndes (1966), for example, has shown how widespread showers in north-westerly airstreams over south-east England in winter may be related to low-pressure perturbations. As yet, there appears to be little possibility of operational numerical forecasts for cloud or fog and Mason (1970) has stated that further consistent improvements in the short-term forecast will depend on a much better representation of physical processes in the atmosphere and finer spatial resolution on the grid networks, together with the provision of more powerful computers. In some cases, sophisticated models are already producing surface forecasts which are barely superior to fairly simple three-level vorticity models and, although the grid networks can be expected to become more and more refined, manual interpolation will presumably always be necessary for the smallest-scale weather features.

### 8.4 Medium- and long-range forecasting

Medium-range forecasting, which covers the period up to about two weeks ahead, is currently based on a combination of numerical prediction and more subjective extrapolation. For example, the US Weather Bureau has issued five-day forecasts twice a week for a number of years using numerical simulation from its six-level model for the first three days and a simple barotrophic forecast for the remainder of the period. With more observations and better models, it may well ultimately be possible to produce entirely numerical forecasts for up to seven days ahead but, much beyond this, the nature of the problem changes.

The immediate limitation to successful long-range forecasting lies in the over-simplified atmospheric simulation models which are currently employed. Generally speaking, these models start from a given initial condition of the atmosphere, with a specified amount of available energy, and then solve a system of simultaneous, non-linear partial differential equations in order to forecast developments through time arising from the originally stated situation. This type of prediction model can go astray either because of shortcomings in the actual observations fed into it or because of a failure to provide an accurate simulation of physical processes operating through the forecast period. Although deficiencies continue to exist in the observational data, the major difficulty relates to a theoretical understanding and adequate numerical representation of the continuing atmospheric processes.

According to Wickham (1970), the early models represented the atmosphere like a mechanical toy which, from its initial wound-up state, would progressively unwind through various predictable phases. Unfortunately, the real atmosphere does not unwind in this way since the energy supply is continuously replaced by solar radiation and beyond a few days such forecasts become much less realistic. In particular, the models do not simulate baroclinic instability to the extent that it exists in the real atmosphere (Corby, 1970). What is required, therefore, is a more detailed appreciation of the thermal relationships between solar energy, the condition of the earth's surface, and the long-term circulation of the atmosphere, as well as an ability to simulate all the processes of radiation, evaporation, condensation, and precipitation. In addition, Lamb (1972) has noted that, as the forecast period is extended, the behaviour of the atmosphere becomes increasingly subject to external influences, such as volcanic eruptions and fluctuations in man-made pollution, which are independent of the pre-existing atmosphere and extremely difficult to predict.

Apart from technical questions about the future adequacy of global observation networks, or doubts as to whether numerical models and their necessary computational support will ever advance sufficiently for operational long-range forecasting purposes, there is the separate issue of the extent to which inherent indeterminism in the atmosphere may impose obstacles beyond which prediction can only be achieved by statistical or probability methods. The range of possible determinism depends largely on the scale of atmospheric phenomena and is much larger for the planetary circulation than for synoptic-scale features, but Robinson (1967) has introduced a note of caution by testing the potential limits of atmospheric predictability as opposed to the practical limits arising from defects in observation and computation. A similar exercise conducted by Lorenz (1969) revealed that small initial errors detected at the present scale of grid resolution doubled in rather less than three days and this suggested that little prospect existed for predicting the positions of migratory cyclones and anticyclones two weeks ahead. Even with a much denser observational network, both authors concluded that, for predicting weather on a particular day, an absolute maximum period of perhaps three weeks ahead was indicated.

On the other hand, Mason (1970) has stated that these estimates may well prove to be too conservative and some of the extended predictions already achieved by the numerical models developed by the Geophysical Fluid Dynamics Laboratory in the US tend to support this latter view. For example, Miyakoda *et al.* (1969) have reported how a nine-level model of the general circulation, incorporating such factors as orography, radiative heat transfer, turbulent exchange of heat and moisture, differences between land and sea, together with differences in the thermal properties between land-ice and sea-ice, was applied to two situations over the northern hemisphere in winter. The model was run for a simulation period of 14 days and the outstanding success came with the prediction of the birth and the subsequent behaviour of second- and third-generation depressions.

Beyond three days, the forecasts for the lower levels were greatly improved by allowing the model to simulate heat exchange with the ocean surface, and although forecast temperatures were persistently too low, fairly high correlation coefficients between some simulated and observed parameters were recorded well beyond the first week. More recently, the model has been run on a seasonal time-scale over the whole globe (Holloway and Manabe, 1971). In this instance, the model simulated precipitation, evaporation, soil moisture, snow depth, and runoff and was run long enough beyond the state of quasi-equilibrium for meaningful climatic statistics to emerge. Thus, the model was able to reproduce the tropical and mid-latitude rainfall zones, the areas of high runoff and the major arid areas of the world such as the Saharan and Australian deserts. The major defects of the model related to the simulation of excessively high pressure in the polar regions and the overestimation of relative humidity near the earth's surface but the general results were encouraging enough to lead the authors to suggest that a possible future application of the model might be in the evaluation of climatic change through human activity.

Although long-range weather forecasting is becoming increasingly dependent on numerical models, Namias (1968) has emphasized how methods in the US Weather Bureau still tend to be weighted towards rather more empirical synoptic and statistical techniques. A similar situation prevails in the British Meteorological Office, where predictions for a month and beyond are based on a combination of a so-called physical discussion and an analogue selection procedure. Both of these approaches focus on the significance of anomalous synoptic patterns and, in particular, on feedback relationships between the earth and the atmosphere associated with abnormal boundary conditions imposed by sea surface temperatures on the presence of a snow and ice cover. Murray (1970) has explained how improvements in operational techniques have been made possible by additional information, mostly of a historical nature, on parameters such as sea ice and surface sea temperature anomalies for the North Atlantic which have then been applied through a number of specific forecasting tools.

In the case of monthly forecasts, for example, Ratcliffe and Murray (1970) have reported a statistically valid feedback relationship between sea-surface temperatures in part of the North Atlantic and atmospheric circulation anomalies one month later over northwest Europe. Figure 8.1 shows that the key area of ocean involved lies to the south and east of Newfoundland and, in general, it was found that the mean pressure anomalies which followed negative and positive variations in the ocean surface temperatures analysed for this area since 1888 were opposite in sign, so that a colder than normal sea surface led to high pressure and blocked atmospheric patterns the following month over north-western Europe, whereas a warm ocean surface allowed the progression

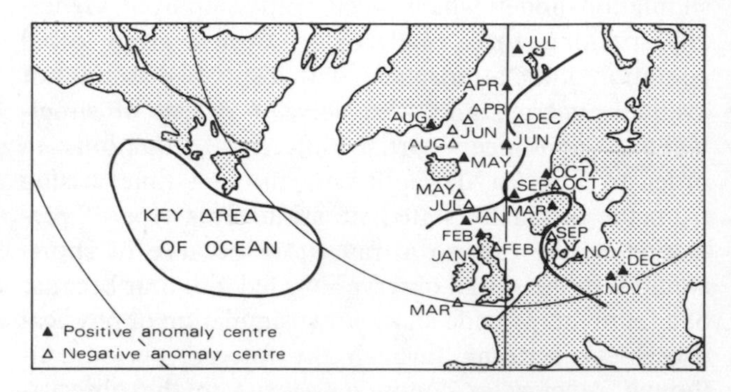


**Fig. 8.1 The positions of positive and negative pressure anomaly centres following warm and cold ocean temperatures in the key area for all months of the year. After Ratcliffe (1970a).**

of depressions and associated low pressure to move into Europe (Ratcliffe, 1970a). Furthermore, the centres of the pressure anomalies were located in approximately the same areas during certain months of the year. Figure 8.1 indicates that from September to December the anomaly centres lie to the east of Britain, from January to March they are close to Britain, and from April to August they are located to the north or northwest.

A closer insight into the physical processes is required but it does appear that high sea-surface temperatures increase both the sensible heat transfer and the exchange of water vapour between the sea and the air, leading to more favourable conditions for the development of Atlantic depressions, whereas low ocean temperatures produce the reverse effect. There is some evidence that comparable mechanisms may be at work elsewhere and Namias (1969) has demonstrated a connection between the presence of anomalously warm water in part of the central Pacific and lower-than-usual pressure south of the Aleutian Islands in winter. It must be stressed that such relationships cannot be used on every occasion and, even when a well defined sea-surface temperature anomaly occurs in the Atlantic, only about 50 per cent of the cases eventually lead to definite probabilities of certain weather types over Britain.

Some other forecasting tools are more direct. For example, Murray and Benwell (1970) have outlined the forecasting potential of synoptic indices, which are based on the statistical relationships of large-scale monthly mean circulation patterns over the north-east Atlantic and western Europe. Murray (1971) has presented several forecasting roles for the month of March deriving from such synoptic-statistical associations that northerly cyclonic Februarys are generally followed by cold Marches, whereas southerly cyclonic Februarys tend to lead to warm Marches. The statistical analysis of past temperature and rainfall data provides an even more direct approach to prediction and there is now a much better appreciation of the persistence and anti-persistence of runs of warm or cold months, or wet or dry months at various times of the year (Murray, 1967a, 1967b, and 1968).

In view of the potential economic benefits to be derived from more extended-range forecasting, the Meteorological Office has started to attack the seasonal-prediction problem (Murray, 1972a). The present, largely empirical method, employs almost 100 years of mean monthly pressure anomaly maps for the northern hemisphere to identify, in the three months preceding the season in question, the main areas where an anomalous circulation might be used as an index of the subsequent seasonal temperature or rainfall. Thus, temperature conditions in spring have been predicted by classifying each spring from 1873 to 1969 as quintiles 1, 2, 3, 4, or 5 in temperature for Central England and then computing mean pressure anomaly maps for each temperature class for the preceding Decembers, Januarys, and Februarys. Figure 8.2 shows the mean pressure anomaly map in February preceding very cold springs compared with the corresponding map for February coming before very warm springs. It can be seen that quite different anomalous circulations are represented and the difference

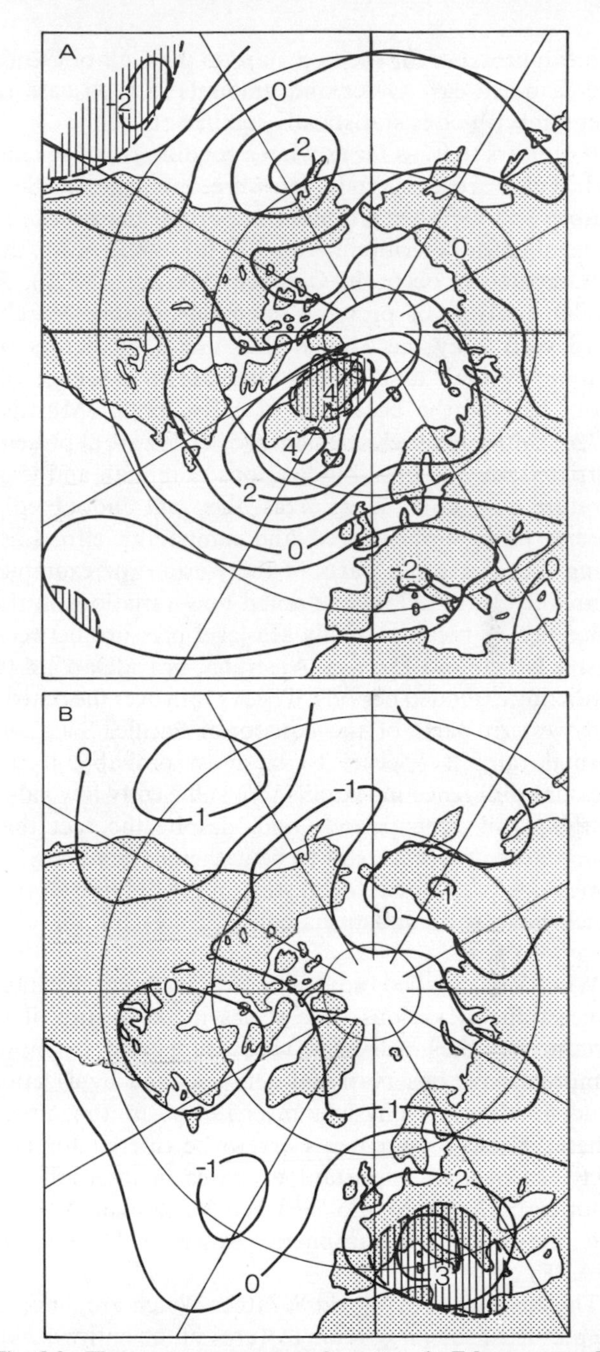


**Fig. 8.2 The mean pressure anomaly pattern in February preceding (A) very cold (quintile 1) springs and (B) very warm (quintile 5) springs over central England. The pressure anomalies are drawn at 1-mb intervals and the shaded areas indicate where anomalies are significantly different from zero at the 5 per cent level according to the t-test. After Murray (1972a).**

in mean pressure for the two maps in the Gulf of Genoa and in north-east Greenland amounts in each case to over 5 mb which is statistically significant.

It appears that, as far as winter conditions in England and Wales are concerned, pressure anomalies in September and October are more important than those in November for determining rainfall; for temperature, the best predictors occur in October (Murray, 1972b). A similar analysis of pressure anomalies during March, April, and May has shown that useful predictors of summer rainfall and temperature are in evidence on about half of the occasions in early spring (Murray, 1972c). Although such seasonal predictions are at present restricted only to broad indications of rainfall and temperature over fairly large areas, they will undoubtedly become more sophisticated, and comparable efforts are being made in other parts of the world. For example, Quin and Burt (1972) have noted how variations in the difference of mean monthly sea-level pressure between Easter Island and Darwin, Australia, provide an aid to predicting extended periods of heavy rain over the central and western parts of the equatorial Pacific. Such abnormal rainfall appears to be most probable if the pressure-difference index falls to a sufficiently low value in the April–June periods and, despite the fact that many more data have still to be gathered, there may be some hope of seasonal forecasting of the El Nino storms which have such disastrous consequences for the economy of Peru.

Whatever detailed advances are made in weather forecasting on various time scales in the future, it is certain that they will have to be supported by more comprehensive observational networks and by a better understanding of physical mechanisms in the atmosphere. These requirements are to be catered for respectively by two important exercises in international co-operation—the World Weather Watch plan (WWW) and the Global Atmospheric Research Programme (GARP).

The basic aims of World Weather Watch are, first, to keep at least the large-scale features of the entire atmosphere under constant surveillance and, second, to provide an integrated, high-speed communications system capable of collecting and disseminating the information received (Sawyer, 1967b; Davies, 1966; Akingbehin, 1966). The main transmission of data will take place between the three World Meteorological Centres, established at Washington, Moscow, and Melbourne, and the larger number of regional centres, which will be concerned with more local conditions. It was estimated by Malone (1968) that only 20 per cent of the global atmosphere was adequately observed during the late 1960s and much of the impetus for the WWW has come from the increased capability that now exists for monitoring the atmosphere by satellites. Mason (1968) and Kerr-Smith (1970) have described some of the numerous ways in which remote sensing of the atmosphere is likely to develop in the immediately foreseeable future. For example, in addition to the provision of new and more accurate sensors on the satellites themselves, it is expected that satellites will play a leading role in the location and interrogation of automatic weather stations sited in remote areas of the world. It is expected that some of the most important benefits of the WWW will be realized in the forecasting field and an early estimate by Thompson (1966) suggested that the prediction of weather several days in advance might lead to savings of 15 to 20 per cent in protectable losses on a world scale.

According to Garcia (1967), the Global Atmospheric Research Programme represents a point of convergence for the developments in numerical weather forecasting and numerical simulation of the general circulation, which have hitherto developed along parallel but separate lines. However, the main objective lies in an improved understanding of the physical and dynamical processes that determine the global circulation up to a height of 30 km and, in some ways, GARP may be considered as the research component of the WWW (Sheppard, 1968). In this context, the Global Atmospheric Research Programme has identified several themes of outstanding significance, including the evolution of convective cloud systems in the tropics and the transfer processes of heat, moisture, and momentum in the atmospheric boundary layer over both land and ocean surfaces (Mason, 1970). All these processes are of major importance to the tropical circulations and a large-scale tropical experiment has been planned for the eastern Atlantic in 1974, in which the US, USSR, UK, Germany, France, and Canada will combine to study the deep convective cloud systems and their relationship to the various physical processes (Mason, 1971). In turn, this will lead to the First Global Observing Year in 1976, when experiments will be conducted to determine the minimum observational and data-processing facilities which will be required for a more detailed understanding of the atmosphere and also to satisfy the fully operational phase of the World Weather Watch.

## 8.5 Weather services and the community

Weather services designed to satisfy particular demands for meteorological information or advice within the community have a long evolutionary history. This is especially true of the forecasting services provided by the various national, government-based meteorological agencies. For example, a daily 'Farmers' Bulletin' was prepared as early as 1876 for consumers in agricultural areas of the US and was distributed to postmasters for display in their offices. Indeed, before the communication of forecasts in newspapers and by telephone became common in the first decade of this century, the US Post Office Department played a major role in the distribution of forecasts by means of a back-stamping system on mail (Doane, 1970). Peters (1955) has outlined the early development of forecasting services in Britain where, somewhat surprisingly, forecasts for long-distance pigeon racing pre-dated by three years the frost warnings for horticulturalists introduced in the spring of 1936. In some cases, the start of a service can be traced back to a specific event, and the warning service for high tides in the Thames estuary was initiated after severe flooding in 1928.

Although it can be shown that meteorological services have grown to meet changing requirements over the years, the process of adaptation has been rather slow in certain respects. Thus, most government services have tended to concentrate on the needs of aviation during the first half of the present century, and in 1966 the provision of a world-wide service for the Royal Air Force occupied nearly one-third of the total staff employed by the British Meteorological Office and was responsible for an annual expenditure of about £3 million out of a budget of £7·2 million (Mason, 1966b). Taken together, military and civil aviation accounted for well over half the annual expenditure, although Mason (1970) has predicted that this pattern is likely to undergo a significant change over the next 10 to 30 years. The chief reasons for this expected change have been summarized by Petterssen (1966), who has foreseen that, as the world population grows, so the importance of meteorological services will increase. The services themselves will expand to fill new roles in weather forecasting and in the better management of environmental resources. Within recent years, most of the national meteorological agencies have become more aware of their responsibility for the general guidance of the community in relation to the whole complex of weather-sensitive decisions which have to be made by the public. Consequently, several attempts have been undertaken to identify the nature and level of consumer demand and to assess the services provided on the basis of a cost: benefit approach.

In the US, the government Weather Bureau has identified no less than 18 different groups, who are frequently involved in weather-sensitive decisions (US Weather Bureau, 1964). After identification, the survey employed a subjective ranking method to determine each user-group's sensitivity to the weather, and, as illustrated in Table 8.1, the general public was ranked first but followed fairly closely by fishing, agriculture, general science and air transportation. Rather more simply, Hibbs (1966) made a fundamental distinction between clients of the US Weather Bureau, who were classified as either *consumers* or *producers*. In this context, consumers were seen as those customers who use the weather services for other than purely economic motives, such as the public trying to optimize leisure time or government departments discharging health, safety, or welfare functions, while the producers were identified by their use of some of the more specialized services to agriculture or industry for the direct enhancement of personal or professional activities.

It is in connection with the latter category that the economic value of weather information becomes most apparent, although McQuigg and Thompson (1966) have claimed that it is still difficult to compute the increased financial return from improved management decisions relating to a particular weather-sensitive operation, either because of the lack of economic data or because of problems involved with conducting actual experiments on the process in question. The management of the flow of natural gas through an urban distribution system was used as an illustration of both the alternative simulation model approach and the basic fact that, to receive the best economic return, the decision-maker must have a very precise knowledge of the inter-relationships between weather events and the operations under his control.

probably justify calling on the services of a private consultant, whereas, when decisions approach the $100 000 level on a continuing basis, a profit-motivated organization may well see fit to appoint a meteorologist to the permanent staff.

Mason (1966b) made a comprehensive assessment of

| *User group* | *(a) Weather sensitivity* | *(b) Decision latitude* | *Product of (a) and (b)* | *Rank* | *Projected 1972 national income* ($\$ \times 10^9$)* |
|---|---|---|---|---|---|
| General public | 2·6 | 2·0 | 5·2 | 1 | † |
| Fishing | 2·3 | 2·3 | 5·3 | 2 | 0·26 |
| Agriculture | 2·0 | 3·0 | 6·0 | 3 | 28·10 |
| General science | 2·7 | 2·3 | 6·2 | 4 | † |
| Air transportation | 2·0 | 3·3 | 6·6 | 5 | 2·30 |
| Forestry | 2·3 | 3·3 | 7·6 | 6 | 1·68 |
| Construction | 2·6 | 3·0 | 7·8 | 7 | 39·30 |
| Land transportation | 2·6 | 3·3 | 8·6 | 8 | 12·90 |
| Water transportation | 3·0 | 3·0 | 9·0 | 9 | 2·05 |
| General welfare | 3·6 | 2·6 | 9·4 | 10 | † |
| Energy production and distribution | 3·3 | 3·0 | 9·9 | 11 | 30·00 |
| Health and safety | 2·9 | 3·6 | 10·4 | 12 | † |
| Resources utilization | 3·0 | 3·6 | 10·8 | 13 | † |
| Merchandising | 3·3 | 3·3 | 10·9 | 14 | 145·30 |
| Water supply and control | 4·2 | 3·0 | 12·6 | 15 | 4·28 |
| Communications | 3·6 | 3·6 | 13·0 | 16 | 22·70 |
| Recreation | 4·0 | 3·3 | 13·2 | 17 | 3·35 |
| Manufacturing | 4·2 | 4·3 | 18·1 | 18 | 262·40 |

*Weather sensitivity*
1 = Great—involves life, very large money values
2 = Important—much money and/or direct health effects
3 = Significant—meaningful costs, secondary health effects
4 = Modest
5 = Small
6 = None

*Decision latitude*
1 = Very flexible
2 = Quite flexible
3 = Some flexibility
4 = Little flexibility
5 = Rigid

*Adapted from information supplied by National Planning Association. All data are in terms of 1962 dollars.
†Not assessed.

**Table 8.1** Ranking of weather influence by user groups in the US. (After US Weather Bureau, 1964.)

In many instances, therefore, the producer type of client will require a very detailed, highly personalized weather service or forecast and, in the US, it has become reasonably common for these demands to be met by private firms of consulting meteorologists. As a rough approximation, Crow (1965) has suggested that any decision-maker, faced with a weather-related problem with financial implications in excess of $1000, could

the economic value of the state meteorological services to the UK, which spends only about 12½p annually per head of the population on the Meteorological Office compared with an expenditure on meteorological services of 30p in Australia, 50p in Canada, and £1·05 in the US. The conclusion was that the economic value of the civil national weather service in Britain was at least £50 million to £100 million per annum which, for a cost of £4 million, represented a crude benefit: cost ratio of around 20. As might be expected, most of the benefits were associated in some way with weather forecasting.

It has been demonstrated (Anon., 1971) that there is a demand from the public for weather information in Britain. In the first place, this demand is met by national and regional bulletins and forecasts distributed through the mass media. All four BBC radio channels and most of the local radio stations broadcast routine weather forecasts and, around breakfast time on weekdays, there is an audience of about 10 million people. Current and expected weather has been described on BBC TV since 1954, with some 9–11 million people watching each evening, and similar presentations are made by most of the independent television companies. A not too dissimilar situation for radio and television applies in Canada as described by McGlening (1968). In addition, weather information and forecasts are printed by most British newspapers and, since 1956, the Post Office Corporation has operated an automatic telephone weather service whereby tape-recorded local forecasts are available to the telephone subscriber. By 1970, this service was available from 49 telephone centres covering areas in which over 30 million people live.

On the other hand, despite the underlying need for general information of this nature, the most rapidly accelerating demand appears to be for a more personal service, whereby anyone with a weather-related problem can discuss it directly with a qualified meteorologist (Anon., 1971). This demand was satisfied initially by a telephone service available from about 40 stations run by the Meteorological Office and serving different parts of the country but, since 1959, these largely aviation-forecasting offices have been supplemented by six Weather Centres, which have been opened in or near major cities to deal with the needs of the community. In the period 1962 to 1970 the total number of inquiries handled annually by both types of Meteorological Office outstation increased from 785 138 to 1 633 478, which represents a rise of 208 per cent. By 1970, the Weather Centres were dealing with 54 per cent of all the enquiries received. An examination of the monthly totals indicates seasonal peaks in mid-winter and during the summer, and notable variations can be attributed to the occurrence of unusual weather, such as the severe snowstorms in February 1969 which generated 203 820 inquiries in that month. Approximately three-quarters of all the inquiries can be allocated to five major areas of interest (Taylor, 1971). As shown in Table 8.2, the largest rate of increase between 1962 and 1970 was experienced by the Holidays and Recreation sector, which has now supplanted industry as the principal category of inquiries.

| *Year* | *Agriculture* | *Industry* | *Holidays and recreation* | *Marine* | *Road transport* |
|---|---|---|---|---|---|
| 1962 | 82 826 | 134 874 | 114 820 | 57 948 | 107 907 |
| 1970 | 148 048 | 323 073 | 329 947 | 141 840 | 190 928 |
| Per cent increase | 179 | 240 | 287 | 245 | 177 |

**Table 8.2** Weather inquiries received by the Meteorological Office relating to the five major areas of interest. (After Taylor, 1971.)

## 8.6 The application of weather information

It will be apparent that one of the main factors influencing the utilization of forecasts and other weather information will be the accuracy of the information provided, but the objective assessment and verification of weather forecasts has proved a difficult and controversial problem for many years (Brier and Allen, 1951). Murphy and Epstein (1967) have stated that verification is meant to serve two quite distinct purposes. First, there is the task of evaluating the forecast in terms of the economic value for the decision-making client; and in this case an *operational evaluation* which measures the utility of the forecast is required. Second, some assessment of the forecasting skill displayed by the meteorologist is wanted and this involves a more *empirical evaluation*. No single method of assessment is suitable for both purposes.

Indeed, Brier (1950) has claimed that, not only may the most useful forecasts not be the most accurate according to some system of arbitrary scores, but also, in this situation, a particular verification scheme can influence the forecaster and encourage him to produce the type of forecast that he thinks will score highly.

Most authorities, however, would accept that the best verification statistic for forecasts is their usefulness to the customer and, following this premise, Glahn (1964) has argued, with others, that forecasts should be issued in terms of probability statements. Thompson and Brier (1955) similarly concluded that probability forecasts were preferable from an economic viewpoint but also noted a need for the better education of the general public with regard to the interpretation and use of such forecasts. In particular, they foresaw a danger that the public would ignore the probability statement and use the forecast as a categorical prediction. Thompson and Brier also considered it desirable to use the climatological expectancy as the basis for the prediction of unfavourable weather conditions. As an example, the San Joaquin Valley raisin crop in California, then valued at $35 million, is exposed to the atmosphere for 4–6 weeks each year on the climatic expectancy that there will be no heavy rain in early autumn. If rain does occur, however, the losses are so large and the costs of covering the crop so small that the operational risk is reduced to a low level. In turn, this means that a decision to issue warnings of heavy rain should be taken at a rather low probability level, perhaps close to the climatological expectancy. Thus, the tendency would be to 'over-forecast' severe weather.

Whatever the methods which are selected to assess forecast accuracy, there is little doubt that weather predictions on most time scales are becoming increasingly reliable. Thus, Mason (1970) has claimed that the 24-h regional forecasts of the general weather situation for the UK are, on average, not seriously in error on more than 3 or 4 days per month. This implies a routine short-term forecast accuracy rather better than 85 per cent and most of the residual inaccuracy may be attributed to timing errors, such as those associated with a frontal passage, which occurs a few hours earlier or later than expected, or to local factors which modify the effects of the regional synoptic situation. The improvements which have been noted in the accuracy of the 30-day forecasts since their introduction in December 1963 are particularly significant. These forecasts, which are issued twice-monthly, deal with the expected pattern of temperature and rainfall in relation to the climatic expectancy for the month in question and provide additional information on the probable frequency of elements such as snowfall, frost, or gales.

| *Category* | *Mean temperature* 1* | 2† | *Rainfall* 1 | 2 | *Additional information* 1 | 2 | *Overall marking* 1 | 2 |
|---|---|---|---|---|---|---|---|---|
| A—no serious discrepancy | 10 | 18 | 10 | 5 | 12 | 20 | 1 | 4 |
| B—good agreement | 24 | 20 | 10 | 18 | 17 | 30 | 24 | 27 |
| C—moderate agreement | 12 | 17 | 20 | 28 | 19 | 21 | 23 | 33 |
| D—little agreement | 8 | 13 | 17 | 17 | 14 | 5 | 15 | 11 |
| E—no real resemblance | 12 | 10 | 9 | 10 | 4 | 2 | 3 | 3 |
| Totals | 66 | 78 | 66 | 78 | 66 | 78 | 66 | 78 |

*Column 1 refers to forecasts from December 1963 to August 1966 inclusive.
†Column 2 refers to forecasts from September 1966 to November 1969 inclusive.

**Table 8.3** Number of 30-day forecasts for the UK in the various categories of success. (After Freeman, 1966 and Ratcliffe, 1970b.)

For the ten forecast regions in the British Isles, the area forecasts have been subsequently checked against 4–6 representative stations and the area mean anomalies of temperature and rainfall have been calculated in an entirely objective manner as described by Freeman (1966). From the scores for each area, an average score was then obtained for the whole country so that each forecast could be assessed nationally in one out of five major categories: A—no serious discrepancy, B—good agreement, C—moderate agreement, D—little agreement, E—no real resemblance. Table 8.3 lists the overall classification of forecast accuracy for the two periods December 1963 to August 1966 and September 1966 to November 1969 based on analyses presented by Freeman (1966) and Ratcliffe (1970b) respectively. It can be seen that, as far as the total marking is concerned, a moderate agreement or better was achieved in 64 out of 78 forecasts (82%) in the latter period compared with only 48 out of 66 (73%) in the case of the earlier sample. This modest improvement appears to be related to both an expansion in the library of data used for analogue-fitting purposes and to the incorporation of continuing research, such as that into Atlantic ocean temperature anomalies and surface-pressure patterns one month ahead, within the forecasting routine. However, it should be noted that some of the success in the more recent period is due to better 'additional information' and there has been much less general improvement in the very best forecasts in categories A and B. Good forecasts of temperature can still be made more frequently than good rainfall forecasts. Similarly, there are seasonal and regional variations in overall forecast accuracy. The forecasts tend to be less reliable in autumn (mid-September) and winter (mid-January) and temperature forecasts are normally better in the northern and western parts of the country than in the south and east.

As weather forecasts become more accurate, their use by all sections of the community is likely to increase. Ellison (1970) has suggested £40 million as the potential financial benefit which may eventually be derived from long-range forecasting alone in the UK, but there are further opportunities for the application of forecast information on all time scales. Some of these opportunities have been demonstrated in earlier chapters of this book and Table 8.4 presents a comprehensive summary of both the present and potential utility of forecasts for various weather-sensitive economic activities. Although accuracy

| *Activity* | *48 h* | *1 week* | *1 month* | *Season* | *Climatic changes* |
|---|---|---|---|---|---|
| Agriculture | Day-to-day operations. Frost protection. | Timing and planning of ploughing, sowing, harvesting, hay-making, crop-spraying, fertilizer application, etc. | Timing of sowing and harvesting. Estimates of demand for fruit and vegetables. | Forward planning of crop schedules. Choice of varieties. Forecasting of crop and milk yields. | Long-term planning of land use. Breeding of new varieties. Investment decisions on buildings and machinery. |
| Aviation | Forecasting terminal and weather en route. Flight planning. Avoidance of special hazards. | | | Seasonal forecasting of traffic patterns, loads, etc. | Long-term planning of new routes and aircraft. |
| Building and construction | Day-to-day operations, avoidance of worst effects of heavy rain, frost, strong winds, etc. | Planning of operations with alternate plans to suit weather. Hiring of plant and machinery. | Planning of work schedules. Weather-proofing of buildings during construction. Hiring of plant and machinery. | Firmer estimates of completion dates, delays, etc. Weather proofing of buildings during construction. | Long-term planning-siting of new towns, industries, motorways. Design of buildings, structures utilities, etc. |
| Electricity and gas | Hourly values, 48 h ahead of weather parameters influencing demand. | Continuously up-dated forecasts of all weather elements affecting demand for as far ahead as possible. Planning of maintenance schedules. | | Forecasts of seasonal demands. Warning of abnormal spells for maintenance schedules and operational stand-by equipment. | Long-term planning, design and location of new power stations, storage and distribution systems. |
| Oil | | Estimation of demand for periods of 1 week to 6 months ahead. Long-term planning. Arrange production and delivery schedules. | | | Long-term planning. |

Contd.

| Activity | 48 h | 1 week | 1 month | Season | Climatic changes |
|---|---|---|---|---|---|
| Manufacturing Marketing Distribution | Assessment of weather in determining demand, delivery of raw materials, delivery of products. | Planning of advertising and marketing. | | Forecast of demands for seasonal products. | Long-term assessments of consumer demand for new and existing products. |
| Road/rail transport | Forecasts of snow, heavy rain, fog, icy roads/rails, rail temperatures in summer. | Forecasts of snow and ice. | Local authority arrangements for supplies of grit and salt. | Estimates of traffic patterns and density. Local authority arrangements for snow-clearing equipment, floodwater drainage, road-heating, etc. | |
| Shipping | Local forecasts for coastal shipping and pleasure craft. Weather routeing of ships. Forecasts for fishing fleets. | | | Forecasts of ice cover, freezing of rivers. | |
| Water resources | Forecasting of precipitation, evaporation, runoff, river flow. Regulation of dams and reservoirs. Irrigation requirements. Flood forecasting. | | | Forecasting of seasonal precipitation and droughts. Estimates of water balance, irrigation requirements. | Long-range planning and estimates of demands. Location and design of dams, reservoirs. Management of water resources. |
| General public | Day-to-day activities. Dress, leisure, sporting activities. Regulation of central heating and air-conditioning plant. | Planning of social and sporting events, outdoor activities, gardening, etc. | | Planning of holidays. | |

**Table 8.4** The utility of weather forecasts on different time scales to weather-sensitive activities. (After Mason, 1970.)

is the prime requirement for all forecasts, it is also important that the information available is relevant to the needs of the user and that the forecast is presented in the best possible way. The following section explores the extent to which these criteria are met for both the general public and some of the more specialist users of weather forecasts.

## 8.7 The application of weather forecasts

A major limitation to the effectiveness of general-purpose, public weather forecasts arises from the gap between the specialist and the layman. Several writers, such as Holford (1964) and Gordon and Bestwick (1969), have indicated that the regional forecasts are often misleading to the public because of the over-general statements which may be impossible to relate directly to local weather events at a specific forecast time. This problem is especially acute when the local weather is strongly influenced by a wide variety of topographic and relief characteristics. At least part of this generality springs from the severe time and space restrictions placed on weather-forecast information by the mass media and Taylor (1972) has criticized the minimal exposure of the television weather map at peak viewing times and the highly compressed verbal summaries of the weather outlook which appear on radio and in newspapers. However, even if provisions are made for an expanded presentation of forecasts, there remains the problem of choosing a vocabulary which is fairly precise without including technical terms which would be incomprehensible to the public.

Thus, Oddie (1964) has instanced the difficulty of

putting a meaningful description to a sky covered with thin stratocumulus or altocumulus cloud since the term 'sunny' is too optimistic and 'cloudy' too depressing. Similarly, common everyday words such as 'fine', 'showers', or 'fog' mean different things to different people and really need some numerical definition. As far as air temperature and frost severity are concerned, the Meteorological Office employs a prescribed series of adjectival phrases which is adjusted seasonally so that, as far as possible, the usage conforms to natural conditions as experienced by the man-in-the-street at that time of year, as shown in Table 8.5. In the case of frost forecasts, the aim is to provide some idea of the amount of damage which could be caused to Items such as garden plants, car radiators, or newly laid concrete, and the incorporation of wind speed as well as temperature enables the probable cooling rate to be expressed on a scale ranging from 'ground frost' through to 'very severe frost'.

| *Temperature departure from average (degC)* | *Spring (mid-March/mid-May) Autumn (mid-Sept./mid-Nov.)* | *Summer (mid-May/ mid-Sept.)* | *Winter (mid-Nov./ mid-March)* |
|---|---|---|---|
| +7 } +6 to +7 } | Very warm | Very hot | Exceptionally mild |
| | | Hot | |
| +4 to +5 | Warm | Very warm | Very mild |
| +2 to +3 | Rather warm | Warm | Mild |
| −1 to +1 | Normal | Normal | Normal |
| −2 to −3 | Rather cold | Rather cool | Rather cold |
| −4 to −5 | Cold | Cool | Cold |
| −5 | Very cold | Very cool or Cold* | Very cold |

*Cold when a marked fall in temperature expected.

**Table 8.5** Terms used to indicate air temperature. (After Oddie, 1964.)

Despite the difficulties of interpretation which surround many routine forecasts, specific storm or hazard warnings are usually clearly understood by the public with the result that a predictable response ensues. For example, White (1966) has described the immediate public reaction to the television and radio bulletins issued when Hurricane Hilda struck the Gulf Coast of the US in the autumn of 1964. To be effective, such warnings necessarily depend on an efficient communications network and, conversely, Arakawa (1966) has pointed out that, when a typhoon struck Japan only 42 days after the historic atom bomb attack on Hiroshima, excessive casualties occurred in the Hiroshima prefecture compared to the neighbouring island of Kyushu where the communications systems were still functioning.

On some occasions, however, the reactions of the general public are sufficiently unpredictable to cause an element of chaos. Such chaos is likely to increase if confusion exists with respect to the forecast information and Stringer (1970) has documented two examples concerning the reactions of commuters to snow in Birmingham, England. Heavy snowfall in the early morning of 9 January 1968 suspended railway operations in the area between 7 and 9 am, because British Rail were relying on an incorrect national forecast which predicted a thaw. On the other hand, Birmingham City Corporation followed an accurate, locally available forecast and cleared the roads ready for the morning journey to work, although most commuters eventually elected to leave their cars at home because of the snow and attempted to travel by the suspended train service. Almost one month later, on 5 February 1968, an accurately predicted afternoon snowfall sent many people home early from work, but these early commuters interfered with salting and gritting operations on the roads, which led to delays, whereas the railways remained fully operational but carried only their normal 30 per cent share of the rush-hour traffic.

Weather forecasting to meet the more specialized needs of agriculture and industry has already been discussed, where appropriate, in previous chapters, but there are certain generalizations which can be made about this aspect of meteorological information. In the first place, the 240 per cent increase in industrial inquiries received by the Meteorological Office between 1962 and 1970 has not been uniformly distributed over the major areas of interest, as shown in Table 8.6.

| *Year* | *Construction* | *Manufacturing* | *Public utilities* | *Total* |
|---|---|---|---|---|
| 1962 | 28 292 | 33 714 | 72 868 | 134 874 |
| 1970 | 97 111 | 82 279 | 143 683 | 323 073 |
| Per cent increase | 343 | 244 | 197 | 240 |

**Table 8.6** Number of industrial weather inquiries, by categories, received by the Meteorological Office in 1962 and 1970. (After Buchanan, 1972.)

Thus, although the public utility sector continues to exert the largest demand on these weather services, the construction industry appears to represent the fastest-growing area and in 1970 accounted for 30 per cent of the demand. Buchanan (1972) has described in some detail the nature of the special forecasting services available for each of these demand sectors and has emphasized the need for a greater integration of the demand for and the supply of meteorological information in order to achieve the greatest benefits. The same point has been urged by Smith (1972) who concluded that the real function of the applied meteorologist was to achieve a better identification of the weather-sensitive problems of the farmer and the businessman, and then effectively communicate the implications of weather forecasts and other meteorological information to these consumers. In terms of the latter objective, the professional meteorologist often fails to market his product adequately and, in a questionnaire survey of selected farmers in Wales and part of western England, Hogg (1972) has revealed the extent to which currently available forecast information failed to meet agricultural requirements as judged by the clients themselves. For example, there was a large unsatisfied demand for more local forecasts and the inclusion of the probability of rainfall in the daily forecasts. Over 90 per cent of farmers thought that weekly forecasts would help them and 81 per cent considered that more accurate monthly weather predictions would be of assistance in agriculture.

## 8.8 Background to weather modification

Man has modified his immediate climatic environment ever since he first began to wear clothing or live in purpose-built shelters but, until recent decades, his capability for consciously altering atmospheric processes remained at the smallest scale. Even today, fully controlled artificial climates exist only within modern air-conditioned buildings and, elsewhere, man's intervention is restricted to temporary actions performed over very small areas often under particular, short-term meteorological circumstances.

These limitations arise partly from the excessive amounts of energy which are required to modify atmospheric mechanisms and partly from an incomplete understanding of how such mechanisms work. For example, Sellers (1965) has indicated that, if the solar energy received by the earth in one day could be collected and stored, it would satisfy the world's industrial and domestic fuel requirements for 100 years; further, it has been estimated that the energy required to increase rainfall by 2 mm over an area of 260 km$^2$ is equivalent to the total output of electrical energy in the US during a 6-day period (Malone, 1967). Not surprisingly, therefore, most attention has been concentrated on the trigger mechanisms which might be manipulated to initiate or retard individual atmospheric processes in a delicately balanced but essentially natural situation, and technological advances over the last 30 years or so have made weather modification experiments an important aspect of applied climatology, especially in North America and the USSR.

Most writers, such as Huschke (1963), regard 1946 as the starting point for deliberate weather experimentation based on scientific principles since, in November of that year, V. J. Schaeffer conducted the first field trial into cloud-seeding by sprinkling washed dry ice from an aeroplane into an altocumulus cloud at an altitude of 4200 m near Schenectady, New York, causing snow to fall for a distance of about 610 m from the seeded cloud before evaporating in the air (Schaeffer, 1946). In the same month, it was discovered by Vonnegut (1947) that minute crystals of silver iodide, liberated in the form of a smoke, acted as highly efficient ice-forming nuclei at temperatures below −5 °C. However, Schaeffer (1968) has stated that the idea that dry ice could be used to produce clouds and rain dates back at least to 1891, although at that time the importance of air colder than 0 °C and supersaturated with respect to ice was not appreciated. In addition, Federov (1967) has claimed that as early as 1932 the Hydrometeorological Service of the USSR was formulating tasks to be solved by the Artificial Rain Institute based on the introduction of hygroscopic substances and electrically charged particles into clouds.

Although cloud-seeding has remained the principal focus of weather-modification activity, and consequently receives most attention in this volume, it should be emphasized that many smaller-scale alterations already exist, and much more ambitious schemes have been suggested as a means of changing climate on the largest spatial scales. Some aspects of the deliberate manipulation of local climates have already been discussed in earlier chapters in connection with, for example, protective agricultural practices or the use of various fog-dispersal techniques at airports, but occasionally rather more fundamental ideas have been put forward.

Thus, Black and Tarmy (1963) have claimed that a thin asphalt covering spread over an area of about 100 km$^2$ would alter the surface energy balance to the extent that greater instability would either induce shower formation or increase natural rainfall. The feasibility of this and similar notions is largely unknown, and there is a need for the accurate prediction of any possible changes through the application of numerical models of the atmospheric boundary layer, as evolved by Estoque (1963), before any such schemes are commissioned. In the case of large-scale atmospheric modification, which is most likely to arise as a result of changes in the planetary wind system, it would clearly be irresponsible to interfere with the balance of natural processes without prior understanding of the detailed consequences of such actions. Therefore, in view of the existing deficiencies in our knowledge of the general circulation and, in particular, of the feedback mechanisms which operate between the earth and the atmosphere, it is to be hoped that plans such as those advanced by Wexler (1958) for changing the direction of ocean currents will remain as purely intellectual exercises for the foreseeable future.

## 8.9 Artificial stimulation of precipitation

According to Mason (1962), experiments designed to increase precipitation are based on the assumption that either some supercooled clouds precipitate inefficiently or not at all because of a deficiency of ice crystals necessary to release snow and rain by the Wegener-Bergeron process, or that some warm clouds at temperatures above 0 °C similarly fail to precipitate efficiently due to a lack of large water droplets which are required to initiate the coalescence process. It is further assumed that the precipitation efficiency can be increased by seeding the clouds artificially with either solid carbon dioxide (dry ice) or silver iodide to produce ice crystals or by introducing water droplets or large hygroscopic nuclei to induce the cloud particles to form rain drops. In practice, however, all these methods are of limited application since, in order to release appreciable rainfall by either natural or artificial means, a cloud must have a sufficiently large liquid content. This means that rain-making experiments are normally restricted to cumulus clouds and deep orographic clouds.

The greatest success has been achieved with the supercooled clouds which extend above the freezing layer. However, although ordinary convective thunderstorms deliver as little as 10 per cent of their total liquid water to the ground and even orographic clouds formed in moist maritime air above coastal mountains naturally precipitate only 25–30 per cent of the available moisture (Wycoff, 1966), the most productive seeding is limited to a fairly narrow range of cloud summit temperatures and is also dependent on the cloud attaining a certain minimum depth. In a summary of recent experimental evidence from the US, Burroughs (1972) concluded that for successful seeding of convective clouds the specific cloud top temperature should lie between −5 °C and −20 °C and the cloud should be at least 3000 m deep with a diameter of approximately 1 km. For orographic winter clouds approximately 100 mb thick and with a cloud base some 1500 m above the mountains, positive results were claimed by observers with cloud summit temperatures in the range from −10 °C to −25 °C. The lower thermal limits exist because, at temperatures of −15 °C and below, seeding experiments tend to become less significant due to the high probability of natural rainfall taking place.

Most convective cloud systems are modified to increase rainfall in semi-arid areas during the growing season but the seeding of winter orographic clouds is often undertaken in order to augment the snowpack so that additional streamflow will be available later in the year to meet irrigation or hydroelectric demands. Ludlam (1955) has described an early experiment of this type in Scandinavia, where it was considered that the cost of increasing winter snowfall by silver iodide seeding would be covered by the enlarged potential for hydroelectricity generation. Less usually, rainfall stimulation has been attempted from stratus clouds and Battan (1969) has reported the results of experiments in the Russian Ukraine, where seeding produced precipitation when the cloud thickness was at least 300 m and the minimum temperature in the cloud layer was below −3 °C. In such circumstances, it was stated that 70 per cent of the liquid water can be made to precipitate to the ground.

In lower latitudes, where cloud top temperatures frequently remain above 0 °C, a different approach has to be adopted. Bowen (1952) introduced the method whereby small water droplets of about 50 μ diameter are sprayed into the lower layers of deep clouds so the growth of cloud particles will be stimulated by coalescence and, on the basis of numerous experiments in Australia, claimed significant results from clouds exceeding 1500 m in depth. Other techniques of warm cloud seeding include the feeding of hygroscopic particles into the lower air layers near the updraft of a growing cumulus cloud and Biswas *et al.* (1967) employed common salt as a seeding agent in three climatically similar areas of north-west India over several monsoon seasons. The data were subjected to three statistical tests and, although the initial results indicated an increase in precipitation of around 41 per cent, it was concluded that 21 per cent may be a more realistic figure in view of uncertainties in the conduct of the experiment.

Some of the progress achieved in the artificial modification of precipitation has been reviewed by Kahan *et al.* (1969), but it is evident that outstanding problems still remain. Many of these problems are related to technical aspects of the seeding operation. For example, although it is widely accepted that silver iodide is the most efficient agent for the promotion of ice nucleation, there is disagreement about the relative merits of ground-based generators and aircraft as methods of delivering the material to the chosen cloud at the right time. With a ground-based generator, which releases smoke impregnated with silver iodide upwind of the target cloud, the operation is entirely dependent on wind diffusion to effect the seeding; although more direct control can be exercised from an aircraft, it is frequently unsafe to fly through storm clouds. In these circumstances there is a need for seeding by reliable and economical ground-launched rocket systems. Furthermore, silver iodide is not only somewhat expensive for large-scale use but also its efficiency tends to deteriorate in sunlight, and there is a continuing search for alternative materials as potential ice nucleants.

Although it is clearly desirable that cloud-seeding should continue in order to perfect the detailed experimental procedures, it appears most unlikely that the continued replication of existing methods will, by itself, resolve the central problem in precipitation modification. This problem remains that of demonstrating conclusively that seeding is capable of producing sufficiently large and reliable increases (or decreases) of precipitation over areas which are extensive enough to justify a permanent, fully operational modification exercise in any one part of the world. Despite hundreds of experiments, the results are still controversial largely because, as in the allied problem of weather prediction, an imperfect understanding of the physical processes involved means that the evidence has to be evaluated on a statistical rather than a scientific basis. Thus, although precipitation undoubtedly occurs after most seeding trials, it is difficult to assess the extent to which artificial stimulation is responsible as opposed to purely natural mechanisms.

Most clouds suitable for seeding are found in a mixed population of clouds, some at least of which will precipitate naturally, and it can be argued that most seeded clouds would eventually have produced some rainfall somewhere. In addition, many experiments have been conducted by commercial operators, who have had a vested interest in claiming over-optimistic results and who have given little attention to the rigorous experimental design which is necessary for valid statistical testing. During the early 1950s, for example, it was not uncommon for cloud-seeding firms to claim that rainfall increases between 50 and 100 per cent could be achieved, depending on the local circumstances (Battan, 1962). Cloud-seeding for research purposes, on the other hand, is now carefully organized and usually conducted according to randomized procedures, which are intended to reduce any bias on the part of the operator in selecting the occasions for seeding trials. In effect, this means that seeding is carried out, on the basis of a predetermined random choice, on only half of the occasions which are judged suitable from a meteorological viewpoint, and the remaining occasions are used to assess natural conditions. A comparison can then be made between the rainfall recorded for each storm or each day in the target area where modification is expected to occur and an adjacent unaffected control area for both seeded and non-seeded events.

The actual delimitation of target and control areas, which should have similar topographic and meteorological characteristics, can prove difficult especially when large-scale seeding with ground generators is undertaken. For example, O'Mahony (1967) rejected results claimed for a seeding experiment in western Victoria, Australia, on the basis of independent tests conducted in a different control area although, in turn, Adderley (1968) refuted these criticisms on the grounds that O'Mahony had applied his tests over areas that were unrepresentative of the seeding operations. Even when the areas have been properly selected, there is still the problem of making accurate precipitation measurements with a specially installed rain-gauge network often combined with radar observations.

Present-day cloud-seeding experiments suggest that increases in precipitation up to about 10 per cent, and possibly up to 20 per cent, can be achieved in widely scattered localities such as the USSR (Battan, 1969) and Puerto Rico (Howell and Lopez, 1966). On the other hand, such experiments are necessarily short-term and increases of this magnitude may be marginal in both economic and physical terms. Thus, Mason (1962) has noted that, even in reasonably humid areas, the natural rainfall in any one year may deviate from a 50-year average by 10 to 15 per cent and in the drier areas, where rainfall stimulation is most needed, the variation may be as much as 50 per cent. The natural variability of rainfall makes it very hard to assess the effects of cloud-seeding on all time scales but it may well be that most of the limitations exist in the longer term when consistent results are required over a number of years.

After a review of ten years of commercial cloud-seeding in California, Dennis and Kriege (1966) interpreted the fact that increases in winter precipitation were confined to periods of convective instability as a confirmation that seeding is unlikely to initiate rainfall in unfavourable meteorological situations. We still know too little about the climatological frequency of favourable seeding situations in different areas, but Huff (1969) has examined the long-period storm pattern in the state of Illinois, with particular reference to the potential implications for rainfall modification during the critical growing season from May to September and the urban water supply replenishment period from October to April. The evidence suggested that increases in seasonal precipitation, especially in the October–April period, would be small, unless seeding intensified existing prolonged heavy rainstorms and produced very large percentage increases in storms of light-to-moderate intensity. Even if such increases were realized, attendant problems of soil erosion, reservoir silting, and flooding may result in more damage than benefits.

Future progress in cloud-seeding appears most likely to come from a trend to fewer but more detailed experiments, which will allow the statisticians to make worthwhile statements, as stressed by Neyman (1967), and, above all, from a better understanding of cloud physics. Some of the priorities for research have been outlined by Mason (1969), who also drew attention to the fact that premature seeding may destroy cloud updraughts before fallout has reached the maximum possible rate so that, on a random selection of days over a 5-year period, silver iodide released from aircraft over Missouri apparently led to a reduction in rainfall by some 20 to 50 per cent over extensive areas. In view of the difficulties involved in monitoring cloud processes in the field, many authorities see the greatest potential in the further evolution of numerical models of cloud growth and development, which will enable meteorologists to predict the behaviour of both seeded and unseeded clouds, although it has been considered unlikely that such models will be accurate enough to detect precipitation modifications as small as 10–20 per cent for some time (Mason, 1970).

Nevertheless, significant progress has already been reported by Simpson and Wiggert (1971) on the numerical modelling of tropical cumulus in Florida, and Weinstein (1972) has employed a one-dimensional steadystate cumulus model to investigate the climatological potential for precipitation augmentation from isolated cumulus over the western US during the summers of 1967 and 1968. After confirming that the principal objective was to seed the cumulus clouds in order to freeze as much supercooled water as fast as possible, and thereby release latent heat of fusion which subsequently initiated dynamic instability within the clouds, Weinstein found that meteorological conditions favourable to the initiation of such thermodynamic instability occurred on approximately 25 per cent of the total days. On these occasions, the model indicated that seeded cumulus could have been induced to precipitate some 50 per cent more rainfall, equivalent to about 8 mm, than their non-seeded neighbours. The model further predicted that, on the basis of the two years involved, the most favourable areas in terms of seeding potential were the southwestern states of Arizona, New Mexico, and Texas compared with the least favourable plains states of Kansas, Nebraska, and South Dakota.

## 8.10 The suppression of severe storms

Although cloud-seeding is principally associated with rainfall stimulation, it has also been used increasingly as a possible means of reducing the impact of severe convective storms. These storms offer a variety of damaging features, including hail, lightning, and strong winds, and modification experiments are normally designed specifically to combat one or other of these hazards. The meteorological background to violent storms is outlined in various texts such as those of Battan (1961) and Dunn and Miller (1960).

*Hail suppression* experiments have a long history in Europe and there is an old tradition in Switzerland, Italy, and Austria of firing cannon and ringing church bells in an attempt to prevent the formation of large hailstones. Since the late 1940s, farmers in northern Italy have fired explosive cardboard rockets loaded with gunpowder into thunderstorms and, in the 1959 hail season, more than 100 000 such rockets were released (Battan, 1962a). The scientific basis for these methods is highly speculative but the rockets do appear to soften hailstones and it has been suggested that the explosion of the rocket may propagate pressure waves, which are capable of cracking and weakening the ice making up the hailstone. On a world scale, however, most experimentation consists of cloud-seeding with ice nucleants. This is because a hailstone grows by collision with supercooled water droplets in the turbulent updraughts of convective clouds and it is believed that, if enough droplets can be converted into ice crystals, an upper limit will be set on the hailstone accretion process.

On the other hand, despite refinements of the hailstorm model proposed by Ludlam (1961) and the laboratory simulation of hailstone growth in vertical wind tunnels as described by List (1960), the exact mechanism of hail formation is still imperfectly understood and many hail-suppression experiments have proved inconclusive. For example, Sanger (1960) reported on the first two years of a randomized project conducted over an area of 3000 km$^2$ of pronounced relief on the border between the Swiss and Italian Alps. In this case, twenty ground-based silver iodide generators were used between May and October but the only result to emerge was an apparent increase in rainfall on the days of seeding. It has often been suggested that such tests have proved unsatisfactory, because current seeding techniques produce insufficient nuclei for an appreciable modification to take place, and it has been claimed that most early trials used between 10 to 100 times less silver iodide than is really required (Battan, 1962). Following a comprehensive review of the evidence for hail suppression by cloud-seeding, Schleusener (1968) concluded that major hailfalls could be reduced only by very high applications of silver iodide equivalent to more than 2000 g/h for every storm.

Within recent years, most anti-hail activity has been concentrated in the USSR, where the area being seeded to reduce the incidence of damaging hail increased from 80 000 hectares in 1964 to about 2·5 million hectares in 1969 (Battan, 1969). According to Benton (1969), operational programmes exist in nine different regions and the effectiveness of the work is evaluated on the basis of a variety of data including claims filed with the State Board of Insurance by state and co-operative farms under mandatory group insurance schemes; estimates of acreage damaged by hail, using insurance data; observations of hail in the protected areas and nearby; and studies of radar echoes and their changes subsequent to seeding.

Most of the seeding is accomplished by the use of artillery shells which inject ice nuclei into the supercooled parts of a developing hail cloud, but in the Transcaucasian region shells containing common salt are also exploded into the warm part of the cloud in order to wash out the precipitation from the lower part. Silver iodide is the most popular seeding agent, although lead iodide is used as well. There is widespread belief in the USSR that these measures have been highly effective in preventing hail over agricultural areas. Typical results may be quoted for the Alazani valley where hail investigations began in 1953 and where, over a seven-year period, measurements at 32 stations showed that, during the seeded period, hail frequency was 70 per cent less within the protected area than during the non-seeded period.

In North America, hail-suppression experiments have been mounted on a smaller scale and restricted to certain areas like central Alberta where the hail season lasts from May to September. Summers and Paul (1970) have shown that hailstorms occur here with maximum intensity about 112–130 km downwind from the Rocky Mountain divide and the fact that preferred zones of hail formation continue to occur at multiples of 56–65 km further east suggests the influence of some airstream wave effect. Over a ten-year period, the average annual incidence of hail was 66 days but there was a June peak with marked persistence effects leading to a clustering of hail-days and no-hail-days into spells. Such persistence means that the probability of a major hail day occurring after a no-hail day is low at around 7 per cent, whereas a major hail day, with a probability of 26 per cent, is almost 4 times as likely to succeed another major hail day. This type of pattern greatly complicates the statistical problem of randomizing the seeding of storms, since the occurrence of hail cannot be assumed to be a random event.

For an appreciation of some of the special problems surrounding the design and evaluation of hail-suppression experiments, reference should be made to the work of Changnon and Schickedanz (1969) and Schickedanz and Changnon (1970). Any effective anti-hail programme in Alberta must be able to cope with the most severe conditions, since an analysis of 33 years of hail insurance data showed that 80 per cent of the total hail damage occurred on the 12 worst days. Another difficulty in Alberta is that Russian techniques with anti-aircraft guns and ground-based rockets cannot be employed because of the high density of air traffic and the large area (almost 200 000 $km^2$) in need of protection; therefore, Summers *et al.* (1972) have developed an alternative cloud-injection system involving a droppable pyrotechnic flare device for use in jet aircraft. This system has not yet been fully tested but it has been claimed that the aircraft's speed is useful in reaching cumulus towers quickly enough to ensure that seeding takes place early enough in terms of cloud evolution, while radio contact with the ground enables appropriate clouds to be positively identified.

*Lightning suppression* is a rather less understood field than that of hail modification. Most experience in the US has come from Project Skyfire, a research programme inaugurated by the US Forest Service to study the characteristics of lightning storms, the physics of lightning discharge, the fire-starting capabilities of lightning storms and the possible effects of weather modification. There is some evidence to suggest that electrical fields appear in clouds when ice crystals are formed and Barrows (1966) has reported that a randomized seeding of clouds with silver iodide over a 2-year period produced a 38 per cent reduction in cloud-to-ground discharges from treated storms. As with hail suppression, however, there is a need for large quantities of silver iodide and it was considered that generator output may have to be increased by at least one order of magnitude. Other methods capable of draining off electrical charges from clouds may emerge and Wycoff (1966) has cited an experiment by the US Army, when millions of minute metallic needles were released from an aircraft into a thunderstorm and apparently caused a significant reduction in electrical field gradient in the vicinity of the charged cloud.

*Hurricane modification* is probably the most desirable of all the goals associated with the suppression of severe storms, largely because of the extremely favourable potential benefit-to-cost ratio of the work as outlined by White and Chandler (1965). In 1961, it was first proposed that some modification of tropical cyclones might be achieved by introducing freezing nuclei into the ring of clouds around the hurricane centre. According to this hypothesis, such seeding would trigger the release of the latent heat of fusion in the eye-wall cloud system which, in turn, would diminish the maximum horizontal temperature gradients in the storm and cause a hydrostatic lowering of the surface pressure. Eventually, this should lead to a weakening of the damaging winds. This suggestion coincided with the early development of pyrotechnic generators, which were capable of delivering large quantities of silver iodide within a few minutes, and in 1962 a joint US Navy and ESSA undertaking called Project Stormfury came into operation. Simpson and Malkus (1964) have indicated that experimentation during the first few years proved inconclusive, although related seeding trials on isolated tropical cumulus clouds have been rather more successful, as reported by Simpson (1967).

Further progress on Project Stormfury has been discussed by Gentry (1969) but perhaps the most convincing evidence for possible hurricane modification came from five seeding trials conducted on Hurricane Debbie at approximately 2-h intervals on both 18 and 20 August 1969 (Gentry, 1970 and 1971). Before the first seeding on 18 August, the maximum wind velocity at 3600 m was 50 m/s (98 kts), as shown in Fig. 8.3, but five hours after the fifth seeding these winds had decreased to 35 m/s (68 kts). A similar result occurred on 20 August when the reduction was from 51 to 43 m/s. These decreases in maximum wind speeds represent a reduction in kinetic energy equivalent to 52 and 28 per cent respectively on the two days. Although it cannot be proved conclusively that these changes were entirely the result of cloud-seeding, such large decreases in wind speed appear to occur only rarely in unseeded storms and computer simulations with numerical-dynamical models of hurricanes have confirmed that a 15 per cent reduction in maximum winds is theoretically possible. It has been stressed by the Rand Corporation Staff (1969) that the prediction of natural atmospheric behaviour through the further development of numerical models is particularly important in the case of hurricanes, since the rarity of, and the lack of suitable controls for, tropical storm systems precludes the evaluation of experiments by statistical means.

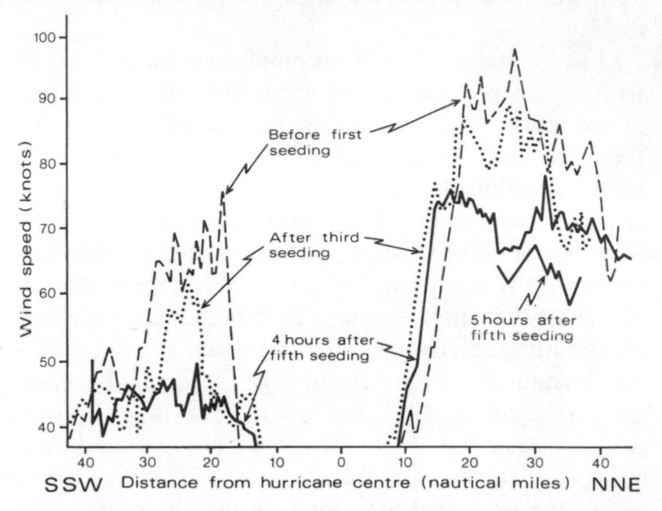


**Fig. 8.3 Wind speed changes with time at 3600 m in Hurricane Debbie on 18 August 1969. After Gentry (1970). Copyright 1970 by the American Association for the Advancement of Science.**

The *tornado* is another type of convective storm which may permit some modification. Vonnegut (1960) drew attention to the fact that tornadoes appear to contain an excess of energy, which may be related to the discharge of lightning through the revolving storm, and Colgate (1967) has postulated that such storms could perhaps be suppressed if the electrical energy could be discharged more widely. However, little field experimentation has been attempted.

## 8.11 The significance of weather modification

In the previous section, emphasis was placed on the difficulties surrounding the scientific assessment of success in weather modification. However, the ultimate significance of weather-modification activities lies in the impact, both real and imagined, which such experiments have on the community at large and it is necessary to examine some of the evidence relating to the implications in the economic and social sphere. The basic aim of weather modification is to optimize the productive capacity of certain weather-dependent enterprises, either by increasing the availability of some essential commodity, such as rainfall, or by providing some measure of protection against atmospheric hazards. It is hardly surprising, therefore, that most evaluations of modification have been limited to largely economic appraisals of the potential benefits which may result in specific areas of interest.

According to White (1966), two basic methods of approach have been adopted. In the first case, an assumed degree of weather modification, such as a 10 per cent increase in precipitation, is studied in terms of its likely impact, while the alternative approach depends on a prior determination of the extent to which a particular activity is influenced by the weather, so that it is then possible to specifiy the degree of weather modification which would be necessary to improve the efficiency of that activity. Most of the existing studies have followed the former pattern, probably because too little is known about the detailed weather-dependency of most economic activities, and have been based on some form of benefit : cost analysis. Although this type of work has yielded valuable results, notably in the case of small-scale projects, there are important limitations to the validity of benefit : cost analysis in certain weather modification situations as detailed by Gutmanis and Goldner (1966). More fundamentally, however, it cannot be assumed that weather modification is always beneficial to everyone. For example, the suppression of hail in one area may possibly lead to flooding through increased rainfall in an adjacent region, and a deliberate increase in precipitation may benefit a farmer in the experimental area only at the expense of another equally needy farmer downwind. Similarly, the dispersal of supercooled fog from the vicinity of an airport may lead to icy roads and increased traffic accidents nearby. This sort of adverse chain reaction means that any realistic economic assessment should take into account such losses as well as the net gains in the activity for which the weather modification operation was conducted. Furthermore, it will be clear that any adverse results deriving from weather experiments raise numerous moral, social, legal, and political issues, which are really quite separate from questions of economic feasibility, and it is probably within this sphere that some of the most complex and intractable problems exist.

The dominating role of experiments designed to stimulate increases in precipitation has, not unnaturally, led to a majority of economic evaluations concerned with the effects of additional rainfall on water supplies, hydroelectric generation, and agricultural production. Even in this general field, there may be some conflict of interest, since any rainfall increase that is limited to the replenishment of soil moisture reserves may benefit agriculture but be of little value for urban water-supply purposes, whereas, conversely, excessive runoff is of no direct use for crop requirements unless it can be stored in an irrigation reservoir. In addition, the critical seasons when rainfall augmentation is most desirable may well be out-of-phase, with agricultural needs concentrated in the summer growing season and the main demand for hydroelectricity located in the winter.

Krick (1952) provided an early assessment of the implications of weather modification for water resources but, as stressed by Crawford and Linsley (1963), the actual increase achieved in precipitation is less important than the net hydrologic effect on the pattern of streamflow. It is possible to analyse the hydrological consequences of any assessed increase in precipitation by means of computer simulation and Crawford (1966) employed the Stanford Watershed Model to estimate the effect of changed rainfall and evaporation conditions on runoff in three different catchments in Australia and the US. It was concluded that streamflow is sensitive to small changes in the catchment water balance and increases in streamflow of about 60 per cent were simulated in some years despite the fact that alterations in precipitation and potential evapotranspiration were only 10 per cent. More recently, Williams (1971) has stated that long-period experiments in California have demonstrated that increases of 5–10 per cent in mean annual runoff can already be achieved and, with increases perhaps as high as 20–50 per cent possible in the future, it appears inevitable that weather modification will become an integral part of river basin management.

A rather more cautious view has been expressed by Eberly (1966) in connection with rainfall stimulation for hydroelectricity generation in the western US. Following cloud-seeding tests in the Lake Almanor area of northern

California, it was found that the cost : benefit ratio for a hypothetical 10 per cent increase in streamflow might vary from 1:0·4 to 1:14·1, depending on the type of watershed and whether the increase occurred in a year of low, high, or normal discharge. For most watersheds, the economic return would reach a maximum in years with average conditions and be of least value in wet years when the reservoirs may well be unable to store the extra runoff.

A complex, interactive response is the main feature which emerges from studies of the effects of increased precipitation on the economics of agricultural production. Some general points have been raised by Castle and Stoevener (1966), who noted that, although the potential benefits could be very large, it is necessary to weigh the costs of obtaining improvements from rainfall stimulation against the costs of achieving the same benefits by alternative procedures, such as the importation of water from elsewhere or weather forecasting. This type of external complication is, of course, applicable to all other fields where weather modification has a role to play.

One of the most detailed agricultural evaluations so far published has been that by Huff and Changnon (1972) of the possible effects of increased rainfall during the growing season on corn and soybeans, which comprise the two major crops in Illinois. After the development of multiple regression equations relating long-period crop yields to weather data in 13 regions of the State with similar yield characteristics, hypothetical seeding models were used to assess the effects of changes in July–August rainfall. It was found that an economic benefit would occur in most regions provided that the seeding operator had the capability to produce rainfall increases of at least 10 per cent. On the other hand, reaction to the potential seeding was estimated to vary considerably with the same seeding model both between regions and from year to year. For example, with seeding costs of the order of 5–15 cents per acre, the optimum model showed a range of economic gain from $1–$10 per acre with a State average of about $5 per acre. These regional variations were largely attributed to differences in soil properties, whereas the year-to-year variations resulted from the temporal variability in rainfall characteristics.

The potential economic benefit to be derived from the modification of severe storms has attracted the attention of a number of authors. Thus, Benton (1969) reported that benefit : cost ratios for the Russian anti-hail programmes have been estimated to vary from 5 : 1 to as much as 17 : 1, and Gentry (1970) has claimed that the US would obtain a more than tenfold return on investment in the Stormfury experiments, if only one major hurricane per decade could be modified to reduce the damage by 10 per cent. Hendrick and Friedman (1966) have presented a summary of potential savings in the American insurance industry which might reasonably arise from the artificial reduction of storm hazards, and, as indicated in Table 8.7, most scope exists with hurricane suppression but the total cost saving at 1966 prices was estimated between $182 and $364 million in an average year.

| *Storm type* | *Current range of annual losses ($ million)* | *Assured reduction in losses* | *Potential range of annual losses with storm controls ($ million)* |
|---|---|---|---|
| Hurricanes | 250–500 | 57% | 108–216 |
| Tornadoes | 100–200 | 15% | 85–170 |
| Thunderstorms | 125–250 | 20% | 100–200 |
| Extra-tropical | 25–50 | 0% | 25–50 |
| Totals | 500–1000 | 36% | 318–636 |

**Table 8.7** Potential impact of storm controls on the American insurance industry. (After Hendrick and Friedman, 1966.)

Such monetary benefits are impressive but have to be set against the dis-benefits which are also associated with weather modification. For example, Rango (1970) has noted that, although precipitation stimulation may appear to be an attractive solution to water shortages in the semi-arid parts of the western US, any significant increase in rainfall would be likely to initiate more vigorous stream channel erosion and sediment yields would rise substantially until, beyond a mean annual rainfall of about 685 mm, the extra growth of vegetation began to reduce sediment production. More fundamentally, Standler and Vonnegut (1972) have conceded that insufficient information is available on the toxic effects of silver iodide used in cloud seeding. There appeared to be little danger of damage to human health caused by the contamination of either air or water supplies within the target areas for experiments but the much higher concentrations observed around the generators could possibly have long-term effects on the seeding operators.

Quite predictably, the multi-facetted nature of weather modification tends to produce varying and sometimes conflicting reactions from the public. Perception studies conducted in the US by Saarinen (1966) and Sewell and Day (1966) have revealed that many people hold serious reservations about both the effectiveness and the value of weather modification, and this uncertainty appeared to be most usually expressed in a desire for more active government participation in the supervision of weather modification operations. More specifically, doubts have been expressed by Morris (1965) about the adequacy of existing legal institutions to cope with the sharp increases in legislation which have been necessary to control weather modification activities, and the need for further research into the legal background has been echoed by Morris (1966) and Johnson (1968). The basic role of government policy has been questioned by Kahan (1968), who has claimed that the research sponsored by the various Federal agencies has concentrated on the scientific and engineering issues at the relative expense of the social implications. On the other hand, it must be admitted that one of the major difficulties in this field is that of evaluating the effectiveness of research programmes concerned with weather modification, particularly in respect of the fact that modification research, as well as that into weather forecasting, inevitably competes with many other areas of possible investment for government funds.

Kates and Sewell (1966) have developed a so-called research and pay-off model but this and similar research and development models normally require information inputs which may not be readily available. Thus, it is always difficult to extrapolate research findings in terms of fully operational techniques in the future but Julian *et al.* (1969) undertook a preliminary questionnaire survey of a sample of research scientists in order to assess the probable potential of existing research into weather forecasting and weather modification. A high degree of optimism was expressed by the scientists about progress over the following decade, during which period 86 per cent of the respondents believed that operational weather forecasts up to 8 days ahead would be achieved. Similarly, 77 per cent anticipated that precipitation modification in non-orographic situations would become feasible and some two-thirds estimated that warm fog dispersal and hail-suppression techniques would also become a practical proposition by the end of the 1970s. This optimism is no doubt an encouraging portent for the future. It is equally clear, however, that such developments in weather technology should not be allowed to outstrip either our broader understanding of the interrelationships between man and his climatic environment or the evolution of the various institutional mechanisms which are necessary to ensure that climatic resources in general, and weather-modification techniques in particular, are managed and utilized for the benefit of all concerned.

## References


Adderley, E. E. (1968). Cloud-seeding in western Victoria in 1966. *Austral. Meteorol. Mag.* **16**:56–63.

Akingbehin, N. A. (1966). World Weather Watch: a means of accelerating development. *W.M.O. Bull.*, **15**:120–31.

Anonymous (1956). *Weather Map. An introduction to weather forecasting*. Met. Office, 4th Edn., HMSO, London, 99 pp.

Anonymous (1971). Special topic—Weather services for the community. *Met. Office, Annual Report 1970*, 1–10.

Arakawa, H. (1966). Usefulness of weather forecasting and storm warnings. *Weather*, **21**:46–7.

Barrett, E. C. (1967). *Viewing weather from space*. Longmans, London, 140 pp.

Barrows, J. S. (1966). Weather modification and the prevention of lightning-caused forest fires. In Sewell, W. R. D. (ed.) *Human dimensions of weather modification*. University of Chicago, Dept. of Geography, Research Paper No. 105: 169–82.

Battan, L. J. (1961). *The nature of violent storms*. Anchor Books, Doubleday & Co. Inc., New York, 158 pp.

Battan, L. J. (1962). *Cloud physics and cloud seeding*. Heinemann, London, 144 pp.

Battan, L. J. (1969). Weather modification in the U.S.S.R.—1969. *Bull. Amer. Met. Soc.*, **50**:924–45.

Benton, G. S. (1969). Some general comments on meteorological and weather modification activities in the Soviet Union. *Bull. Amer. Met. Soc.*, **50**:918–22.

Benwell, G. R. R. and Timpson, M. S. (1968). Further work with the Bushby-Timpson 10-level model. *Q. Jl. R. Met. Soc.*, **94**:12–24.

Biswas, K. R., Kapoor, R. K., and Kanuga, K. K. (1967). Cloud seeding experiment using common salt. *J. Appl. Met.*, **6**:914–23.

BLACK, J. F. and TARMY, B. L. (1963). The use of asphalt coatings to increase rainfall. *J. Appl. Met.*, **2**:557–64.

BOWEN, E. G. (1952). A new method of stimulating convective clouds to produce rain and hail. *Q. Jl. R. Met. Soc.*, **78**:37–45.

BRIER, G. W. (1950). Verification of forecasts expressed in terms of probability. *Mon. Weath. Rev.*, **78**:1–3.

BRIER, G. W. and ALLEN, R. A. (1951). Verification of weather forecasts. In Malone, T. F. (ed.) *Compendium of meteorology*. Amer. Met. Soc., Boston, 841–8.

BUCHANAN, R. A. (1972). Weather forecasting for industry. In Taylor, J. A. (ed.) *Weather forecasting for agriculture and industry*, David and Charles, Newton Abbot, 115–25.

BURROUGHS, W. J. (1972). Weather modification in the USA. *UK Scientific Mission*. Report No. 72/16, Department of Trade and Industry, London, 7 pp.

BUSHBY, F. H. and TIMPSON, M. S. (1967). A 10-level atmospheric model and frontal rain. *Q. Jl. R. Met. Soc.*, **93**:1–17.

CASTLE, E. N. and STOEVENER, H. H. (1966). The economic evaluation of weather modification with particular reference to agriculture. In Sewell, W. R. D. (ed.). *Human dimensions of weather modification*, University of Chicago, Dept. of Geography, Research Paper No. 105:141–58.

CHANGNON, S. A. and SCHICKEDANZ, P. T. (1969). Utilisation of hail-day data in designing and evaluating hail suppression projects. *Mon. Weath. Rev.*, **97**:95–102.

COLGATE, S. A. (1967). Tornadoes: mechanism and control. *Science*, **157**:1431–4.

CORBY, G. A. (1970). The future of extended range forecasting. *Q. Jl. R. Met. Soc.*, **96**:339–40.

CRAWFORD, N. H. (1966). Hydrologic consequences of weather modification: case studies. In Sewell, W. R. D. (ed.) *Human dimensions of weather modification*. University of Chicago, Dept. of Geography, Research Paper No. 105:41–57.

CRAWFORD, N. H. and LINSLEY, R. K. (1963). Estimate of the hydrologic results of rainfall augmentation. *J. Appl. Met.*, **2**:426–7.

CROW, L. W. (1965). *Economic values of weather-related decisions*. Weather Bureau, ESSA, 4 pp.

DAVIES, D. A. (1966). World Weather Watch. *New Scientist*, **31**:546–9.

DENNIS, A. S. and KRIEGE, D. F. (1966). Results of ten years of cloud seeding in Santa Clara County, California. *J. Appl. Met.*, **5**:684–91.

DOANE, E. R. (1970). Early weather forecasts sent by U.S. mail. *Weather*, **25**:529–40.

DROESSLER, E. G. (1966). Atmospheric sciences and society. *Bull. Amer. Met. Soc.*, **47**:518–21.

DUNN, G. E. and MILLER, B. I. (1960). *Atlantic hurricanes*. Louisiana State University Press, Baton Rouge, 326 pp.

EBERLY, D. L. (1966). Weather modification and the operations of an electric power utility: the Pacific Gas and Electric Company's test program. In Sewell, W. R. D. (ed.) *Human dimensions of weather modification*, University of Chicago, Dept. of Geography, Research Paper No. 105:209–26.

ELLISON, T. H. (1970). Introductory remarks at Discussion Meeting on 'Long-range weather forecasting'. *Q. Jl. R. Met. Soc.* **96**:326–8.

ESTOQUE, M. A. (1963). A numerical model of the atmospheric boundary layer. *J. Geophys. Res.* **68**:1103–13.

FAWCETT, E. B. (1962). Six years of operational numerical weather prediction. *J. Appl. Met.*, **1**:318–32.

FEDOROV, E. K. (1967). Weather modifications. *W.M.O. Bull.*, **16**:122–30.

FREEMAN, M. H. (1966). The accuracy of long-range forecasts issued by the Meteorological Office. *Met. Mag.*, **95**:321–5.

GARCIA, R. V. (1967). The global atmospheric research programme. *W.M.O. Bull.*, **16**:212–18.

GENTRY, R. C. (1969). Project Stormfury. *Bull. Amer. Met. Soc.*, **50**:404–9.

GENTRY, R. C. (1970). Hurricane Debbie modification experiments. *Science*, **168**:473–5.

GENTRY, R. C. (1971). To tame a hurricane. *Sci. J.*, **7**:49–55.

GLAHN, H. R. (1964). The use of decision theory in meteorology —with an application to aviation meteorology. *Mon. Weath. Rev.*, **92**:383–8.

GORDON, I. E. and BESTWICK, N. (1969). Understanding weather forecasts. *New Soc.*, **14**:898–9.

GUTMANIS, I. and GOLDNER, L. (1966). Evaluation of benefit/cost analysis as applied to weather and climate modification. In SEWELL, W. R. D. (ed.) *Human dimensions of weather modification*, University of Chicago, Dept. of Geography, Research Paper No. 105:111–25.

HENDRICK, R. L. and FRIEDMAN, D. G. (1966). Potential impacts of storm modification on the insurance industry. In Sewell, W. R. D. (ed.) *Human dimensions of weather modification*, University of Chicago, Dept. of Geography, Research Paper No. 105:227–46.

HIBBS, J. R. (1966). Evaluation of weather and climate by socio-economic sensitivity indices. In Sewell, W. R. D. (ed.) *Human dimensions of weather modification*. University of Chicago, Dept. of Geography, Research Paper No. 105: 91–109.

HOGG, W. H. (1972). The weather forecasting requirements of specific types of agriculture and horticulture. In Taylor, J. A. (ed.) *Weather forecasting for agriculture and industry*, David and Charles, Newton Abbott, 69–85.

HOLFORD, I. (1964). The public and the weather. *Weather*, **19**:12–14.

HOLLOWAY, J. L. and MANABE, S. (1971). Simulation of climate by a global general circulation model: 1. Hydrologic cycle and heat balance. *Mon. Weath. Rev.*, **99**:335–70.

HOUGHTON, D. M. (1965). Current forecasting practice. *Q. Jl. R, Met. Soc.*, **91**:524–6.

Howell, W. E. and Lopez, M. E. (1966). Cloud seeding in southern Puerto Rico, April–July 1965. *J. Appl. Met.*, **5**:692–6.

Huff, F. A. (1969). Climatological assessment of natural precipitation characteristics for use in weather modification. *J. Appl. Met.*, **8**:401–10.

Huff, F. A. and Changnon, S. A. (1972). Evaluation of potential effects of weather modification on agriculture in Illinois. *J. Appl. Met.*, **11**:376–84.

Huschke, R. E. (1963). A brief history of weather modification since 1946. *Bull. Amer. Met. Soc.*, **44**:425–9.

Johnson, R. W. (1968). Weather modification and legal research. In U.S. National Science Foundation. *Human dimensions of the atmosphere*. NSF. 68–18, Washington, D.C., 87–98.

Julian, P. R., Kates, R. W., and Sewell, W. R. D. (1969). Estimating probabilities of research success in the atmospheric sciences: results of a pilot investigation. *Water Resour. Res.*, **5**:215–27.

Kahan, A. M. (1968). The place of government programs in weather modification. *Bull. Amer. Met. Soc.*, **49**:242–6.

Kahan, A. M., Stinson, J. R., and Eddy, R. L. (1969). Progress in precipitation modification. *Bull. Amer. Met. Soc.*, **50**:208–14.

Kates, R. W. and Sewell, W. R. D. (1966). The evaluation of weather modification research. In Sewell, W. R. D. (ed.) *Human dimensions of weather modification*, University of Chicago, Dept. of Geography, Research Paper No. 105: 347–62.

Kerr-Smith, M. W. (1970). The role of meteorological satellites. *Weather*, **25**:244–53.

Knighting, E. (1958). Numerical weather forecasting. *Weather*, **13**:39–50.

Krick, I. P. (1952). Increasing water resources through weather modification. *Amer. Wat. Wks. Ass. J.*, **44**:996–1020.

Lamb, H. H. (1972). Problems and practice in longer-range weather and climate forecasting. In Taylor, J. A. (ed.) *Weather forecasting for agriculture and industry*. David and Charles, Newton Abbot, 34–43.

List, R. (1960). Design and operation of the Swiss hail tunnel. In Weickmann, H. (ed.) *Physics of precipitation*. Geophys. Monograph 5, Amer. Geophys. Union, Washington, D.C., 310–16.

Lorenz, E. N. (1969). Three approaches to atmospheric predictability. *Bull. Amer. Met. Soc.*, **50**:345–9.

Lowndes, C. A. S. (1966). The forecasting of shower activity in airstreams from the north-west quarter over S.E. England in October to April. *Met. Mag.*, **95**:248–52.

Ludlam, F. H. (1955). Artificial snowfall from mountain clouds. *Tellus*, **7**:277–90.

Ludlam, F. H. (1961). The hailstorm. *Weather*, **16**:152–62.

Ludlam, F. H. (1966). *The cyclone problem: A history of models of the cyclonic storm*. Inaugural Lecture delivered at Imperial College of Sci. and Tech., University of London, 49 pp.

McGlening, K. (1968). Canadian weather services for radio and television. *Bull. Amer. Met. Soc.*, **49**:377–8.

McQuigg, J. D. and Thompson, R. G. (1966). Economic value of improved methods of translating weather information into operational costs. *Mon. Weath. Rev.*, **94**:83–7.

Malone, T. F. (1967). Weather modification: implications of the new horizons in research. *Science*, **156**:897–901.

Malone, T. F. (1968). New dimensions of international cooperation in weather analysis and prediction. *Bull. Amer. Met. Soc.*, **49**:1134–40.

Mason, B. J. (1962). *Clouds, rain and rainmaking*. Cambridge University Press, 145 pp.

Mason, B. J. (1966a). Weather forecasting by computer. *New Scientist*, **32**:96–8.

Mason, B. J. (1966b). The role of meteorology in the national economy. *Weather*, **21**:382–93.

Mason, B. J. (1968). The role of satellites in observing and forecasting the global behaviour of the atmosphere. *Proc. Roy. Soc. (London), Ser. A*; **308**:157–72.

Mason, B. J. (1969). Some outstanding problems in cloud physics—the interaction of microphysical and dynamical processes. *Q. Jl. R. Met. Soc.*, **95**:449–85.

Mason, B. J. (1970). Future developments in meteorology: an outlook to the year 2000. *Q. Jl. R. Met. Soc.*, **96**:349–68.

Mason, B. J. (1971). Global Atmospheric Research Programme. *Nature*, **233**:382–88.

Maunder, W. J. (1969). The consumer and the weather forecast. *Atmosphere*, **7**:15–22.

Miles, M. K. (1961). The basis of present-day weather forecasting. *Weather*, **16**:349–63.

Miyakoda, K., Smagorinsky, J., Strickler, R. F., and Hembree, G. D. (1969). Experimental extended predictions with a nine-level hemispheric model. *Mon. Weath. Rev.*, **97**:1–76.

Moorman, T. (1968). The forecasting problem. *Bull. Amer. Met. Soc.*, **49**:12–15.

Morris, E. A. (1965). The law and weather modification. *Bull. Amer. Met. Soc.*, **46**:618–22.

Morris, E. A. (1966). Institutional adjustment to an emerging technology: legal aspects of weather modification. In Sewell, W. R. D. (ed.) *Human dimensions of weather modification*, University of Chicago, Dept. of Geography, Research Paper No. 105:279–88.

Murphy, A. H. and Epstein, E. S. (1967). Verification of probabilistic predictions: a brief review. *J. Appl. Met.*, **6**:748–55.

Murray, R. (1967a). Sequences in monthly rainfall over England and Wales. *Met. Mag.*, **96**:129–35.

Murray, R. (1967b). Persistence in monthly mean temperature in central England. *Met. Mag.*, **96**:356–63.

Murray, R. (1968). Sequences in monthly rainfall over Scotland. *Met. Mag.*, **97**:181–3.

MURRAY, R. (1970). Recent developments in long-range forecasting in the Meteorological Office. *Q. Jl. R. Met. Soc.*, **96**:329–36.

MURRAY, R. (1971). Forecasting March temperature and rainfall for England and Wales. *Met. Mag.*, **100**:161–73.

MURRAY, R. (1972a). On predicting seasonal weather for England and Wales from anomalous atmospheric circulation over the northern hemisphere. *Weather*, **27**:396–402.

MURRAY, R. (1972b). An objective method of foreshadowing winter rainfall and temperature for England and Wales. *Met. Mag.*, **101**:97–110.

MURRAY, R. (1972c). Prediction of summer rainfall and temperature for England and Wales from anomalous atmospheric circulation in spring. *Met. Mag.*, **101**:253–264.

MURRAY, R. and BENWELL, P. R. (1970). PSCM indices in synoptic climatology and long-range forecasting. *Met. Mag.*, **99**:232–45.

NAMIAS, J. (1968). Long range weather forecasting—history, current status and outlook. *Bull. Amer. Met. Soc.*, **49**:438–70.

NAMIAS, J. (1969). Seasonal interactions between the north Pacific ocean and the atmosphere during the 1960's. *Mon. Weath. Rev.*, **97**:173–92.

NEYMAN, J. (1967). Experimentation with weather control. *J. Roy. Stat. Soc.*, **130**:285–326.

ODDIE, B. C. V. (1964). The language of forecasts. *Weather*, **19**:138–43.

O'MAHONY, G. (1967). Cloud-seeding in Wimmera-Mallee, Victoria, 1966. *Austral. Meteorol. Mag.*, **15**:133–47.

PETERS, S. P. (1955). Forecasting and public services. *Met. Mag.*, **84**:192–8.

PETTERSSEN, S. (1966). Recent demographic trends and future meteorological services. *Bull. Amer. Met. Soc.*, 950–62.

POTHECARY, I. J. W. and RATCLIFFE, R. A. S. (1966). Satellite pictures of an old occluded depression and their usefulness in analysis and forecasting. *Met. Mag.*, **95**:332–9.

QUIN, W. H. and BURT, W. V. (1972). Use of the Southern Oscillation in weather prediction. *J. Appl. Met.*, **11**:616–28.

RAND CORPORATION STAFF (1969). Weather modification progress and the need for interactive research. *Bull. Amer. Met. Soc.*, **50**:216–46.

RANGO, A. (1970). Possible effects of precipitation modification on stream channel geometry and sediment yield. *Water Resour. Res.*, **6**:1765–70.

RATCLIFFE, R. A. S. (1966). A case illustrating the value of satellite photographs in forecasting for the British Isles. *Met. Mag.*, **95**:257–61.

RATCLIFFE, R. A. S. (1970a). Sea temperature anomalies and long-range forecasting. *Q. Jl. R. Met. Soc.*, **96**:337–8.

RATCLIFFE, R. A. S. (1970b). Meteorological Office long-range forecasts: six years of progress. *Met. Mag.*, **99**:125–30.

RATCLIFFE, R. A. S. and MURRAY, R. (1970). New lag associations between North Atlantic sea temperature and European pressure applied to long-range weather forecasting. *Q. Jl. R. Met. Soc.*, **96**:226–46.

RICHARDSON, L. F. (1922). *Weather prediction by numerical process.* Cambridge University Press, Cambridge, 236 pp.

ROBINSON, G. D. (1967). Some current projects for global meteorological observation and experiment. *Q. Jl. R. Met. Soc.*, **93**:409–18.

SAARINEN, T. F. (1966). Attitudes towards weather modification: a study of Great Plains farmers. In Sewell, W. R. D. (ed.) *Human dimensions of weather modification*, University of Chicago, Dept. of Geography Research Paper No. 105: 323–8.

SANGER, R. (1960). The Swiss randomized hail suppression project in the Tessin. In Weickmann, H. (ed.) *Physics of precipitation.* Geophysical Monograph 5, Amer. Geophys. Union, Washington, D.C., 388–94.

SAWYER, J. S. (1967a). Weather forecasting and its future (Part 1). *Weather*, **22**:350–60.

SAWYER, J. S. (1967b). Weather forecasting and its future (Part 2). *Weather*, **22**:400–11.

SCHAEFER, V. J. (1946). The production of ice crystals in a cloud of supercooled water droplets. *Science*, **104**:457–9.

SCHAEFER, V. J. (1968). The early history of weather modification. *Bull. Amer. Met. Soc.*, **49**:337–42.

SCHICKEDANZ, P. T. and CHANGNON, S. A. (1970). The design and evaluation of hail suppression experiments. *Mon. Weath. Rev.*, **98**:242–51.

SCHLEUSENER, R. A. (1968). Hail fall damage suppression by cloud seeding—a review of the evidence. *J. Appl. Met.*, **7**:1004–11.

SELLERS, W. D. (1965). *Physical climatology.* University of Chicago Press, 272 pp.

SEWELL, W. R. D. and DAY, J. C. (1966). Perception of possibilities of weather modification and attitudes toward government involvement. In Sewell, W. R. D. (ed.) *Human dimensions of weather modification*, University of Chicago, Dept. of Geography, Research Paper No. 105:329–44.

SHEPPARD, P. A. (1968). Global Atmospheric Research. *Weather*, **23**:262–83.

SIMPSON, J. (1967). An experimental approach to cumulus clouds and hurricanes. *Weather*, **22**:95–114.

SIMPSON, J. and WIGGERT, V. (1971). 1968 Florida cumulus seeding experiment: numerical model results. *Mon. Weath. Rev.*, **99**:87–118.

SIMPSON, R. H. and MALKUS, J. S. (1964). Experiments in hurricane modification. *Sci. Amer.*, **211**:27–37.

SMITH, C. V. (1972). The organisation of meteorological advice for agriculture and industry. In Taylor, J. A. (ed.) *Weather forecasting for agriculture and industry.* David and Charles, Newton Abbot, 19–33.

STAGG, J. M. (1971). *Forecast for Overlord.* Ian Allan, London, 128 pp.

STANDLER, R. B. and VONNEGUT, B. (1972). Estimated possible effects of AgI cloud seeding on human health. *J. Appl. Met.*, **11**:1388–91.

STRINGER, E. T. (1970). The use of cost-benefit studies in the interpretation of probability forecasts for agriculture and industry: an operational example. In Taylor, J. A. (ed.) *Weather economics*, Pergamon Press, Oxford, 83–91.

SUMMERS, P. W., MATHER, G. K., and TREDDENICK, D. S. (1972). The development and testing of an airborne droppable pyrotechnic flare system for seeding Alberta hailstorms. *J. Appl. Met.*, **11**:695–703.

SUMMERS, P. W. and PAUL, A. H. (1970). Some climatological characteristics of hailfall in central Alberta. In Nelson, J. G., Chambers, M. J., and Chambers, R. E. (eds.) *Weather and climate*. Methuen, Toronto, 193–211.

SUTCLIFFE, R. C. (1952). Principles of synoptic weather forecasting. *Q. Jl. R. Met. Soc.*, **78**:291–320.

SUTCLIFFE, R. C. (1964). Weather forecasting by electronic computer. *Endeavour*, **23**:27–32.

TAYLOR, J. A. (1971). Curbing the cost of bad weather. *New Scientist*, **50**:560–3.

TAYLOR, J. A. (1972). The revaluation of weather forecasts. In Taylor, J. A. (ed.) *Weather forecasting for agriculture and industry*. David and Charles, Newton Abbot, 1–18.

THOMPSON, J. C. (1966). The potential economic and associated values of the World Weather Watch. *World Weather Watch Planning Report* No. 4, WMO, Geneva, 35 pp.

THOMPSON, J. C. and BRIER, G. W. (1955). The economic utility of weather forecasts. *Mon. Weath. Rev.*, **83**:249–54.

U.S. WEATHER BUREAU (1964). *The national research effort on improved weather description and prediction for social and economic purposes.* Interdepartmental Committee on Atmospheric Sciences, Federal Council for Science and Technology, 84 pp. (Quoted by Maunder, 1970.)

VONNEGUT, B. J. (1947). Nucleation of ice formation by silver iodide. *J. Appl. Phys.*, **18**:593–5.

VONNEGUT, B. (1960). Electrical theory of tornadoes. *J. Geophys. Res.*, **65**:203–12.

WEINSTEIN, A. I. (1972). Ice-phase seeding potential for cumulus cloud modification in the western United States. *J. Appl. Met.*, **11**:202–10.

WEXLER, H. (1958). Modifying weather on a large scale. *Science*, **128**:1059–63.

WHITE, G. F. (1966). Approaches to weather modification. In Sewell, W. R. D. (ed.) *Human dimensions of weather modification*. University of Chicago, Dept. of Geography Research Paper No. 105:19–23.

WHITE, R. M. (1966). Broadcasting the weather. *Bull. Amer. Met. Soc.*, **47**:21–4.

WHITE, R. M. and CHANDLER, R. A. (1965). Project Stormfury: status and prospects. *Bull. Amer. Met. Soc.*, **46**:320–2.

WICKHAM, P. G. (1970). *The practice of weather forecasting*. HMSO, London, 187 pp.

WILLIAMS, M. C. (1971). Status of weather modification in watershed management. *J. Irrig. Drainage Div. (Amer. Soc. Civ. Engrs.)*, **97**:585–600.

WYCOFF, P. H. (1966). Evaluation of the state of the art. In Sewell, W. R. D. (ed.) *Human dimensions of weather modification*. University of Chicago, Dept. of Geography Research Paper No. 105:27–39.

# Author Index

# Subject Index